Solar Air Systems –
A Design Handbook

Published by James & James (Science Publishers) Ltd,
35-37 William Road, London NW1 3ER, UK

© 2000 Solar Heating and Cooling Executive Committee of the
International Energy Agency (IEA)

All rights reserved. No part of this book may be reproduced in any form or by any means electronic or mechanical, including photocopying, recording or by any information storage and retrieval system without permission in writing from the copyright holder and the publisher.

A catalogue record for this book is available from the British Library.

ISBN 1 873936 86 9

Printed in the UK by Hobbs the Printers

This book is dedicated to George Löf, Professor Emeritus, Colorado State University, USA

Contents

Foreword		*v*

I INTRODUCTION — 1
I.1 Introduction — 3
Robert Hastings
I.2 Design process and system selection — 7
Ove Mørck with Christer Nordström and Charles Filleux

II SYSTEMS — 15
II.1 System 1: Solar heating of ventilation air — 17
Søren Østergaard Jensen
II.2 System 2: Open collection loop with radiant discharge storage — 38
Giancarlo Rossi and Gianni Scudo
II.3 System 3: Double envelope systems — 49
Christer Nordström and Torbjörn Jilar
II.4 System 4: Closed collection loop with radiant discharge system — 69
Charles Filleux and Peter Elste
II.5 System 5: Closed collection loop with open discharge loop — 91
Hans Erhorn
II.6 System 6: Closed collection loop with heat exchange to water — 102
Ove Mørck

III ALTERNATIVE USES OF SYSTEMS — 113
III.1 Domestic hot water heating — 115
Ove Mørck
III.2 Cooling applications and additional use of air systems — 120
Karel Fort
III.3 Hybrid photovoltaic/heating system — 128
John Hollick

IV COMPONENTS — 131
IV.1 Flat-plate air collectors — 133
Ove Mørck and Hubert Fechner
IV.2 Window air collectors — 146
Charles Filleux
IV.3 Perforated unglazed collectors — 150
John Hollick
IV.4 Double facades and double-shell facades — 158
Thomas Zelger, Matthias Schuler and Alexander Knirsch
IV.5 Spatial collectors — 172
Kevin Lomas
IV.6 Hypocaust and murocaust storage — 175
Karel Fort
IV.7 Rockbeds — 185
Sture Larsen
IV.8 Phase change materials (pcm) for heat storage — 193
Gerhard Zweifel

V ACCESSORIES — 199
V.1 Fans — 201
Johann Reiss
V.2 Air-to-water heat exchangers — 212
Ove Mørck
V.3 Controls — 215
Joachim Morhenne
V.4 Fire protection — 221
Christer Nordström

APPENDICES — 223
A Glossary — 225
Robert Hastings
B Properties of materials — 227
Aninja Grilc and Alex Knirsch
C Meteorological data — 229
Frank Heidt
D Climate impact on performance — 240
Frank Heidt
E Economic evaluation — 246
Frank Heidt
F Collector testing — 250
Hubert Fechner
G Trnsair: thermal simulation program for solar air systems — 255
Alex Knirsch
H Nomogram assumptions — 258
Alex Knirsch
I Survey of PC programs — 269
Matthias Schuler
J Bibliography — 273
Heike Kluttig
K IEA Solar Heating and Cooling Programme — 282
Robert Hastings

Foreword

Using air to transport heat from a solar collector to its end use is a very old and, until recently, largely forgotten idea. In Europe and North America this concept has enjoyed a renaissance, finding applications in climates from Norway to Italy and in building types ranging from houses to large industrial structures. Solar air systems have proven reliable for heating spaces, ventilation air and even domestic hot water in summer. An air system, unlike a water system, needs no freeze protection, nor are leaks damaging to the building structure or its contents. In contrast to passive design, active air systems provide better heat distribution, improved comfort and fuller use of solar gains. They are a natural fit to mechanically ventilated buildings and mechanical ventilation is increasingly common, not only in commercial and institutional buildings, but also in very low energy residences.

In order to pool the experience in designing such systems, the Solar Heating and Cooling Programme of the International Energy Agency initiated a five-year project. Twenty experts from nine countries working in collaboration have produced:

- a book illustrating 33 exemplary buildings with diverse solar air systems
- a catalogue of manufactured components and guidelines for selecting them
- a PC-based, easily used program to predict energy performance and comfort
- and now, at the end of the collaboration, this handbook for designing a system.

This handbook is intended as a practical tool for selecting a system from the six types of systems presented, optimizing the system using nomograms and curves, and finally dimensioning the components of the system. Tips are offered regarding the construction and how to avoid problems.

The goal in the 1980s was to demonstrate the capabilities of such systems. The goal in the 1990s was to demonstrate reliability and improved cost effectiveness. The purpose of this series of publications is to share these experiences in order to achieve a wide market penetration as we enter a new century, a century of transition from fossil fuels to renewable energy sources.

Robert Hastings
January 2000

I. Introduction

I.1 Introduction

WHY A SOLAR AIR SYSTEM?

Solar air systems have found use in single and multi-family residences, institutional buildings, sports facilities and industrial buildings. The reasons for this widespread use are that air is a reliable and economical heat carrier for heating a building space, ventilation air or even domestic hot water in summer. Some solar air systems can also be reversed to extract air out of a space by the chimney effect. Advantages and limitations of solar air systems are listed below.

Advantages

- Improved cost effectiveness when:
  - the collector is an integral part of the building's skin, and the building structure itself provides heat storage and distribution;
  - more than one energy function is served, i.e. space heating in winter, domestic water heating in summer;
  - non-energy uses are provided, such as an acoustic barrier to street noise;
- fast responding: solar air collectors heat up quickly as soon as the sun shines;
- thermal buffering: the collectors create a warm envelope for the building in periods of low radiation;
- easily controlled: fan-powered solar air systems are easy to control and can be integrated into conventional heating, ventilation and air conditioning systems (HVAC);
- durability: no frost or corrosion problems, leaks are not as serious compared with water systems and no anti-freeze chemicals are needed.

Limitations

- Heat transport by air needs large channel cross-sections to move the needed volume;
- potential noise problems must be designed out (to avoid too high air velocities);
- electric consumption by fans must be minimized (to minimize pressure drops in air channels);
- dust and moisture in open systems must be planned for.

A SHORT HISTORY*

The first patented solar air heater was designed and built in the USA. E S Morse built the prototype in 1881 which was then marketed in 1890. Mounted on a south-facing wall, it was essentially a vertical sheet of black metal or slate supported inside a shallow, glass-covered box. Air was warmed by contact with the black, sun-heated surface and would rise by natural convection. Cold air entered an opening in the lower end of the panel, and heated air passed from a top opening through the south wall of the building directly into the room behind the wall. Several such solar air heaters were installed in Massachusetts buildings, but there was no further development or commercialization.

Responding to the threat of energy shortages in the Second World War in the USA an American, K W Miller designed and patented a solar air collector, the rockbed system. It was constructed and tested by George Löf. Heat storage in a rock-filled bin was evaluated in a full-size laboratory installation, and, in 1945, a small house in Colorado was retrofitted with a solar air heating system, comprising collector, heat storage rock bin, air ducts and automatic controls. The system was able to cover approximately one-third of the heating demand, according to monitoring results.

Prior to the research on solar air heating in Colorado, design, construction and testing of space heating with liquid systems was initiated in Massachusetts by Hoyt Hottel. Three residential-type structures were provided with heating systems in which water was used in the collecting, transport and storage of solar heat. The results of this work provided the technical basis for the performance of all types of flat-plate solar collectors.

Another solar air heating development took place in Massachusetts during the 1946–8 period, introduced by M Telkes. A residence was designed with south-facing vertical collectors from which solar-heated air was delivered to interior chambers in which heat was stored by melting several tons of Glauber's salt, crystalline sodium sulphate ($Na_2SO_4 \bullet 10H_2O$), in 20-litre containers at temperatures of 40–50°C. Air was circulated back to the collector for re-heating and to the rooms as needed. Full reversible melting and re-crystallizing of the salt in the stagnant containers apparently did not

---

*Based on material provided by George Löf, Colorado State University.

occur, resulting in decreased heat storage capacity and the termination of the experiment.

A second vertical solar air heater and phase-change storage system was built in about 1950, in a New Jersey building which served as a solar testing laboratory. The 60 m² collector was provided with double glass covers, and air was circulated either behind the metal absorber plate or between the plate and the inside glass pane. Approximately 10 m³ of phase-change heat storage was provided. After limited testing, the system was dismantled.

A simple type of solar air heater was built into the roof of an Arizona school building in 1948. Designed by Arthur Brown, the collector was fabricated on site by covering the nearly horizontal aluminium roof channels with additional blackened aluminium sheets to form rectangular passages through which air was warmed by 5 to 10 K and forced into the building. Daytime heating was thus provided in sunny weather. In this mild climate, over 80% of the heating requirements were met by solar energy. Although a low-cost system, it was not adopted elsewhere.

A considerably different concept was developed for a rural house in Arizona by Donovan and Bliss. The site-built collector consisted of a single glass cover and a loosely woven black fabric absorber through which air was circulated. The heated air was conducted to a rock-filled bin for heat storage and/or to the house rooms. Although the concept was validated, the durability of the collector material proved inadequate.

Interest in solar space heating languished in the 1950s until 1957 when a US window glass manufacturer funded the design, construction and evaluation of a solar air heating system on a new Colorado residence. Employing the same collector design and system concept retrofitted into the 1945 Colorado house, this installation demonstrated the practicality and durability of solar air heating (Figure I.1.1). Today, more than 40 years later, it is still faithfully providing heat for the residence. Extensive performance data have been obtained and published. Although functioning satisfactorily, this site-built collector was not suitable for factory production. Numerous improvements were identified during the many years of observing the system operation. Performance measurements after 17 years of use showed that heat output had declined by about one-fourth; after cleaning and refurbishing the system delivered solar heat at its original level.

These half-dozen solar air heating projects were all that occurred before a major US initiative in 1973 resulted in extensive development and use of solar heating of various types. In a time of rising fuel prices and threats of shortages, energy department officials in the USA and Canada launched research and incentive programs to develop solar energy technology and use. Many projects were established, primarily in universities and their affiliated laboratories, ranging from fundamental research to full-scale system development and testing. In addition to government funding, heating equipment manufacturers supported extensive development work between 1973 and 1985.

Almost simultaneously, commercialization began, with old and new manufacturing companies developing, producing and selling complete solar air systems. Commercial activities depended heavily on government subsidies.

During this era the commercial manufacture and sale of solar air heating systems grew, in parallel with the growth of the solar water heating manufacturing industry. A 1980 survey of solar air collector manu-

Figure I.1.1. A 1945 solar air heated home in Colorado

facturers identified 85 US companies offering solar collectors for commercial and residential use. Many of the companies also provided complete systems and installation.

In 1985, the federal subsidies that had been in place since 1975 were discontinued. In some states, tax credits (subsidies) continued, but the effect of the termination of the federal subsidy was sudden and severe. Most manufacturers of solar heating systems of all types closed down their business in 1986 because the price to the customer of a system was too high without the 40% subsidy by the government. The manufacture of all types of solar space heating systems virtually ceased. Research and development in solar air systems also came to a halt. In contrast to the period from 1977–85 when over 20 technical papers on the subject were published in the journal *Solar Energy*, only three papers were published in the following three years.

In Canada, a similar scenario occurred when federal subsidies ended in 1988. Only one solar air company survived, Conserval Engineering, and it had focused on low-cost building-integrated designs where south walls were converted to solar heaters using metal facades as the solar absorber. With Canadian government support, a new unglazed panel was tested and since 1990 virtually every solar air system installed in Canada has been of the unglazed perforated absorber type described in chapter IV.3. Since 1992, most USA solar air installations have also used the unglazed perforated panel (or transpired panel as it is commonly referred to by the Department of Energy).

Another slow renaissance occurred in the 1990s, this time in Europe. Without the back-up of research institutions, architects began building solar air heated residences and schools. The buildings demonstrated very good performance and interest in this alternative way of transporting and storing solar heat revived. While many systems were architect-planned and site-built, all the same the solar air collector industry was reborn, targeting market niches. Inexpensive package collectors came on the market to temper and dehumidify vacation cottages from Sweden to Spain. Mass-produced, high performance collectors found applications in sports halls, commercial buildings and residences. The largest commercial success has occurred in the marketing of unglazed collectors for heating ventilation air for large buildings.

The promise of this technology gained the attention of governments increasingly motivated by environmental and energy concerns. The members of The Executive Committee of the International Energy Agency agreed to hold a series of small workshops in the Netherlands, Denmark and Switzerland. The workshops provided a forum for exchanging experience and looking for ways to improve system performance. It became apparent that solar air systems showed great promise, but if the technology were to develop, help was needed. Neither the architects designing individual systems, nor the fledgling manufacturers could finance the effort needed. A research proposal and work plan were written and approved by the committee, and funding was obtained from nine countries to carry out this project under 'Task 19'. This handbook is one of the outputs of this effort, presenting six types of solar air systems.

TYPE 1: SOLAR HEATING OF VENTILATION AIR

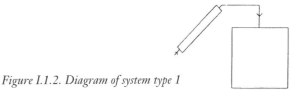

Figure I.1.2. Diagram of system type 1

This is one of the most economical applications (Figure I.1.2) and accordingly is the most popular form of solar air heating with the most square metres of solar panels sold worldwide.

Outside air is circulated through an unglazed or glazed collector directly into the space to be ventilated and heated. This system can achieve very high efficiencies because cool air is supplied to the collector. In the summer the collector can be vented to the outside.

Appropriate applications range from keeping unoccupied vacation cottages from becoming damp and musty to ventilating schools, offices and large industry halls.

TYPE 2: COLLECTOR/ROOM/COLLECTOR

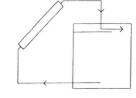

Figure I.1.3. Diagram of system type 2

This so-called Bara Costantini System, named after its Italian inventor, circulates room air into the collector where it is heated, rises and returns via a thermal storage ceiling back into the room, all by natural convection (Figure I.1.3). The storage radiates heat after sunset. In summer the collector can be vented at the top to the outside, extracting room air, which can then be replaced by cooler air from an earth register or open north-facing windows. This system has been used most for apartment buildings.

TYPE 3: COLLECTOR-HEATED AIR CIRCULATED THROUGH CAVITY IN THE BUILDING ENVELOPE

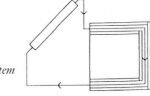

Figure I.1.4. Diagram of system type 3

By circulating collector-warmed air through a hollow building envelope, heat losses through the building envelope can be drastically reduced (Figure I.1.4). Because the air returning to the collector from the building envelope is relatively cool, the collectors operate at a high efficiency. In summer, a bypass from the collectors directly to an air-to-water heat exchanger can allow heating of domestic hot water as well. Key components of this system are a relatively simple construction collector (since it can operate at low temperatures), one or more fans, dampers, a hollow building envelope and the control system to determine when the fan should run. Switching to the summer water heating mode can be a manual operation. The system is especially appropriate for retrofitting poorly insulated existing apartment buildings.

TYPE 4: CLOSED LOOP COLLECTOR/ STORAGE AND RADIANT DISCHARGE TO BUILDING SPACES

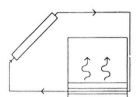

Figure I.1.5. Diagram of system type 4

In this most classical type of solar air system, collector-warmed air is circulated through channels in a massive floor or wall which then radiate the heat into the room four to six hours later (Figure I.1.5). This system has the advantage of large radiating surfaces providing comfort. Fan-forced circulation provides the best overall system efficiency and output. Applications include all building types which have large surfaces available for the radiant surfaces.

TYPE 5: OPEN SINGLE LOOP COLLECTOR TO THE BUILDING SPACES

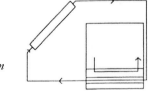

Figure I.1.6. Diagram of system type 5

This system is similar to system type 4, but separate channels in the storage allow a controlled active discharge (Figure I.1.6). Since the storage can be better insulated, it can be charged to higher temperatures and not discharged until heat is desired. Further, the storage can be located remotely from the rooms to be heated. Relatively few buildings exist with this system because of the expense involved.

TYPE 6: COLLECTOR-HEATED AIR TRANSFERRED TO WATER VIA AN AIR/WATER HEAT EXCHANGER

By this system hot air from the collectors passes over an air-to-water heat exchanger (Figure I.1.7). The hot

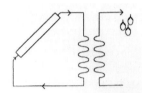

Figure I.1.7. Diagram of system type 6

water can then be circulated to conventional radiators, radiant floors or walls or to a domestic hot water tank. Applications where the heat must be transported over a distance are particularly suited for this system. Retrofit of existing buildings is simplified since the heating distribution systems often use water. Because the air-to-water heat exchanger is typically located within the heated building envelope, anti-freeze can normally be omitted. A tight frost damper is essential, however, to prevent back circulation from the collector to the heat exchanger at night.

STRUCTURE OF THIS HANDBOOK

This handbook is structured to follow the decision-making process for designing a solar air system. Chapter I.2 describes a process for using this book to select and design a system. The system chapters include a graphic means to estimate the energy performance, allowing for modifications in the basic design parameters. For help in selecting and designing the individual components of the system the reader is referred to the appropriate components chapter. Of special interest are multiple uses of a given component, such as the use of a photovoltaic panel to generate electricity and provide hot air. Finally, appendices provide data needed to use the tables and nomograms of this handbook. When the configuration is found to be more or less acceptable, more detailed analyses can then be carried out using the PC tool Trnsair.

Neither the experts nor the IEA SHAC can assume any liability for information provided in this document.

Chapter author: Robert Hastings

I.2 Design process and system selection

INTRODUCTION

The design of a solar air system needs to be linked to the design of the building itself. Double exploitation of building components for constructional and solar system use is essential to obtain cost-efficient applications. A solar air system is in fact as much an architect's business as that of an engineer, so the process must be carried out in close co-operation between both parties.

The design process starts with establishing building constraints (i.e. the site, the climate and the user profile) and goals (i.e. indoor environment, the energy use, the economics and sustainability of the design). The designer must consider the external conditions of the building and the possible orientation of solar collecting surfaces. Then comes the selection and layout of a solar system. Finally, the detailed design of the selected solar system and its components can be made.

CONDITIONS AND GOALS (INPUT PHASE)

Building programme

Identify unbroken external wall and roof surfaces by tilt and orientation, usable for solar collectors, and heavy building parts with a large thermal mass. The auxiliary heating and ventilation system should be described at this stage.

Determine the building heat losses, the internal gains and the (net) heating demand of the building without the solar air system.

The input phase should establish the use of the solar air system. Should the system be planned for domestic hot water heating (DHW), or future extensions of the building? Also, any regulations affecting the building design and solar system selection should be clarified.

Climate

Important features are the location (latitude), climate (solar radiation, ambient temperatures) and microclimate (topography, vegetation, surrounding buildings, shading and prevailing winds). In Appendix C some key figures for a number of climates have been tabulated. One of these figures is an index showing the relation between global solar radiation and degree-days. The idea behind the calculation of this index is that climates with similar climate index may provide similar conditions for solar heating systems.

The four climate zones shown in Table I.2.1 and representative locations have been selected for the calculations in this book. Determine the climate similarity index for your climate situation and find the best matching climate type in Appendix C, using the nomograms for performance estimations and for computer simulations.

Calculation period for simulations
In the systems chapters II.1–II.6 the results of all six system types are presented. The systems have been calculated for all building insulation levels and all climates for the period September–May. This was decided to ensure comparable results, and to avoid the errors involved in fixing heating periods consisting of whole months for the different building types and insulation levels. The fact that the building's heating load is also simulated assures that only useful solar contributions from the solar systems are counted for. The domestic hot water system was calculated for the three-month period June–August.

Orientation of collectors
The calculations in this book have been produced for a limited number of angles of orientation of the solar

Table I.2.1. *Climates used for the calculations presented in this handbook*

Climate type	Representative location	Period	Global horizontal radiation kWh/m²	Degree days Kd	Climate similarity index kWh/Kd m²
Sunny/temp	Brig (CH)	Oct–Apr	520	3465	0.150
Cloudy/temp	Copenhagen	Oct–May	450	3815	0.119
Sunny/cold	Denver	Oct–Apr	800	3560	0.225
Sunny/mild	Rome	Nov–Mar	375	1610	0.233

collectors, that is a slope of 45° or vertical and directed towards south. Often it is not possible to place a solar collector at an optimum position towards the sun, which means the energy performance of the solar system will be reduced. The isolines for orientation and collector slope presented in Appendix C indicate the impact of installing a solar collector at different orientations and slopes. Table I.2.2 shows the solar insulation at different orientations for the four climates selected for the creation of the performance estimation nomograms for each system.

Table I.2.2. Radiation on 45° tilted surfaces

kWh/m²a	SW	S	SE
Copenhagen	1060	1160	1130
Brig	1435	1525	1440
Rome	1630	1795	1740
Denver	1985	2205	2050

Economic/energetical goal
At the start of a project the design team will define together with the client/builder what fraction of the heat demand should be covered by the solar system. This discussion is important, because the result will influence the costs of the saved energy. Basically there are three options:

- high solar contribution;
- high specific output of the solar system;
- optimum costs of saved energy.

A high solar contribution will cover a significant portion of the heat demand, say 50% in a sunny/temperate climate, and is obtained by a large collector area per floor area. A high specific output of the system will be obtained with a small collector area per floor area, and covers say 15%, in a sunny/temperate climate of the heat demand. Finally, optimum costs of saved energy will be obtained by a compromise between both options. Figure I.2.1 shows that optimum costs of saved energy occur at a lower fraction of collector area per heated floor area for large systems, than for small systems.

Calculation of saved energy costs
Several methods of economic analysis are simple payback, advanced payback (taking into account interest rates and inflation), present worth value, internal interest rate and annuity. Three of these methods are described in detail in Appendix E. The most intuitively direct method is probably to compare the cost of the energy produced or saved by the solar energy system (calculated by the annuity method) to the cost of conventional energy. Table I.2.3 shows the main input data.

Table I.2.3. Typical economic calculation for a solar system

Conventional energy costs	
Electricity	[€/kWh]
Heat	[€/kWh]
Solar air system costs	
Investment per system unity	[€/m²]
Lifetime	[a]
Operation and maintenance costs	[%/a]
System gain per unity	[kWh/m²a]
Interest rate	[%]
Annuity investment costs	[€/a]
Yearly maintenance costs	[€/a]
Total costs per year	[€/a]
Investment costs	[€/kWh]
Operation and maintenance costs	[€/kWh]
Total energy costs	[€/kWh]

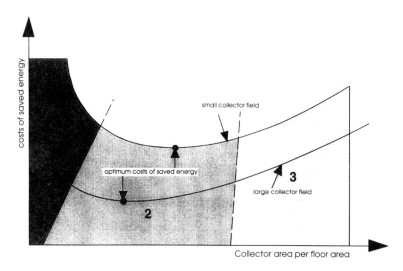

(1) High specific output
(2) Optimum costs of saved energy
(3) High solar contribution

Figure I.2.1. Optimum costs of saved energy

Environmental protection aspects

Life Cycle Assessments (LCA) study the environmental impact throughout a product's life. Some parts of the LCA are still at an early stage of development. The current state of the art is described in:

- ISO 14040: Environmental management – Life cycle assessment – Principles and framework
- ISO 14041: Environmental management – Life cycle assessment – Goal and scope definition and life cycle inventory analysis
- ISO 14042: Environmental management – Life cycle assessment – Life cycle impact assessment (Draft)
- ISO 14043: Environmental management – Life cycle assessment – Life cycle interpretation (Draft)

Impact potentials
Some substances strongly contribute to environmental imbalance through:

- enhancement of the greenhouse effect;
- depletion of ozone layer;
- photochemical oxidant formation;
- acidification;
- nitrification.

Parameters have been developed to 'score' the contribution of a product to the effect under consideration. For example, the Global Warming Potential (GWP) is a relative parameter which uses CO_2 as a reference: the extent to which a mass unit of a given substance can absorb infrared radiation compared with a mass unit of CO_2. If a LCA study of 'steel production' leads to a GWP of steel (low-alloy) of $2.9\,kg\,CO_{2equ.}/kg$ this means that throughout the product's life various greenhouse gases have been emitted, the average total contribution of which to the greenhouse effect is the same as that of $2.9\,kg\,CO_2$.

The potentials and reference substances of the other effects mentioned above are:

- ODP (Ozone Depletion Potential) – CFC-11
- POCP (Photochemical Ozone Creation Potential) – C_2H_4
- AP (Acidification Potential) – SO_2
- NP (Nitrification Potential) – PO_4^{3-}

The greenhouse effect of unglazed and glazed collectors
The total GWP of the solar wall, air collector materials and the fan can be considered as 'GWP investment costs', and likewise the GWP of the annual electricity consumption as 'annual GWP operating costs' and the GWP of the fossil fuels as 'annual GWP fuel savings'. Therefore the ratio in Figure I.2.2 can be regarded as a simple GWP payback period (Tables I.2.4, I.2.5 and I.2.6 and Figure

$$\frac{\text{GWP investment costs}}{\text{annual GWP fuel savings} - \text{annual GWP operating costs}}$$

Figure I.2.2

Table I.2.4 Collector GWP investment costs

Materials	$[kg/m^2]$	GWP* $[kgCO_{2equ.}/kg]$	$[kgCO_{2equ.}/m^2]$
Unglazed Solar Wall Collectors[1]			
steel	5.0	2.9	14.5
aluminium (100% recycl.)	2.7	1.3	3.5
zinc	0.3	5	1.5
black paint (polyester)	0.05	1.6	0.1
			19.6
Glazed Air Collectors[2]			
aluminium (30% recycl.)	7.12	9.1	64.8
steel sheet (galvanized)	11.45	0.59	6.8
glass	10.0	1.1	11.0
mineral wool	6	1.4	8.4
Supplies:			
aluminium (30% recycl.)			
supporting elements	3.15	9.1	28.7
			119.7
Fan			0.2

1 Conserval Eng. Inc., Bruce Hanson + SOLAR WALL International
2 Grammer Solarlufttechnik

*Source: Baustoffdaten – Ökoinventare [Hg. v. Institut für Industrielle Bauproduktion (ifib), Universität Karlsruhe; Lehrstuhl Bauklimatik und Bauökologie, Hochschule für Architektur und Bauwesen (HAB) Weimar; Institut für Energietechnik (ESU), Eidgenössische Technische Hochschule (ETH) Zürich; M.Holliger, Holliger Energie Bern: Karlsruhe/Weimar/Zürich, 1995]

Table I.2.5 Annual GWP Fuel savings

	Annual heat production $[kWh/m^2a]$	Natural gas GWP** $[kgCO_{2equ.}/kWh_{calorific value}]$	GWP savings $[kgCO_{2equ.}/m^2a]$
Unglazed Collector (best case)	755	0.2628	198.4
Unglazed Collector (worst case)	250	0.2628	65.7
Glazed Collector (best case)	450	0.2628	118.3
Glazed Collector (worst case)	120	0.2628	31.5

** including fuel production, delivery and incineration in typical furnaces (natural gas, low NO_x < 100 kW)

Table I.2.6 Annual GWP operating costs

	Annual electricity consumption of the fan [kWh/m²a]	GWP electricity UCPTE*** [kgCO$_{2equ.}$/kWh]	GWP [kgCO$_{2equ.}$/m²a]
Unglazed collector (best case)	38	0.54	20.5
Unglazed collector (worst case)	13	0.54	7.0
Glazed collector (best case)	30	0.54	16.2
Glazed collector (worst case)	15	0.54	8.1

***UCPTE = Union pour la coordination de la production et du transport de l'électricité (Belgium, Germany, France, Greece, Italy, Luxembourg, Netherlands, Austria, Portugal, Switzerland, Spain)

Unglazed collector (best case): $\dfrac{19{,}6 + 0{,}2}{198{,}4 - 20{,}5} = 0{,}1$

Unglazed collector (worst case): $\dfrac{19{,}6 + 0{,}2}{65{,}7 - 7{,}0} = 0{,}3$

Glazed collector (best case): $\dfrac{119{,}6 + 0{,}2}{118{,}3 - 16{,}2} = 1{,}2$

Glazed collector (worst case): $\dfrac{119{,}6 + 0{,}2}{31{,}5 - 8{,}1} = 5{,}1$

Figure I.2.3. Example calculations of simple GWP payback periods

I.2.3). Thus for all four cases the GWP payback periods are much shorter than the financial payback periods.

Thermal comfort

Open loop systems

Here, heated ambient air is injected directly into the rooms of the building. Comfort should be critical.

The inlet of the system in the rooms should contain diffusers, which reduce the speed of the incoming air. The air speed should not exceed 0.25 m/s in the comfort zone. In systems where ambient air is always circulated through the solar air collector the temperature of the incoming air may, at times, be lower than room temperature. Therefore, it should be possible to further heat the incoming air. For many diffusers a temperature of the incoming air of 3 K below room temperature will create draught problems, however some diffusers allow a temperature difference of up to 10 K.

Warm air heating, if not well designed, can result in uncomfortable draughts. Air velocities at entrance grills must not exceed generally accepted levels, and warm air should be supplied at a sufficient number of points to provide uniform conditions in the room. The incoming air may sometimes be too warm. In this case it should be possible to bypass the solar collector, especially in the summer.

Closed loop systems

These systems, for example a double envelope solar air system or a system with a hypocaust storage, improve comfort in winter, since the building envelope will be warmer than that of an ordinary building. Comfort can be achieved at a lower room temperature which will save energy. It is important that envelope surfaces, not covered by the double envelope (i.e. windows) have low U-values. Otherwise, they will increase the required temperature and could cause draught.

Hygiene

Open loop systems

An open solar air system is integrated into the building's ventilation system, and should therefore be designed following the same hygienic rules and standards.

As the system delivers heated fresh air to the building it is excellent for solving most of the problems associated with 'sick building problem'. However, it is very important that fresh air is delivered to the building. The inlet to the solar air collector should be located in areas of low pollution.

A filter located in the inlet to the solar air collector should trap dust, pollen, etc. A filter is especially important when using collectors where the airflow is above or through the absorber of the solar air collector, as dust etc. on the cover and the absorber will decrease the efficiency and change the appearance of the solar air collector. Occupants (especially those with allergies) will furthermore appreciate the clean air.

Cleaning or replacement of filters belong to routine maintenance. The air may be contaminated in the solar heating system if the materials used are toxic in some way, contain loose dangerous fibres or give off smell. Many types of plastic may not produce toxic gases but may give off an odour. Mineral wool should be encapsulated in order to prevent fibres from becoming airborne. Some paints may also cause problems at the high temperatures at which the systems operate. The performance of the chosen materials should not only be investigated at traditional temperature levels, e.g. at typical room temperatures below 25°C but also in the whole range of conditions to which the materials will be exposed.

Dust filters remove only solid and liquid particles from circulating air. Gaseous contaminants, cooking odours, noxious vapours, and humidity build-up can be reduced by ventilation, i.e. replacing contaminated indoor air with fresh outdoor air.

Fresh air is naturally incorporated in solar air systems by outdoor air leaking into solar collectors through joints and connections, mixing with solar heated recirculated air, and being delivered to rooms during the direct solar-to-rooms cycle. In some systems the fresh air fraction is increased by the intentional introduction of outdoor air into the suction side of the solar fan and/or distribution fan.

Experience with many rock bed installations does not indicate any problems with growth of moulds, algae, insects, fungi or other life forms. The highly fluctuating temperature in the rocks, often with variations as high as 60–70 K, and the extremely dry conditions at those temperatures, prevent contamination. In hypocaust storages connected to open systems access to allow cleaning should be provided. For open discharge systems it is important to check the origin of the materials used in the storage, i.e. if the rocks used emit radon.

Closed loop systems
In a closed loop, the air in the solar system does not pass through the occupied space, so there are no special hygienic considerations. Filters are recommended between storage and air collector to avoid dust in the collector. This may not be necessary if the storage is kept clean during construction.

Semi-closed systems
If the system is open or 'semi-closed' (i.e. if a sunspace is used), the duct system must allow regular cleaning and higher humidity must be planned for.

Acoustics

Forced warm air heating, whether conventional or solar-supplied, generates low-level sound. At grille openings into rooms, air friction against metal grids can be audible if velocities are excessive. Quiet operation can be achieved by proper design. The collector fan and distribution fan can be located in an equipment room separated from living areas by walls and floors. The acoustic problems of solar air systems may be solved using conventional HVAC knowledge.

A special problem occurs in systems where the heated air is admitted to an adjacent room with the fan situated in the wall between the collector and the room (System 1). Here it is not possible to decrease the noise by traditional noise-reducing ducts since the ducting is nearly non-existent. In this case special low noise fans should be used or the fan should run at a low speed. Fan manufacturers provide nomograms to assess noise levels. The noise should not exceed 30 dB 1 m from the inlet to the room in dwellings. The noise in larger volume buildings may be higher, e.g. 60 dB.

In closed loop systems the circulated air does not pass through the building, reducing low frequency fan noise. Noise will be further reduced if the fans and the duct system are kept from direct contact with the building's structure. Since noise causes most problems in housing during the night, it is advantageous if the fans do not run at night (i.e. passive discharge system).

If a system requires quite high air flow rates (e.g. System 6) noise can be generated in the air ducts, dampers and heat exchanger. Also the fan will generate noise. Occupied spaces must be isolated from this noise. By increasing the cross-section of the ducts, the air speed can be reduced. To avoid noise, the air speed should be kept lower than 4–5 m/s.

SYSTEM SELECTION (LAYOUT PHASE)

During this phase system layouts are developed using rules of thumb and energy performance curves.

Integration of the solar system in the building construction should be evaluated to doubly exploit building material and labour. Such considerations early on can improve the economy of the solar system.

The six generic system types described in this handbook differ significantly in thermal as well as building integration opportunities. Not all are suitable for a given application. Each system chapter has a section on suitable applications.

SYSTEM DESIGN (DESIGN PHASE)

Here the solar air system is designed in more detail. It is important that the function of the system including day/night and seasonal modes are calculated. This is also the moment to make an economic estimation for the solar project and the building project, separately and together.

The final system design may be simulated, using a dynamic computer simulation program (i.e. Trnsair). The simulation will show the dynamic behaviour and the energy performance of the solar system and also indicate the need for corrections.

INSTALLATION, COMMISSIONING, OPERATION AND MAINTENANCE

Installation

During the system installation there can be problems involving contractors and services. The energy producing and distribution systems are usually installed by a ventilation company. There may occur coordination problems between the contractors responsible for the building envelope and the ventilation system. If the heat storage is controlled by the heating system, it may be unclear who is responsible for which work. Therefore it is recommended to clearly define the role of each company before construction begins. Otherwise there may be a risk that the contractors will deny responsibility when later it becomes clear that there are problems and consequently nobody will rectify the defects. It is also advisable to make sure that all involved contractors fully understand the concept of the solar air heating system. This will increase their interest and improve the quality of their work.

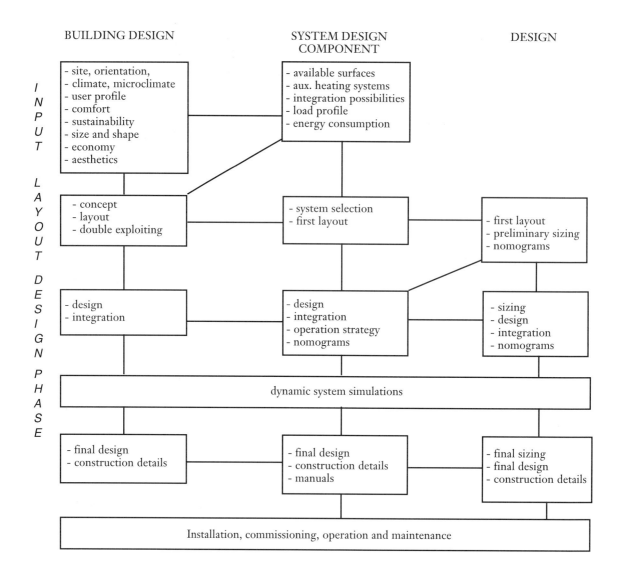

Figure I.2.4. Building design system flow chart

During construction, precautions should be taken to prevent dust from getting into the air cavities of the components. Do not perform dusty work like mounting plaster boards, cutting tiles or insulating with mineral fibres if the solar air components are still open to the ambient. An open air collector will draw in large amounts of dusty air on a sunny day. Also, hypocausts and rockbeds may accumulate dust during construction.

Commissioning

Follow a traditional HVAC commissioning procedure and use a handing-over certificate. The test procedure should check:

- mounting and connections of the collector field;
- the resulting air flow rate, important because very often the pressure drop is larger than obtained from calculations;
- the air-tightness of critical parts of the system, i.e. solar air collectors;
- the on-off sequences of the different thermostats;
- the noise level of the fans and possible noise from the air flow in the ducts;
- at the conclusion of the commissioning, the owner should be given an instruction manual including operation and maintenance instructions from the component suppliers.

Often the actual load is different from the design load. Temperatures or the total load might differ. It is therefore highly advisable to monitor the system, the load and the performance for the first year of operation. Adjustments of the system controls and air mass flow rates may be necessary for optimal performance.

Operation and maintenance

The operation and maintenance of solar air systems is similar to that of traditional systems. Filters should be changed, the duct works should be cleaned and the operation of dampers and fans should be checked, all at regular intervals, described in the user manual.

During summer some high performance systems might prove to be oversized for the domestic hot water load and the temperatures of the storage may reach 80–95 °C. To avoid excessive electricity consumption for the fan, it should be switched off, and the collector left idle. In a closed loop system there is no need to vent the collector, as it should be constructed to withstand stagnation temperatures without problems and emissions. Collectors and supporting construction should be made of non-combustible materials. Open systems might just be ventilated with ambient air.

Collectors should be checked annually for cracks in the cover, dried out joint sealants, dust on the absorber, etc. Likewise the insulation on the back of the collector and on the duct system should be checked, if possible. The covers of glazed solar collectors should from time to time be cleaned, more for aesthetic than for performance reasons. If the collector is roof-mounted the rain will probably keep the cover clean.

In an open system, the solar collector air channels will become dirty after some years and need to be cleaned. This is very important, especially if the system is connected to the intake of fresh ventilation air. In this case, the solar collector should be regarded as a conventional ventilation system with the same requirements for maintenance. Filtration of incoming air is desirable. It is also necessary to check the ducts for moisture which can create problems with odours and rust.

In a closed system, the air will be used only to transport heat from the collector to the air space and there will be no mixing with the ventilation air. Since the air inside the collector will have no contact with the outside air, it will stay clean and the need for maintenance will thus be very low.

The functioning of fans, dampers (including frost dampers), the control system and sensors should periodically be checked. On-site inspection is needed. For example, the control system might report that a fan is running, without this actually being the situation. Maintenance is normally limited to periodic lubrication of fans and damper motors, cleaning or replacing dust filters. Fire detectors must be cleaned according to the manufacturer's instructions.

DESIGN PROCESS - STEP BY STEP

This chapter concludes with a flow chart (Figure I.2.4, facing page) describing the phases of solar air system design.

BIBLIOGRAPHY

Fanger PO (1972) *Thermal Comfort*, McGraw-Hill, New York.

Labhard E, Binz A and Zanoni T (1995) *Renewable Energies and Architecture - Relevant Aspects in the Design Process (Erneuerbare Energien und Architektur [Fragestellungen im Entwurfsprozess - Ein Leitfaden])*. Bundesamt für Konjunkturfragen Bern, EDMZ Best. Nr. 724.251D; also available in Italian.

Chapter author: Ove Mørck
Co-authors: Christer Nordström and Charles Filleux

II. Systems

II.1 System 1: Solar heating of ventilation air

INTRODUCTION

Solar air heating of ventilation air is a simple and cost-effective application where fresh air to a building is heated in a solar air collector. The concept has proven to be very successful. Many commercial products are available including 'off the shelf' systems. Available systems range from small units for dehumidification/heating of holiday cottages to more than 50,000 m² installed in industrial buildings to provide solar heated fresh air. The payback time, especially for large applications e.g. industry, is very short, often a couple of years. If the installation of the system is carried out as part of the renovation of a wall or a roof the extra cost of a solar heating system compared to a traditional facade or roof will be small.

The performance of such systems is high, as solar heated air below room temperature is also valuable. The systems will thus save energy even on cloudy days. Because the systems deliver heated fresh air to the buildings they may eliminate problems often associated with 'sick building syndrome'.

System description

Figure II.1.1 shows the reference configuration of System 1. It consists of a flat-plate wall-mounted solar air collector connected to the building via a duct system. There may or may not be a fan in the system. The system consists of the following components:

- collector
- duct system
- diffusers
- optional fan, filters, dampers and noise reducers.

System variations

Several variations of the systems exist according to the type of collector, where the heated air is delivered to and if the system includes storage.

Collectors
Flat-plate collectors, window collectors, perforated collectors, double facades and spatial collectors may be used in this type of system.

Distribution
The ventilation system connected to the collector may be identical for all types of collectors. The air from the solar air collector may be:

- Delivered directly to an adjacent room: the collection may either be 'active' i.e. fan-driven or 'passive' i.e. with no fan, but e.g. driven by the underpressure created by exhaust ventilation of the building.
- Ducted to a remote room. Heated air is delivered by ductwork to spaces further away from the collector, in particular to north-facing colder rooms. This is a good solution if the direct solar gain is high through the windows of the room behind the solar air collector. This variation will in most cases be fan-driven.
- Distributed to a central mechanical ventilation system. Heated air is delivered to all the rooms connected to the mechanical ventilation system of the building.

Storage
The solar air system may be with or without storage:

- Without storage: The heat from the system is delivered to the building without delay. The air flow may be active- or passive-driven.
- With storage: The delivery of heat to the building is delayed and the delay increases with increasing

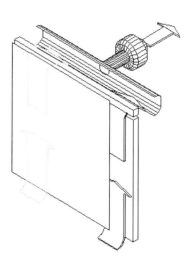

Figure II.1.1. Reference system of System 1

storage. This is beneficial if there is a large direct gain through the windows of the room(s) to which the air is delivered and heat is, therefore, not needed before the evening/night. The storage may be with either active or passive discharge. In active discharge an air stream through the storage discharges the stored heat in the same way as it is charged. In passive discharge the stored heat is mainly discharged by radiation and convection from the external surfaces of the storage.

Storage is often not beneficial if the room temperature is deliberately lowered at night (setback) or when the building is unoccupied and fresh air is not required.

Control

The way the system is controlled is not a system variation as such, but may strongly influence how the systems perform. Control options include:

- continuous operation
- temperature control
- solar cell control
- timer control.

Storage and control

Not all combinations of storage and control strategies are reasonable. In order for active discharge storage to be useful the building must have a heat demand in the evening and there must be an air flow through the storage at least in the evening. The latter is not the case in systems with temperature and solar control, but may be the case with time control, if the ventilation system is allowed to run in the evening.

Passive discharge storage may be combined with all the control strategies. However, as with active discharge storage the building must have a heat demand during the evening/night in order to profit from the stored heat (Table II.1.1).

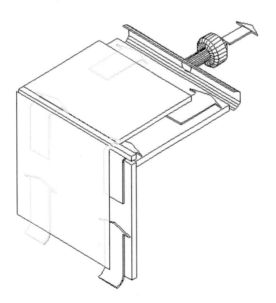

Figure II.1.2. System 1 variation with a hypocaust

Table II.1.1. Reasonable combinations of control strategy and storage type

Control strategy	Storage	
	active discharge	passive discharge
Always running	possible	possible
Temperature or solar control	not recommendable	possible
Time control	maybe	possible

Advantages and limitations of the system

ADVANTAGES

- Very simple and inexpensive especially if installed as a replacement to existing ventilation systems. In that case the ducting and fans as well as the electricity are already paid for.
- When heating ventilation air, temperatures below room temperature are also useful. A principal difference between System 1 and Systems 2–6 is that small temperature increases and low absolute temperatures are adequate to preheat fresh air for a building. There is no demand that specific temperature levels have to be reached i.e. >20°C for space heating. The building has to be supplied with fresh air, so if the ambient temperature is –10°C and the temperature increase across the collector is 5 K (leading to an absolute temperature out of the collector of –5°C) the gained heat is still useful. If the incoming solar heated air, however, is warmer than the room temperature, part of the building transmissions heat losses will also be covered.
- As small temperature increases are valuable, there is generally no restriction on the collector types. Covered and uncovered, simple (e.g. perforated collectors) and advanced collectors but also spatial collectors (sunspaces, atria or roof space collectors) may serve as a collector in a solar air heating system for heating of ventilation air. A solar air collector may be integrated into the facade or the roof as part of a renovation of the building.
- The system consists of standard ventilation components (the engineering is close to 'business as usual' for the HVAC engineer, allowing less expensive and more reliable systems). Problems of noise, filtration of the air, discomfort due to draught, etc. can be solved in the same manner as for traditional ventilation systems.

LIMITATIONS

- A solar heating system coupled to a ventilation system with heat recovery will decrease the performance of the heat recovery unit, as the inlet air to the heat recovery unit will be warmer. This, on the other hand, means that the heat recovery unit may be smaller (less expensive) and that freezing of condensed water will occur less often. The performance of the total system will be higher than the performance of a ventilation system with only a solar air collector or only a heat recovery unit.

- As fresh air to the building is heated directly in the solar air collector the materials of the collector should be non toxic, contain no dangerous fibres, be odourless and inflammable at stagnation temperatures (see Chapter V.4 and local fire codes).

Suitable applications

INDUSTRY

Many industry buildings have a large demand for ventilation air. They have the highest potential for heating ventilation air by solar energy. The solar heating system may be connected as a preheater to a central mechanical ventilation system and controlled by this or possibly a separate system in combination with exhaust ventilation, and will most often be controlled by a timer. Over- and under-pressurization of industrial buildings will probably not create serious problems. The solar heating system may be controlled by temperature either on/off or variable flow, or solar cell control, or by a timer or even permanently running if necessary. Systems for industries are generally very simple and cost effective.

DWELLINGS – APARTMENTS, ROW HOUSES AND SINGLE FAMILY HOUSES

When combined with a *central mechanical ventilation system* the solar air collector acts as a preheater to the ventilation system and will most often be running at all times. However, a timer may reduce the air flow rate at times when less ventilation air is required. Storage may be integrated in the system to delay the heat injection until the evening.

Exhaust ventilation

The solar heating system has its own fan or no fan in the case of a passive system. Where a fan is present the solar heating system will most often only be running during the day. At night when the fan of the solar air system is off, the exhaust ventilation system may maintain a lower flow rate of air through the solar heating system. The system may be time-, temperature- solar cell-controlled or solar cell driven. An active discharge storage is not sensible due to the low or no flow rate during the night.

Dehumidification

These buildings typically have no other form of ventilation system. In order not to force humidity into the construction of the building, the solar air system only runs when it is able to deliver warm and dry air. The system is, therefore, controlled by temperature, a solar cell or is solar cell driven. Storage is not recommended.

OFFICES, SCHOOLS, ETC.

Central mechanical ventilation system and exhaust ventilation

If a ventilation system is present it will typically only be running during daytime. There is most often a demand for a constant flow rate of ventilation air in order to maintain a good air quality. The control of the solar heating system is thus either via the central mechanical ventilation system if such is present or by a timer. As the buildings are normally only occupied during the day storage is not profitable.

SPORTS HALLS

The operation is similar to offices, schools, etc., but storage may be considered, as many sports halls are often occupied until late evening. However, the room temperature is often low.

SWIMMING HALLS

The operation is similar to offices, schools, etc. Due to higher air change rates because of the humid conditions and a higher room temperature the benefit of the systems is not limited to the heating season. The performance of the systems is, therefore, high.

Energy performance

To give a first impression of the performance of the system, the performance of reference systems for large volume buildings (industry, warehouses and sports halls) and dwellings (apartments, row houses and single family houses) are shown here.

Reference systems

The reference system of System, heating of ventilation air, has the following characteristics:

- tilt of collector: 90°
- azimuth of collector: 0° (south)
- control strategy: the air is always drawn through the collector except if the temperature of the building exceeds 23°C
- there is no storage in the system
- the heat from the system is delivered equally to all parts of the building
- the heat loss from the ductwork is considered as useful, i.e. the ductwork is located inside the thermal envelope of the building
- recovered heat loss through the wall on which the collector is mounted is not included in the shown savings
- fan power dissipated in the air stream is not included in the savings
- no heat recovery

Specifically for large volume buildings:

- wall-mounted perforated collector (see Chapter IV.3)
- flow rate of air through the collector: 21 m^3/m^2_{floor} h (equal to three air changes per hour) for industry and 3.5 m^3/m^2_{floor} h (equal to 0.5 air changes per hour) for sports halls/warehouses

Specifically for dwellings:

- wall-mounted high performance collector, see Chapter IV.1

- flow rate of air through the collector: 1.2 m³/m²$_{floor}$h (equal to 0.4 air changes per hour)

Energy performance of reference systems
Figure II.1.3 shows the saved energy for large volume buildings and Figure II.1.4 shows the saved energy for dwellings – both as a function of the collector area per heated floor area. Nomograms appear later in this chapter for different collectors, air flow rates through the collector and heat demands of the building. The influence of fan power, storage in the system and the saved heat loss through the construction on which the collector is mounted may also be evaluated.

Control system variation performance
The performance of the systems using different control systems may be evaluated using the nomograms. Here is a brief ranking of the performances:

- *Always running* will have the highest performance, but should, however, only be applied if night ventilation is required. Storage may in some cases increase the performance further.
- *Time control* may have the same high yield of solar energy as the always running system if the system is allowed to run from sunrise to sunset. A later start and earlier stop will decrease the performance.
- *Solar cell control: On/off control* is designed for operation only under sunny conditions and does not make use of low level solar irradiation since the systems are typically switched on and off at irradiation levels in the range of 100 to 300 W/m². The performance will thus be lower than that of an always running system. *Solar cell driven fan:* variable fan speed with high airflow at high solar radiation. The flow rate of air through the collector at low irradiation levels leads to a lower efficiency of the solar air collector, while having a higher efficiency at high radiation levels.
- *On/off temperature control* performance depends on the chosen set point for starting and stopping the airflow through the collector. Increasing starting and stopping temperatures lead to decreasing performance, as preheating during cloudy conditions is lost. This control may be preferred if comfort is the issue.

In order to maintain a certain temperature of the air out of the collector a variable speed fan (or a mixing damper) may be used, i.e. low air flow through the collector at low

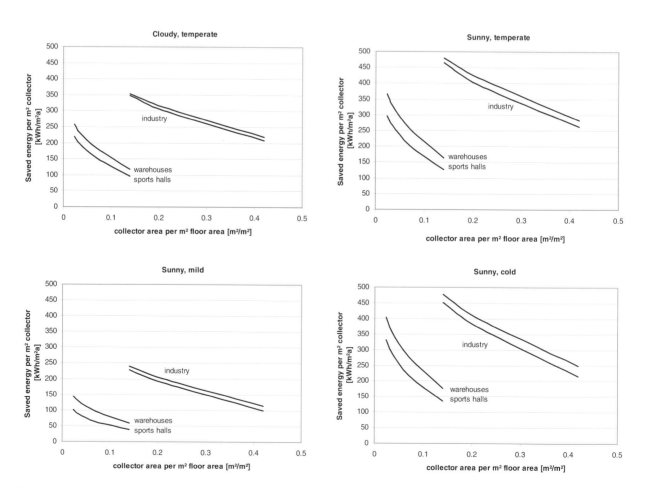

Figure II.1.3. Energy performance of the reference system for large volume buildings

irradiation level and high air flow at high radiation level leading to saved electricity to the fan. Variable speed fans are similar to solar driven fans.

Energy performance of built examples
Table II.1.2 shows the energy performance of several built examples[1] of System 1. The energy performance data in Table II.1.2 should only be used as indicative as the values have been taken out of their context. The performance of the systems is strongly influenced by the heat and ventilation demands of the buildings, the control strategy of the systems and the air speed in the collectors.

Design

System design

The main design parameter for System 1 is not the heat demand of the building, but the necessary ventilation rate. The heat demand of the building influences the performance of the system but does not really need to be known in order to design the system.

The air change in a building with any kind of mechanical ventilation system consists of air change due to the ventilation system and air change due to infiltration

Table II.1.2. Energy performance of built examples of System 1

Building type	Collector type	Collector area m²	Location	Climate	Energy performance kWh/m²$_c$
Industry	Glazed, non-selective	126	Helsinge DK	cloudy, temperate	76
	Unglazed, perforated	611	Montreal C	sunny, cold	705
	Unglazed, perforated	725	Colorado Springs USA	sunny, cold	810
Residential	Glazed, non-selective	5	Struer DK	cloudy, temperate	215
	Glazed, fibre cloth	1.28	Slette-strand DK	cloudy, temperate	227
	Unglazed, perforated	335	Windsor C	sunny, cold	584
Schools	Roof space collector	112	Newport Pagnell UK	cloudy, temperate	50
	Glazed, perforated	177.5	Koblach A	cloudy, temperate	254
Offices	?	59	Karlsruhe D	cloudy, temperate	85

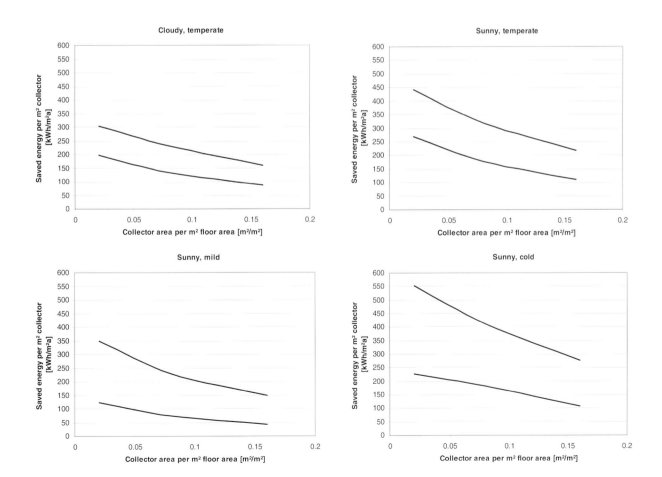

Figure II.1.4. Energy performance of the reference system for dwellings

(cracks in the building envelope induced by wind and stack effects and opening of doors and windows). The solar air heating system can only heat the ventilation air.

The maximum air flow through the solar air collector of a System 1 is thus the total air change of the building minus infiltration.

SYSTEM 1 DESIGN PROCEDURE

1. Define necessary basic data about the building and climate

- collect building data
- define the net heat load of the building
- determine the climate
- define the energy goals of the system

See also Chapter I.2.

2. Determine if it is possible to obtain enough collector area
Determine how large a collector it is possible to locate on the building. Unbroken and south-facing facades and roofs are preferable. Choose unshaded location. From Figures II.1.3 and II.1.4 it is possible to roughly size the collector.

Consider if part of the buildings may serve as a collector, i.e. a sunspace, an atrium or a double facade, or if a construction can easily be converted into a collector, i.e. a facade or the roof. If the facades or the roof have to be renovated anyway the extra cost of a solar air collector may be negligible. Minimize the mounting cost and length of ducts.

3. Determine ventilation rate through the solar air collector
Determine the necessary ventilation rate of the building. Determine if all or only a part of the ventilation rate of the building may be heated by the solar heating system. Further, determine to where the heated air should be directed: to a room just behind the solar collector, to remote rooms or to a central mechanical ventilation system. Determine the flow rate through the solar air collector accordingly.

4. Determine whether there are restrictions on the inlet temperature from the ventilation system
Does comfort dictate a minimum or maximum inlet temperature? In the former case the air flow through the solar air collector may first be started when the desired temperature level is reached, leading to decreased performance of the system. Or the air temperature may be raised to the desired level either by the exhaust air in a heat recovery unit or by an auxiliary heater.

High ventilation inlet temperatures may be prevented by mixing ambient air with heated air or by stopping the air flow through the collector.

5. Investigate if it is appropriate to include storage in the system to delay heat delivery to the building
Does the delivered energy from the solar heating system compete with large direct gains through windows? Is heat from the system useful during the evening/night? Storage is normally not feasible in a building with night setback of the temperature or unoccupied buildings.

Determine the storage capacity using the information/nomograms in Chapters IV.6–IV.8. The main sizing parameter is the length of the time delay.

The appropriateness of storage in the system may be determined assessing the cost of storage vs. the increase in energy savings. The latter may be estimated using the nomograms in Nomograms below, or may be calculated using Trnsair.

6. Define the needed control strategy
Define the control strategy based on the decisions made in steps 3, 4 and 5. Guidelines for choosing components for the control system may be found in Chapter V.3.

Control strategies:

- Air is always directed through the system.
- The system is controlled on the inlet temperature.
- The system is controlled on the irradiation level.
- The system is controlled by a time switch.

7. Choose a solar collector
Choose the collector by use of the information in Chapters IV.1–IV.5. Choose the collector based on the temperature level from 4, the performance of the system and the price of the collector. The performance may be evaluated using the nomograms in Nomograms below or by e.g. Trnsair simulations or simulations with Swift in the case of large volume buildings (Appendix I).

8. Investigate if the system may serve other purposes
Consider if preheating domestic hot water may improve the payback of the system.

Heated air may be directed to an air-to-water heat exchanger as in Chapter II.6. The air-to-water heat exchanger may be sized using the information in Chapter V.2. Alternatively a liquid absorber may be integrated in the solar air collector serving both as absorber for heating air and liquid. The latter configuration will have a higher performance than the solution with a heat exchanger. The liquid loop may be designed using techniques for traditional liquid-based solar air heating systems.

9. Determine the collector area
Determine the necessary collector area based on the chosen collector in step 7, the flow rate of air through the collector from step 3, any restrictions on the inlet temperature from step 4, the performance from the nomograms given in Nomograms below (or calculations with e.g. Trnsair or Swift) and the cost of the collector.

An increase in area will lead to a higher total yield, but a decrease in performance per m² collector area, an increase of the air temperature out of the collector and a higher cost for the system. Determining the optimal collector area is thus an iterative process.

If a simple, inexpensive collector is chosen the extra price of the collector may be negligible compared to traditional external wall cladding. In this case the collector should be as large as possible with respect to the required air flow. If, on the other hand, an expensive collector has been chosen, it is very important precisely to find the optimal collector area.

10. Dimensioning of ducts
Dimension the ducting using traditional HVAC procedures based on the decisions made under step 3 about where the air should be delivered. The duct system should preferably contain a filter and noise reduction. The dimension of the ducts is based on the flow rate of air in order to obtain minimal pressure drop across the duct system seen in relation to the cost of the ducts. The air speed in the ductwork should normally not exceed 4–5 m/s.

If the ducts are located in unheated rooms where the heat loss from the ducts may not be useful or the ducts are located in rooms where the heat loss is unwanted, the ducts have to be insulated. The level of insulation should be two or three times more than for traditional ventilation ducts; 100 mm is recommended.

11. Choose a fan
The fan is located on the hot side of the solar collector. The fan must withstand the maximum outlet temperature of the collector. This may be evaluated using the information in Chapters IV.1–IV.5.

Consider if the system may be connected to an existing mechanical ventilation system and thus share the fan with this system. The fan should be dimensioned based on the needed flow rate and the pressure drop across the system (collector + filter + ducts + maybe storage); see Chapter V.1. A fan with a low power demand should be chosen.

A quiet fan is essential for systems where the air is blown directly into the room behind the collector. The noise should in dwellings not exceed 30 dB 1 m from the inlet. In large volume buildings, e.g. industry, the noise may be considerably higher but should not exceed other noise sources.

12. Choose diffuser
Choose diffusers of the inlet based on the flow rate and any restrictions on maximum/minimum temperatures of the inlet air in order to avoid any discomfort from draughts. This is traditional HVAC knowledge.

Component design/specific design advice
COLLECTOR
The main difference between the operation of collectors for heating of ventilation air and collectors for space heating is that the room air is not recirculated through the collector. The inlet temperature of the collector is the ambient temperature. This means that the efficiency is to be found close to the y-axis on the traditional curves (based on ΔT) shown in the collector chapters, Chapters IV.1 and IV.2. The efficiency of the collectors is, therefore, mainly dependent on the air flow rate through the collector, see Chapters IV.1–IV.3.

The characteristics of solar air heating of ventilation air allow more freedom in selecting the type of collector. The heat loss from the collector is less important because of the rather low temperature level, while the solar transmittance of the cover and the heat transfer between the absorber and air are of major importance. Accordingly inexpensive, uncovered and poorly insulated solar collectors may be of interest for these kinds of systems. Perforated collectors without a cover, where the air is drawn in through the absorber, are an example. The incoming air (see Chapter IV.3) captures the heat loss from the absorber. This type of collector generally uses higher air flow rates per m² than covered collectors in order to be efficient.

Traditional building components may also be used as collectors. Warm air may be taken from atria and sunspaces, through window collectors, from attics (with a traditional or transparent roof) or from beneath the roof elements (especially if the roof is made of metal sheets), etc. In this case the cost of the 'collector' is negligible given its other, primary use.

When using spatial collectors, such as sunspaces, atria or double facades as heat sources for a solar air heating system, one should realise the limiting constraints for these types of 'collectors'. This is further discussed in Chapters IV.4 and IV.5.

STORAGE
Storage may be useful if the heated air is delivered to a room with considerable direct solar gain through windows. It can decrease overheating by delaying the heat release until the evening when the room starts to need heat. The heat from the collector may always be delivered to the storage or may first be delivered directly to the room and later on to the storage if the room temperature reaches a predefined level.

Suitable types of storage include hollow core floor slabs, walls with ducts, rock beds, latent heat storage or the room itself. If the walls of the room are thermally heavy they may passively absorb so much heat that this smoothes out the temperatures of the room.

Storage may be designed with passive discharge as in System 4, or it may be designed with active discharge. In the latter case it is necessary that the ventilation system is always running. If active discharge storage is designed to have a delay so that the heat is delivered in the evening, there has to be an air flow through the storage not only during the day, but also during the night.

Storage may also be integrated in the collector. This is often the case in ventilated solar walls for heating fresh air. Here a simple transparent cover is mounted in front of the existing wall, which is painted black and acts in this way as the absorber. If the existing wall is concrete this will decrease the maximum outlet temperature of the collector and delay the peak value a few hours after the peak solar radiation. As the storage in this type of solar air collectors is only insulated to the ambient by the cover much of the stored heat may be lost to the ambient.

Storage may be dimensioned using the guide-lines in Chapters IV.6–IV.8.

DISTRIBUTION

The distribution system normally uses components from traditional HVAC systems. The solar air heating systems may in fact share the ducting including fans with the ventilation system of the building. On the other hand, the solar heated air may also be distributed via the rooms of the building or by building constructions, e.g. via hollow floor slabs which then also will act as storage in the system.

The ducts do not need to be insulated if the heat loss from them is useful. There is no point in insulating the ducts if they are located in the room to which the heated air is admitted. If insulation is necessary, e.g. if the ductwork goes through unheated attics, the insulation should be two to three times better (100 mm is recommended) than in traditional HVAC systems.

The fans will traditionally be located on the hot side of the system. This means that the fan must withstand the maximum temperatures of the system. If the fan is located after storage in the sequence there will normally be no risk of the fan overheating.

The fan may be shared with a ventilation system, in which case the electricity consumption of the fan will only have minor influence on the economy of the solar air heating system. The pressure drop of a solar air collector will normally be marginal compared to the pressure drop of a ventilation system. A larger fan will, therefore, often not be required.

In solar cell driven systems the fan should preferably have a low voltage DC motor since the conversion to 220 VAC requires an inverter. The installation of an inverter will increase the cost of the installation and decrease the amount of electricity available for the fan from the solar cells due to the loss in the inverter.

The system should admit ambient air directly to the building, bypassing the collector in the summer, if mechanical ventilation is needed then. The heat from the collector may instead be directed to an air-to-water heat exchanger for heating of domestic hot water. In this case the system is a variation of System 6 (see Chapter II.6).

If the system is not continuously running it may be necessary to include dampers in order to prevent any unwanted draughts through the system.

The distribution in large volume buildings may be rather simple, only consisting of a fan and a perforated duct, which also serves as diffuser. Some examples are shown in Construction below. If the perforated ducts are located just under the ceiling in high ceilinged rooms the warmer air under the ceiling will preheat the cold incoming air and in this way prevent draught. This further leads to a lower air temperature under the ceiling and thereby a decrease in the heat loss through the ceiling. The savings in auxiliary energy due to this destratification may be as much as the energy saved by the solar collector.

CONTROL

Rather simple controls can be used in solar air heating systems for heating ventilation air.

No control

This is the case in a 'passive' system where the solar collector or spatial collector is located in front of a room with no fan in the system. The collector will then decrease the infiltration losses of the room. If there is exhaust ventilation of the room the benefit of the collector will be increased. Due to under-pressurization of the building exhaust ventilation will increase the flow rate of air through the collector compared to pure 'passive' systems.

Always running

The only control that may be necessary is a collector bypass to prevent overheating; this may either be a manual or an automatically operated damper. Where a heater is installed in the ducting in order to obtain a certain minimum temperature, the heater power should decrease with increasing solar air temperature.

Temperature control

Thermostats: When using thermostats the system is started and stopped at certain temperatures. For example, the system may start at a collector temperature of 20°C and stop at 18°C. Furthermore it should be possible to stop the system or bypass the collector if the temperature of the room exceeds a certain maximum temperature. Thermostats may further be used to change flow rate through the collector as a function of the collector temperature in order to decrease the maximum temperature of the air out of the collector or to maintain a constant temperature out of the collector. Temperature control decreases the performance compared to an always running system, as small temperature differences in cold periods are not made use of, but may be preferred when comfort is an issue.

Differential thermostats: If the room temperature is constant the use of a differential thermostat is identical to the use of a thermostat except that the system starts and stops on certain temperature differences between the collector and the room.

Solar cell control

The system may either be started and stopped at a certain radiation level or the fan may be driven directly by a PV-module. In the latter case the air flow through the collector will vary over the day and be at zero during the night. The advantage of a solar driven fan is that the outlet temperature of the collector has a smaller amplitude than with a fixed flow rate through the collector. The temperature at low radiation levels will thus be higher, leading to smaller problems with draughts. A DC fan may directly be connected to the solar cell panel. If, however, a fixed ventilation rate of a room is needed solar driven fans are not a good choice unless backup power from the grid is available.

Timer control
The system is switched on at a certain time in the morning and switched off at a predefined time in the evening. A timer may also be used to introduce a low flow rate during times where less ventilation is required. This will save energy, both space heating and electricity to the fan.

Combination
In an always running system there is often a need for a thermostat controlled bypass of the collector. If a solar air collector is combined with a ventilation system with a heat recovery unit it should preferably be possible to bypass the heat recovery unit if the temperature of the air from the solar air collector is higher than the exhaust air from the building, as heat otherwise would be lost to the exhaust air. The bypass should be controlled by a thermostat.

Variable air flow through the collector and additional heating of the fresh air may be achieved by a mixing damper located between the collector and the fan. The purpose of the mixing damper is to mix fresh air with room air. The more room air released into the stream of fresh air (recirculation of room air) the lower the flow rate of fresh air. This may be desirable in periods when the collector delivers fresh air much below room temperature. Check with the local authorities if recirculating of room air is allowed. The flow rate of fresh air may of course never be less than what is necessary to maintain sufficient air quality inside the building.

Do's and don'ts
Do's

- Set the controls to allow for longer runtime (yields higher performance).
- Use high-temperature resistant and low-power fans
- Decrease the noise generated by the fans.
- Use diffusers and additional heating of the incoming air to decrease draft problems.

Don'ts
- Don't increase the air exchange rate of the building to increase the efficiency of the collector. This most often increases the heat loss from the building beyond the benefit of the increased performance of the solar heating system.
- Storage is generally not profitable in buildings with night setback.
- Don't locate the inlet to the collector in polluted or noisy areas.

Construction

The construction may differ from a traditional ventilation system, only in that the inlet of fresh air to the ventilation system is taken through a solar air collector. Simple solutions have been developed for large volume buildings as illustrated in the following examples.

Large volume buildings – industry, warehouses, sports halls, etc
Large volume buildings seldom need sophisticated ventilation systems. These buildings often have exhaust ventilation and high spaces where the warm air tends to accumulate under the ceiling leading to high heat losses.

The above situation may easily be improved by introducing a very simple version of System 1 consisting of a solar air collector, a bypass, a fan and a diffuser as shown in Figure II.1.5. Solar heated air is blown into the room via a perforated duct or textile tube. The incoming air enters the space at low speed and thereby avoids discomfort due to down-falling cold air, as fresh air is mixed/heated by the warm air located under the ceiling. The bypass is used during the summer period where heating of the fresh air is undesirable.

Small systems will often be single systems while large systems may consist of several subsystems as shown in Figures II.1.6 and II.1.7.

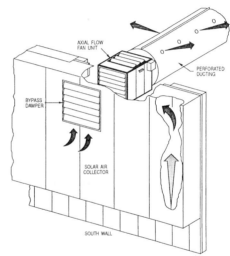

Figure II.1.5. A simple but efficient solar air heating system for preheating of fresh air to workshops, production facilities, warehouses, sports halls, etc.[2]

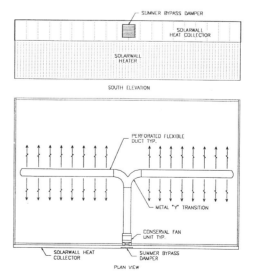

Figure II.1.6. An example of the system installed in a small building[2]

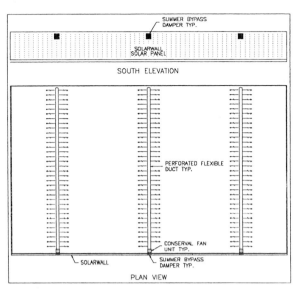

Figure II.1.7. An example of the system installed in a large building[2]

Nomograms

Two sets of nomograms have been developed: one set for large volume buildings like industry buildings, warehouses and sports halls and one set for dwellings, apartments, row houses and single family houses.

The performance values shown in the monograms in Figures II.1.8–18 were computed on the basis of the specific building and system configurations described in Appendix H. The diagrams are normalized to facilitate their wider application. However, if the imput values to the diagrams diviates too far from the values used to generate the curves, results may be misleading. For such cases, a full simulation is essential, i.e. using Trnsair.

Assumptions
The following simplifications were necessary, apart from the general assumptions in Appendix H.

GENERAL
- Heated air is delivered equally to both the south and north side of the building. The solar collector is connected to a central ventilation system, or an exhaust fan is located in the north-facing part of the building, and will in this way draw air from the south-facing part of the building and the collector due to under-pressure of the building.
- Any heat loss from the ductworks is considered useful, i.e. the ductwork is located inside the thermal envelope of the building.
- The ventilation system has no heat recovery unit.
- Only solar collectors with small heat capacity are considered, i.e. not solar walls where, e.g., a concrete wall may act as an absorber.

SPECIFIC FOR LARGE VOLUME BUILDINGS
- Two building types: light industry and warehouses/sports halls.
- Ventilation rates: 21 m^3/hm$^2_{floor}$ (equal to three air changes per hour) for light industry and 3.5 m^3/hm$^2_{floor}$ (equal to 0.5 air changes per hour) for warehouses and sports halls.
- A fan of 3.5 and 0.6 kW for light industry and warehouses/sports halls respectively is assumed to be included in all the system variations. The motor is situated outside the duct. Only 50% of the power to the fan is, therefore, transferred to the air as heat.

SPECIFIC FOR DWELLINGS
- A fan of 25 W is assumed to be included in all the system variations. The motor of the fans is situated inside the duct so all of the power to the fan is transferred to the air as heat.

Nomograms
The nomograms are developed for systems with at vertical collector facing due south.

LARGE VOLUME BUILDINGS
The nomograms for large volume buildings assume one control strategy: the air is always taken through the collector unless the temperature in the building exceeds 23°C. In that case the collector is bypassed and the air is taken directly from ambient.

The nomograms in Figures II.1.8–11 assume that the buildings are occupied every day. If an actual building, is occupied for five weekdays, (setback of the room temperature and the ventilation system is shut off during the weekend), the saving is easily corrected for this by dividing the saving from the nomograms by 7 and multiplying it with the actual number of days a week the building is occupied.

Three graphs are shown in the figures. The graph for light industry and warehouses/sports halls allows for interpolation to other ventilation rates, as the main difference between light industry and warehouses/sports halls is the ventilation rate. The building is identical, the only further difference is the internal gains (see Appendix H). The two other graphs show close-ups for light industry and warehouses/sports halls respectively for more detailed dimensioning.

Dissipated energy from the fan is not included in the saving shown in the nomograms.

The collector areas in Figures II.1.8–11 correspond to the flow rates in the collectors shown in Table II.1.3.

Table II.1.3. Relation between collector area and flow rate through the collectors in Figures II.1.8–11

Light industry	Warehouses/sports halls	Air flow rate through the collector
m$^2_{collector}$/m$^2_{floor}$	m$^2_{collector}$/m$^2_{floor}$	m^3/m$^2_{collector}$h
0.14	0.023	150
0.21	0.035	100
0.42	0.07	50
–	0.14	25

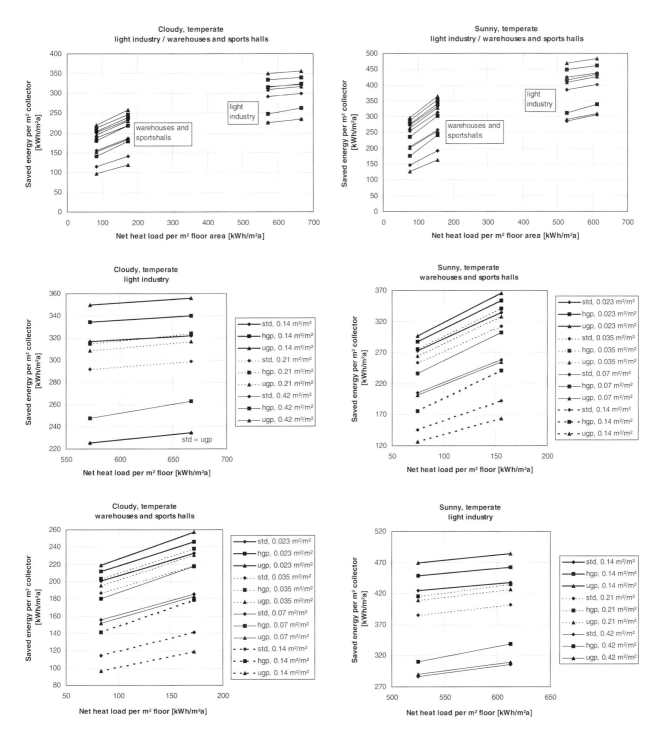

Figure II.1.8. Energy performance of System 1 in large volume buildings under cloudy, temperate weather conditions. Key: hgp = high performance collectors; std = standard collectors; ugp = unglazed perforated collector. The unit of the collector areas in the legend box is $m^2_{collector}/m^2_{floor}$

Figure II.1.9. Energy performance of System 1 in large volume buildings under sunny, temperate weather conditions. Key: hgp = high performance collectors; std = standard collectors; ugp = unglazed perforated collector. The unit of the collector areas in the legend box is $m^2_{collector}/m^2_{floor}$

Dwellings

Nomograms for dwellings have been developed for three control strategies:

- Strategy 1: the fan of the system is always running. The collector is bypassed if the temperature in the building exceeds 23°C. Only applicable if an always running ventilation system is necessary or already existing.
- Strategy 2: the fan of the solar heating system starts at a collector temperature of 18°C and stops at a collector temperature of 16°C. The fan also stops if the temperature of the building exceeds 23°C.
- Strategy 3: the fan of the solar heating system starts at an irradiation level of 300 W/m² and stops at 250 W/m². The fan also stops if the temperature of the building exceeds 23°C.

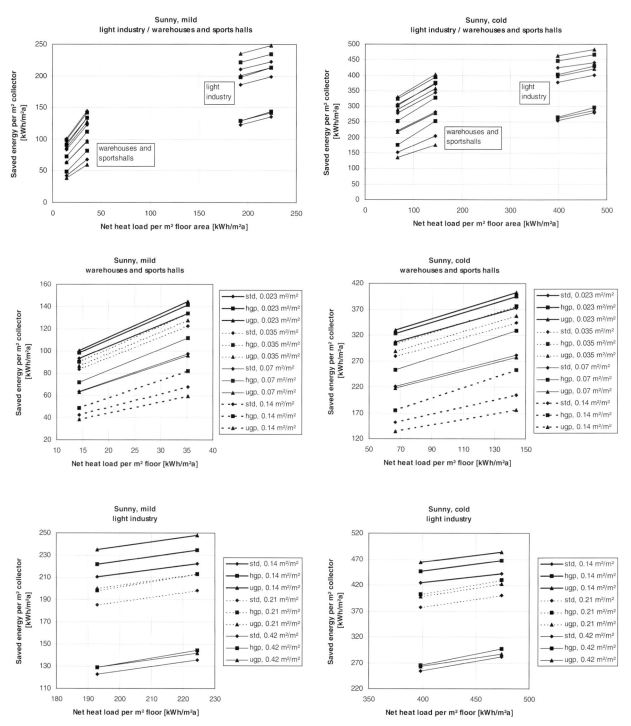

Figure II.1.10. Energy performance of System 1 in large volume buildings under sunny, mild weather conditions. Key: hgp = high performance collectors; std = standard collectors; ugp = unglazed perforated collector. The unit of the collector areas in the legend box is $m^2_{collector}/m^2_{floor}$

Figure II.1.11. Energy performance of System 1 in large volume buildings under sunny, cold weather conditions. Key: hgp = high performance collectors; std = standard collectors; ugp = unglazed perforated collector. The unit of the collector areas in the legend box is $m^2_{collector}/m^2_{floor}$

The nomograms for the three control strategies are shown in Figures II.1.12–15. The nomograms are divided in two graphs: a graph showing the normalized saving (at a collector area of 0.08 $m^2_{collector}/m^2_{floor}$) as a function of the heat demand of the building, the collector type and the flow rate through the collector, plus a graph with a correction factor (as a function of the flow rate through the collector) for the actual area of the considered collector. In order to obtain the performance of the considered system the normalized saving should be multiplied with the correction factor f_c.

Dissipated energy from the fan is not included in the saving shown for control strategy 1, while it is enclosed for control strategy 2 and 3 (see Dissipated energy from the fan, below).

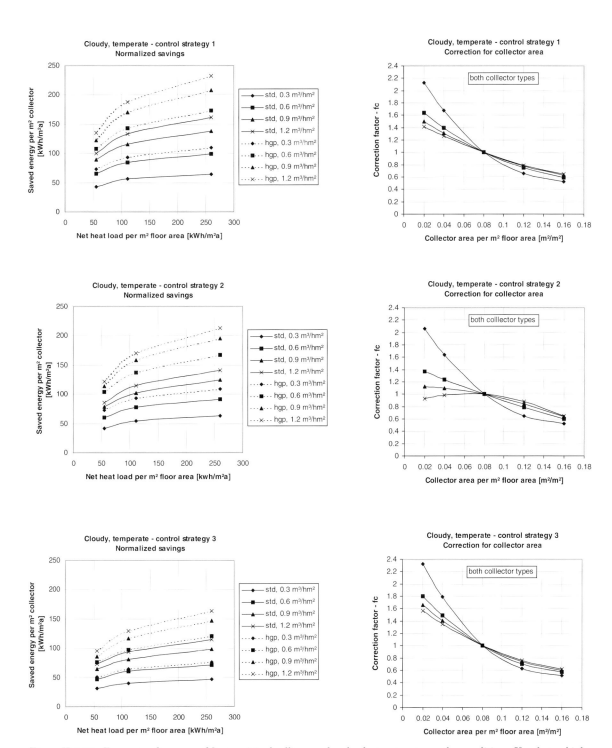

Figure II.1.12. Energy performance of System 1 in dwellings under cloudy, temperate weather conditions. Key: hgp = high performance collectors; std = standard collectors. The unit of the collector areas in the legend box is $m^2_{collector}/m^2_{floor}$

GENERAL

The efficiency curves of the two collectors, standard and optimized, are shown in Chapter IV.1 while the unglazed perforated collector is discussed in Chapter IV.3. Use these graphs for defining the actual collector compared to the collectors used in the nomograms and for interpolation in the nomograms in case of a different actual collector.

In order to use the nomograms the annual auxiliary demand has to be known for the building with a night setback of the room temperature from 20 to 15°C in the period 22:00– 6:00 h. However, night setback does not influence the performance of the system (without storage). If the annual auxiliary demand only is known without night setback the nomograms may still be applied, but the obtained saving will be 3–5% too high.

Dissipated energy from the fan

LARGE VOLUME BUILDINGS

Only the energy delivered from the collector is shown in Figures II.1.8–11, i.e. no energy dissipated from the fan

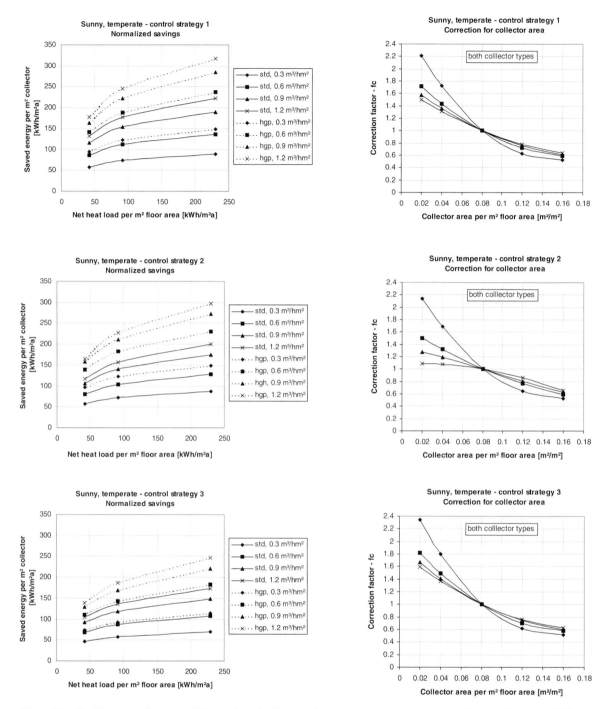

Figure II.1.13. Energy performance of System 1 in dwellings under sunny, temperate weather conditions. Key: hgp = high performance collectors; std = standard collectors. The unit of the collector areas in the legend box is $m^2_{collector}/m^2_{floor}$

is included in the normalized saving. This is because this type of System 1 will often be installed in connection with a necessary ventilation system, where the energy demand for a fan already is present and not caused by the solar air heating system.

However, the useful dissipated energy from the fan (during collection of solar energy) may easily be calculated using the values given in Table II.1.4.

As a fan of 3.5 and 0.6 kW with 50% of the power dissipated in the air has been used for industry and warehouses/sports halls respectively it is easy to determine the run-time for the system (when the dissipated energy from the fan is useful for the building). The run-time is found by dividing the dissipated energy from Table II.1.4 by 3.5/2 or 0.6/2. If a larger fan is required in the system the energy delivered from the system including the dissipated energy from the fan is the saved energy from Figures II.1.8–11, minus the dissipated energy from Table II.1.4, divided by the collector area, plus the dissipated energy from Table II.1.4 divided by 3.5/2 or 0.6/2 and multiplied by the actual fan power in kW (corrected for the amount of the power dissipated in

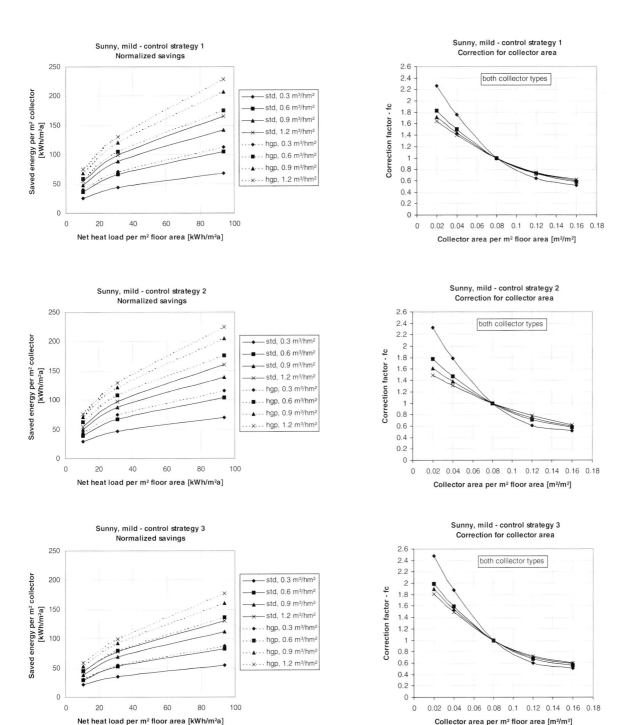

Figure II.1.14. Energy performance of System 1 in dwellings under sunny, mild weather conditions. Key: hgp = high performance collectors; std = standard collectors. The unit of the collector areas in the legend box is $m^2_{collector}/m^2_{floor}$

the air stream), and divided by the collector area. See the examples below for further information.

The dissipated energy in Table II.1.4 is given as total values and not, as in Figures II.1.8–11, as values normalized based on the collector area.

DWELLINGS

The energy dissipated from the fan is treated differently in the nomograms for control strategy 1 compared to control strategies 2 and 3. For control strategy 1 only the energy delivered from the collector is shown, i.e. no

Table II.1.4. The energy from the fan dissipated (during collection of solar energy) in the air stream by the fan

Building	Cloudy, temperate kWh/a	Sunny, temperate kWh/a	Sunny, mild kWh/a	Sunny, cold kWh/a
Light industry	7500	7250	4500	4700
Warehouses & sports halls	1270	1250	800	840

energy dissipated from the fan is included in the normalized saving. This is because this type of System 1 should only be installed in connection with a necessary (central

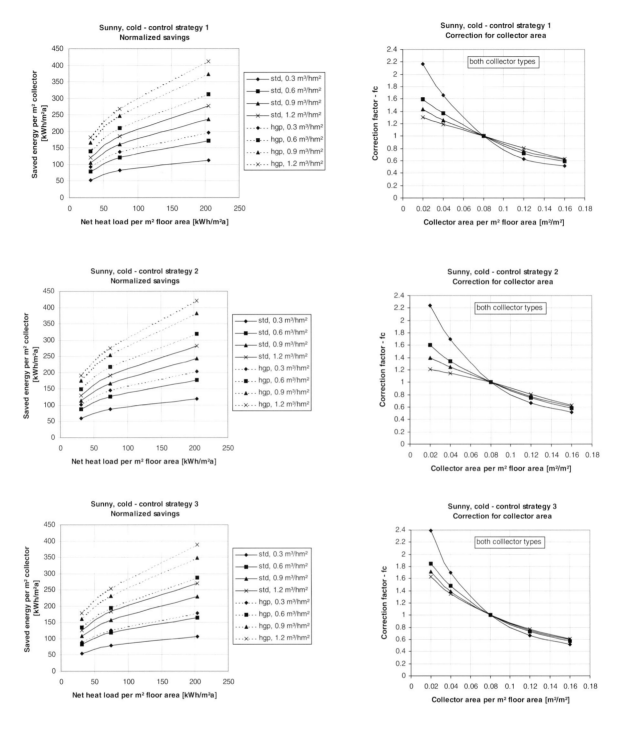

Figure II.1.15. Energy performance of System 1 in dwellings under sunny, cold weather conditions. Key: hgp = high performance collectors; std = standard collectors. The unit of the collector areas in the legend box is $m^2_{collector}/m^2_{floor}$

or exhaust) ventilation system, where the energy demand for a fan is already present and not caused by the solar air heating system.

For control strategies 2 and 3 the dissipated energy (= all the energy needed by the fan) is included in the savings as this type of system normally needs a fan dedicated to the solar air heating system.

The energy dissipated from the fan for control strategies 1 and 3 is nearly independent of the size of the collector as the fan is either always running or starts and stops at a certain level of irradiation. The dissipated energy in control strategies 1 and 3 is mainly dependent on the insulation level of the building. At high insulation level the room temperature more often exceeds 23°C. The fan is, thus, more often switched off. The dissipated energy (during collection of solar energy), equal to the necessary energy demand of the fan, is shown in Table II.1.5. Of the two values, the higher is for poor and medium-level insulated buildings while the lower is for high-level insulated buildings.

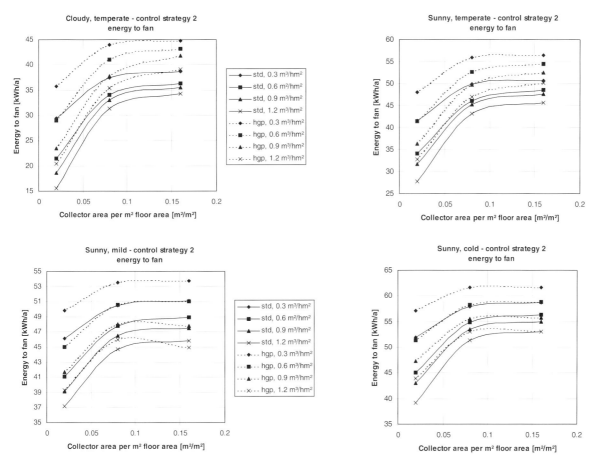

Figure II.1.16. The energy from the fan dissipated in the air stream equal to the necessary energy demand of the fan for control strategy 2 Key: hgp = high performance collectors; std = standard collectors. The unit of the collector areas in the legend box is $m^2_{collector}/m^2_{floor}$

As control strategy 2 is temperature dependent the dissipated energy from the fan will vary more due to the influence of the collector size and collector type. A large efficient collector will more often be capable of obtaining sufficient high temperatures than a small, poor collector for the same flow rate through the collector. The latter collector will thus have shorter run-time than the former. The run-time is here nearly not dependent on the insulation level of the building. The dissipated energy (= energy demand of the fan) for control strategy 2 is shown in Figure II.1.16.

As a fan of 25 W has been used in all cases it is easy to determine the run-time for the system. The run-time is found by dividing the dissipated energy from Table II.1.5 and Figure II.1.16 by 0.025 (25 W = 0.025 kW). If a larger fan is required the energy delivered from the system, including the dissipated energy from the fan, is the saved energy (only for control strategies 2 and 3) from Figures II.1.12–15, minus the dissipated energy from Table II.1.5 or Figure II.1.17, divided by the collector area plus the dissipated energy from Table II.1.5, or Figure II.1.16 divided by 0.025 and multiplied by the actual fan power in kW (corrected for the fraction of the power dissipated in the air stream) divided by the collector area. See the examples below for further information.

The dissipated energy in Table II.1.5 and Figure II.1.17 is given as total values, and not, as in Figures II.1.12–15, as values normalized based on the collector area.

Orientation and tilt of the collector
The influence of the orientation and tilt of the collector is dealt with in Chapter I.2 and Appendix C. Please consult this chapter if the collector is not oriented due south and does not have a tilt of 90° for which the nomograms have been developed.

However, a change in orientation within ±30° from south and a change in tilt from 90° down to 45° will, for most systems, only lead to a change in performance of up to ±10%.

Storage
LARGE VOLUME BUILDINGS
No storage has been considered for large volume buildings. This is because the direct gain through the windows

Table II.1.5. The energy from the fan dissipated in the air stream (during collection of solar energy) equal to the necessary energy demand of the fan for control strategies 1 and 3

Control strategy	Cloudy, temperate kWh/a	Sunny, temperate kWh/a	Sunny, mild kWh/a	Sunny, cold kWh/a
Control strategy 1	164	150–164	100–153	127–159
Control strategy 3	19	26–32	16–32	30–44

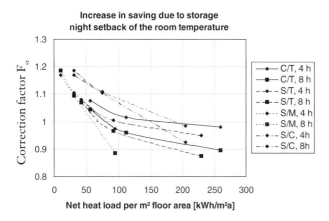
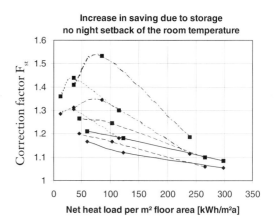

Figure II.1.17 The possible increase in saving when introducing storage in the system as a function of the delay introduced by the storage and if night setback of the room temperature is applied in the building. Key: C/T: cloudy, temperate; S/T: sunny, temperate; S/M: sunny, mild; S/C: sunny, cold; 4 h: a delay of 4 hours; 8 h: a delay of 8 hours

and the internal gains are small compared to the ventilation flow rate. The heat from the collector can most of the time be utilized as it comes, also from large collector areas. The solar fraction is normally low for large volume buildings compared to dwellings.

Even if a surplus of energy from the solar heating system is present, the night setback of the room temperature will make storage unprofitable as shown under Dwellings below.

DWELLINGS
The nomograms have been developed for systems without storage. Storage in the system will influence the performance of the system. Storage postpones the time when heat is delivered to the building. This may be useful if there are large direct gains through the windows of the building.

Figure II.1.17 allows evaluation of the benefit of storage in a system. Two sizes of storage are shown: storage where the peak value of delivered heat is postponed four and eight hours from the peak value of heat out of the collector. The two graphs further show the benefit of storage in a building with and without night setback of the room temperature.

The graph for buildings with night setback shows that it normally is not a good idea to introduce storage here. This is because the heat is stored to periods with no space heat demand. The storage may even decrease the saving of the system. Only in highly insulated buildings may storage be beneficial, as this type of building often has nearly no heat demand during daytime. However, the absolute savings in these building may often be too low to repay the cost of storage.

The correction factor f_{st} from Figure II.1.17 should be multiplied by the saving found in the nomograms (after any correction for fan power and tilt/orientation).

Recovered heat loss
The collector may lead to a decrease in the heat loss of the building if it is located on a structure directly connected to the heated rooms of the building. The heat loss is reduced due to the extra cover on the wall/roof formed by the cover and absorber of the collector and any additional insulation of the collector.

The saving due to the reduced heat loss of the building may be evaluated using the graphs in Figures II.1.18–19. The saving is showed as a function of the U-value of the structure on which the collector is mounted. Two cases are shown: with and without rear insulation in the collector, i.e. the collector consists only of a cover and absorber mounted on the wall (e.g. when installed during the erection of the building) or the collector has its own insulation of 50 mm mineral wool (e.g. when a collector is mounted on an existing building).

In Figures II.1.18–19 it is assumed that the absorber is thermally light, i.e. no heat is stored in the collector. A thermally heavy collector will lead to a higher saved heat loss but also to a lower direct performance in the form of heated fresh air. Together this will often lead to a smaller overall performance for systems with a thermally heavy collector.

Please be aware that the shown savings in Figures II.1.18–19 are only indicative, e.g. thermal bridges in the structure or the collector may drastically change the saved heat loss.

The saving due to reduced heat loss should be added to the saving found in the nomograms (after any correction for fan power, tilt/orientation and storage).

LARGE VOLUME BUILDINGS (FIGURE II.1.18)

DWELLINGS (FIGURE II.1.19)
GENERAL
Mounted in front of well-insulated walls or roofs with a U-value below 0.5 W/K the collector will only lead to marginal recovery of heat loss through the wall or roof. This is because the heat loss through such constructions is already rather low.

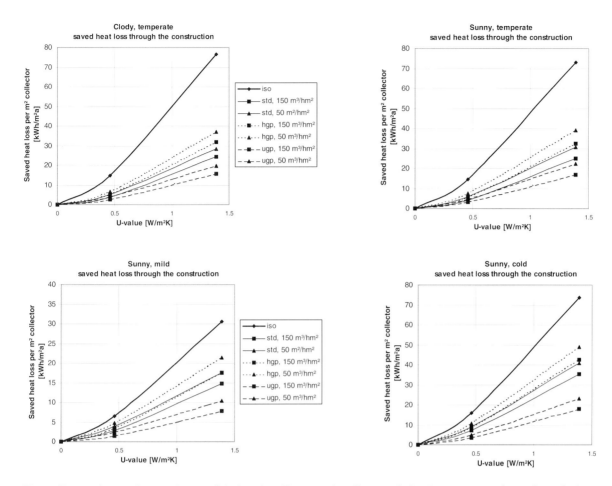

Figure II.1.18. Saving due to reduction of the heat loss. Key: iso = the collector includes 50 mm rear insulation; hgp = high performance collectors; std = standard collectors; ugp = unglazed perforated collector. The units of the flow rate through the collector in the legend box are $m^3/m^2_{collector}$

Example of dwelling calculations

In order to show how to use the nomograms a calculation example for a dwelling is given below.

Location:	Denmark
Size of building:	100 m²$_{floor}$
Tilt and orientation:	tilt 45°, 15° from south towards east
Collector:	optimized collector: 5 m², i.e. 0.05 m²/m²$_{floor}$
Air flow through collector:	100 m³/h, i.e. 1 m³/hm²$_{floor}$
Control strategy:	temperature controlled – control strategy 2
Fan power:	100 W with 50% dissipated in the air stream
Space heating demand:	12,500 kWh/year, i.e. 125 kWh/m²$_{floor}$a
U-value of roof:	0.75 W/m²K

The way to calculate the annual saved energy is shown here. Copenhagen has cloudy, temperate weather conditions so the values for the calculation of the annual saving due to the solar air heating system are found in the graphs and tables under cloudy, temperate climate.

SYSTEM PERFORMANCE

The normalized annual saved energy is found in the left graph of Figure II.1.12, while the correction factor for the collector area is found in the right graph. The method is shown in Figure II.1.20.

Normalized annual saved energy:	168 kWh/m²a
Correction factor for collector area:	1.05
Annual saved energy:	168 · 1.05 = 176 kWh/m²a

DISSIPATED FAN ENERGY

The annual fan energy is found in Figure II.1.17, as shown in Figure II.1.21. The annual fan energy is found to be 30 kWh/a. As the fan power in Figure II.1.17 is 0.025 kW the run-time of the fans is 30/0.025 = 1200 h.

The real fan power of the fan in the example is 0.1 kW with 50% dissipated in the air stream. The dissipated fan energy is thus: 50% of 1200 · 0.1 = 60 kWh/a.

The annual saved energy including dissipated fan energy is thus, as the collector area is 5 m²: 168 + (60 – 30)/5 = 182 kWh/m²a

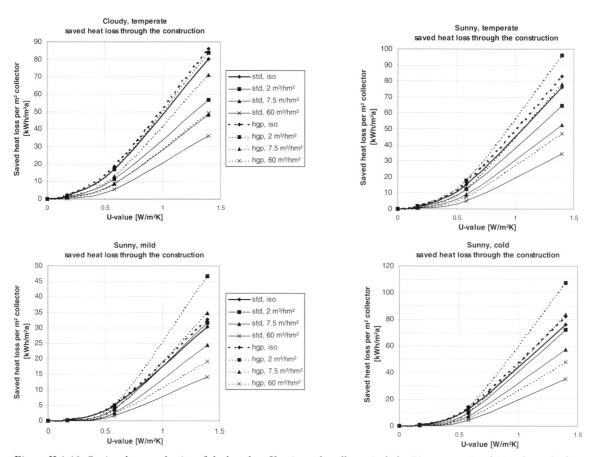

Figure II.1.19. Saving due to reduction of the heat loss. Key: iso = the collector includes 50 mm rear insulation; hgp = high performance collectors; std = standard collectors. The units of the flow rate through the collector in the legend box are $m^3/m^2_{collector}$

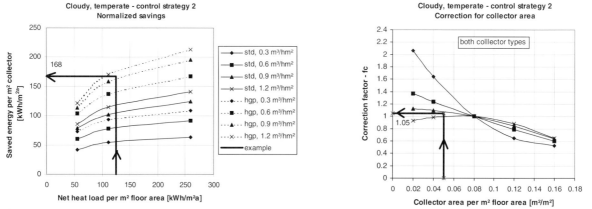

Figure II.1.20. Example: normalized saving and correction factor for actual area. The value is found by interpolation between two curves for a high performance collector (hgp) with a flow rate of 0.9 and 1.2 $m^3/h\ m^2_{floor}$

STORAGE

There is no storage in the system. If there had been, this may be evaluated using Figure II.1.16.

RECOVERED HEAT LOSS

The recovered heat loss through the roof is found in Figure II.1.19, as shown in Figure II.1.22. The recovered heat loss is equal to 22 kWh/m²a. This means that the total annual saved energy due to the solar air heating system is 182 + 22 = 204 kWh/m²a.

ORIENTATION AND TILT

From Orientation and tilt it can be seen that an orientation of 15° towards east and a tilt of 45° will lead to a change in the performance of less than ±10%. A tilt of 45° will lead to slightly higher performance than a tilt of 90°, while an orientation of 15° towards east will lead to slightly lower performance (see also Appendix C). It is judged that this leads to nearly no change in the performance. A change of ±10% is further within the uncertainty of the calculation. So no correction is introduced here.

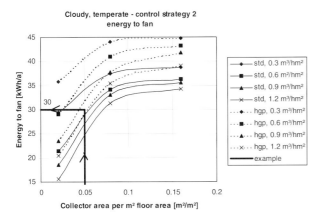

Figure II.1.21. Example: energy delivered from the fan to the air stream. The value is found by interpolation between two curves for a high performance collector (hgp) with a flow rate of 0.9 and 1.2 $m^3/h\ m^2_{floor}$

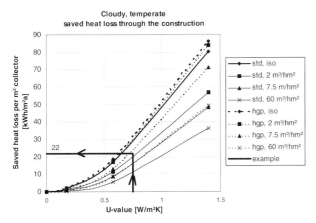

Figure II.1.22. Example: saved heat loss through the roof. The flow rate through the collector is 100 m^3/h and the collector area is 5 m^2. This gives a flow rate through the collector of 20 $m^3/h\ m^2_{collector}$. The value is found in the graph by interpolation between a flow rate of 7.5 and 60 $m^3/h\ m^2_{collector}$ for a high performance collector

References

1. Hastings R (ed) (1999) *Solar Air Systems, Built Examples*, James & James (Science Publishers) Ltd, London.
2. Hollick J (1990) *Solarwall – Air Heating System Design Manual*. Conserval Engineering Inc., 200 Wildcat Road, Downsview, Ontario M3J 2N5, Canada.

Further reading

Gramkow L (ed) (1999) *2nd Generation Solar Walls – Building Integrated Solar Energy* (in Danish). COWI, Parallelvej 15, DK-2800 Lyngby, Denmark.

Jensen S Ø (1992) *Tests on an Uncovered Solar Collector with Decra Roof Sheets*. Institute for Buildings and Energy, Technical University of Denmark, Building 118, DK-2800 Lyngby, Denmark. Report no. 92–5.

Jensen S Ø (1994) *MF-demonstration Project at Wewer's Brickyard in Helsinge*. Institute for Buildings and Energy, Technical University of Denmark, Building 118, DK-2800 Lyngby, Denmark. Report no. 267.

Jensen S Ø (1994) *Results from Tests on a Multi-Function Solar Energy Panel*. Institute for Buildings and Energy, Technical University of Denmark, Building 118, DK-2800 Lyngby, Denmark. Report no. 213.

Jensen S Ø (1994) *Test of the Summer House Package from Aidt Miljø*. Institute for Buildings and Energy, Technical University of Denmark, Building 118, DK-2800 Lyngby, Denmark. Report no. 94-1.

McClenahan D (1994) *Performance of the Perforated-Plate/Canopy Solarwall at GM Canada, Oshawa*. Energy Technology Branch/CANMET, Department of Natural Resources Canada, 580 Booth Street, 7th Floor, Ottawa, Ontario K1A 0E4, Canada.

Morhenne J and Barthel H (1997) *Solare Luftkollektoren für kommunale Gebäude am Beispiel von Turnhalle (Solar Air Collectors for Communal Buildings – Demonstration Project Karls Gymnasium Munich)*. Abschlußbericht BMFT 329016C. Ingenieurbüro Morhenne GbR, Schülkestrasse 10, D-4227 Wuppertal, Germany.

Rasmussen N B K and Jensen S Ø (1997) *Solvarmeanlæg med luft som transportmedium (Solar Heating System with Air as Transport Medium)*. Solar Energi Centre Denmark, DTI Energy, Gregersensvej, P.O.Box 141, DK-2630 Taastrup, Denmark.

US Department of Energy (1998). *Transpired Collectors (Solar Preheaters for Outdoor Ventilation Air)*. NTDP Program Manager, Federal Energy Program, U.S. Department of Energy, 1000 Independence Ave. SW, EE-92, Washington DC 20585.

Chapter author: Søren Østergaard Jensen

II.2 System 2: Open collection loop with radiant discharge storage

INTRODUCTION

System description and principles

In this system air circulates, by natural or mechanical driving forces, through the collector, the ceiling channels, the room space and back to the collector. The mass of the ceiling discharges heat with a time lag of a few hours, mainly by radiation. A large fraction of heat is supplied to the rooms by hot air from ceiling storage channels. In summer collector hot air is vented outdoors through upper and lower dampers, open or closed seasonally.

The system consists of the following elements:

- thermosyphon or fan-driven air collector;
- ducts inside a massive ceiling;
- ceiling duct outlets to room and return air inlet to the bottom of the collector;
- seasonal and daily dampers;
- daily control system.

Characteristics
ADVANTAGES
- proven solutions for residential buildings including multi-storey apartment blocks;
- no fans or mechanical controls required in the passive version, therefore reduced maintenance and running cost;
- ventilating operation modes in summer;
- no manifolds are needed (difficult to integrate into the building structure).

DISADVANTAGES
- low efficiency (thermosyphon systems);
- diverse dampers required (leakage potential).

System variations

Fan-driven collector connected to radiant discharge storage and space
Air from collectors is fan-forced through the ceiling, enters the room and goes back to the collector through floor ducts (semi-open loop). A thermostat in the collector activates the fan and a damper, which avoids reverse circulation in the night.

CHARACTERISTICS
Advantages
- proven solutions for residential building;
- integration of fan in the heating unit (customised fan-coil) below the window;

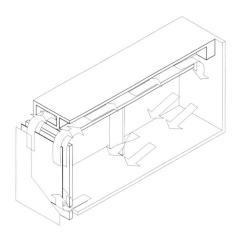

Figure II.2.1. Basic elements of the open loop system: thermosyphon collector–radiant discharge storage–room space, integrated into building structure

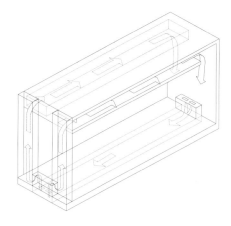

Figure II.2.2. Fan-driven collector with hypocaust as storage and radiant discharge to space

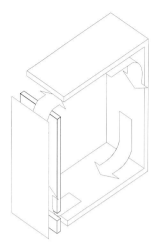

Figure II.2.3. Open loop without storage: wall-integrated collector

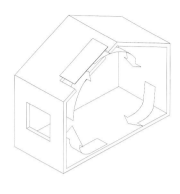

Figure II.2.4. Open loop without storage: roof-integrated collector

- possibility of using night air flush for summer radiant comfort.

Disadvantages
- higher cost.

Air collector without storage
Air circulates from the collector directly into the heated space without any primary storage structure. In wall-integrated solutions air can circulate by natural (a) or forced convection, while in a roof system circulation must be forced (b).

CHARACTERISTICS
Advantages
- suitability for retrofit as an integrated building element or as a panel applied over building cladding;
- fast response;
- simple direct use of solar heated air.

Suitable applications

The standard configuration system is particularly suitable for residences where heat stored in the day can be used during the night. Without storage it can be used in schools, small offices and warehouses, providing that mechanical ventilation does not interfere with the thermosyphon loop of the collector.

Characteristics, advantages and limitations

System
No manifolds are required. Manifolds are difficult and expensive to integrate into a building structure.

Collector
Thermosyphon collectors can be site assembled or factory delivered. Flat plate collectors are on the market in Europe. A thermosyphon collector (called also a 'Solar Chimney') has been produced in Italy as a part of the building envelope lowering the cost of the solar system. When the collector is a building element, special attention must be given to thermal stress, wind and weather tightness. One of the main problems in the Italian experience was heat loss and leakage from seasonal dampers.

Storage
Ceiling storage offers large surface areas for radiant heat discharge and provides an alternative to floor heating from hypocaust storage. It is strongly recommended that internal duct surfaces be accessible for cleaning.

ENERGY PERFORMANCE

Standard configuration of solar system

To assess the energy performances of a thermosyphon collector with radiant discharge, a reference system has been defined:

- thermosyphon collector, integrated in south facade, tilt 90°, azimuth 0 (Figure II.2.1);
- thermal ceiling storage type D3 with material properties given in Appendix G; duct section: 0.5×0.15 m; duct number equal to the integer value of collector width divided by 0.7; channel length: 10 m; outlet per each duct: 0.3×0.3 m;
- the building is divided into four thermal zones: ground floor south, ground floor north, first floor south, first floor north. This was necessary because otherwise it was not possible to calculate the different performance of the internal and external hypocausts.
- the hypocaust area is 14 m² (2×1.4 m $\times 5$ m) for each zone, so four hypocausts are used in this system; other areas are built up as a 'normal' ceiling or roof with the same layers as the hypocaust (see hypocaust D typology, Appendix G):

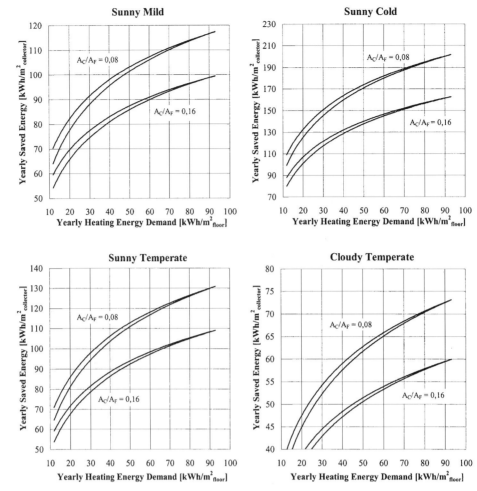

Figure II.2.5. Performance of a thermosyphon system

- air loop resistance is R2 (Figure II.2.8, Table II.2.2);
- automatic control;
- if the collector outlet temperature is lower than north zone temperature, the system doesn't operate;
- if the collector outlet temperature is 4 K above the north zone temperature then the system starts working. If the temperature difference between the window collector outlet and the average north zone is below 1 K the system stops.

For more details on System 2 and building data, see Appendix G.

Energy performance of standard configuration

The energy performance of the reference system is shown in Figure II.2.5, illustrating a residential building in four different climates types: sunny mild, sunny cold, sunny temperate, cloudy temperate. Trnsair was used to generate the data.

The energy performance, on the vertical axis, is expressed as saved energy, per heating season, by a unitary collector surface (kWh/m$^2_{coll}$). The horizontal axis input is the 'yearly net heating demand of unitary floor surface' (kWh/m$^2_{floor}$) of the building or dwelling.

The parameter is the ratio of A_{coll} to A_{floor}. Each pair of curves shows data for a heavy mass building (lower) and for light mass (higher). The heating season has been set from September until May for all climates to achieve a better comparison.

The best energy performance is in high specific heat demand buildings. This means that System 2 works better in a low insulated residential building if the climate is sunny temperate and in a medium-high insulated one if the climate is sunny cold.

The open collection loop system is a good application, but mainly for residential buildings, not commercial buildings.

Figure II.2.5 presents the system energy performance of a thermosyphon collector in a standard configuration. Units are saved energy, per heating season, by a unitary collector surface as a function of yearly net heating demand of unitary floor surface of a building or dwelling. Each pair of curves shows data for a heavy mass building (lower) and light mass.

Energy performance of system variations

The overall performances related to the reference case can be estimated as follows:

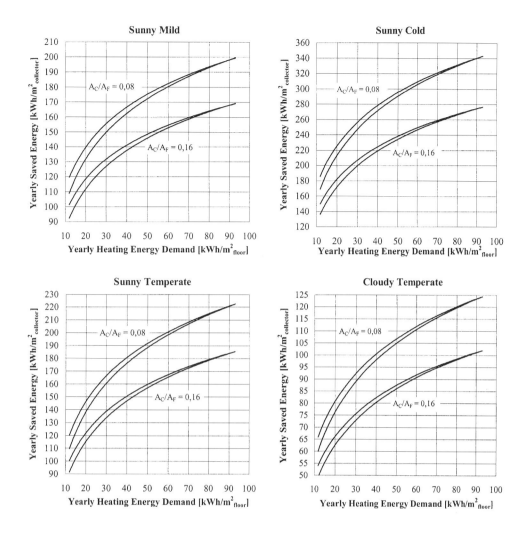

Figure II.2.6. System performance of a fan-forced system

- standard configuration: 100%;
- fan-driven collector with storage (Figure II.2.2): roughly 180%; more details on energy performance are given by the diagrams in Figure II.2.6 that apply to a system similar to standard configuration but fan-driven. Yearly saved energy is computed with mass flow rate equal to 50 m^3/(h m$^2_{coll}$).
- wall-integrated thermosyphon collector without storage (Figure II.2.3): 60%; for a best evaluation, calculate requested input data, enter in Figure II.2.6 and multiply output data by 0.33.
- roof, fan-driven collector without storage (Figure II.2.4): 130% (south azimuth, low tilt); for a best evaluation, calculate requested input data, enter in Figure II.2.6 and multiply output data by 0.7 –0.8 for south orientation and tilt near to 25– 40°.

Figure II.2.6 presents the system energy performance of a fan-driven collector configuration. Units are in terms of saved energy, per heating season, by a unitary collector surface as a function of yearly net heating demand of unitary floor surface of a building or dwelling, with mass flow rate equal to 50 m^3/(h m$^2_{coll}$).

Each couple of curves shows data for a heavy mass building (lower) and for light mass.

Energy performance of built examples

Data of Table II.2.1 were obtained from long-term in situ measurements.

SYSTEM DESIGN

Design advice

For the reference solar air system and its variations the basic factors to be considered are:

Air collector
- The system can use both forced or natural air collectors.
- The efficiency can be high (in the range of 60–80%) or intermediate (range 40–60%) depending on material properties of the transparent cover, absorber and insulated back.
- The efficiency of the thermosyphon collector is generally lower (in the range 25–35%).

Table II.2.1. Example of built systems

Case (bibliog. ref.)	Type	Climate	Collector tilt	Collector azimuth	BLC[1] [W/m²$_{floor}$]	A_{coll}/A_{floor} [—]	Yearly saved energy [kWh/m²$_{coll}$]
Salisano (IT) (4.7.3)	natural	sunny mild	90°	0°	2.1	29%	90
Marostica (IT) (4.7.6)	natural	sunny mild	90°	0°	1.8	16%	100
Luino (IT) (4.7.10)	forced	sunny mild	90°	0°	1.6	15%	150

[1] building load coefficient (transmission + infiltration)

To increase efficiency of the thermosyphon collector:
- use a double flow, single glazing cover;
- maximize the inlet opening at the bottom of the collector to keep air in contact with upper and lower absorber surfaces and to lower air flow resistance;
- optimize the air gap (in mild, sunny climates each air gap is about 5 cm).

- The air flow rate with high radiation values (in the range of 300–600 W/m²) approximates 30–40 m³/(h m²)$_{collector}$ with a maximum temperature rise in the range of 20–25 K, as shown in Figure II.2.7.

Air distribution (ducts)
- Ducts in the ceiling provide both thermal storage and air distribution.
- The cross-section of the ducts in the thermal ceiling and of the inlet/return air openings have to be similar to the cross-section of the thermosyphon collector to minimize pressure drops in the loop.

Thermal storage
- Geometry and dimensions are often determined by load-bearing functions.
- Ceiling storage is not insulated towards the room, though thin insulation can be added to slow the heat release. Thermal mass in residential buildings is usually 200–250 kJ/(m²$_{coll}$ K) (see Chapter IV.6 on hypocaust storage).

Dampers and fan
- Well-insulated dampers manually operated are generally used for seasonal control.

- Back draught dampers (light plastic film) integrated into the return air openings were formerly used to prevent inverse circulation during the night. These had inadequate thermal and acoustic resistance, restricted air flow and leaked as the plastic films aged. It is therefore strongly recommended to use automatically operated dampers driven by temperature sensors and electromechanical logic that switch dampers to the open state when the absorber temperature exceeds 30°C and closes dampers when the absorber plate goes below 20°C.
- In the fan-driven variation, the fan is usually part of the air-heating unit integrated in the south facade.
- Seasonal dampers are operated a few times a year, and are manually operated dampers at the bottom and upper external part of the air panel (see Figure II.2.8).

Auxiliary heating
- The auxiliary heating must shut off quickly when the solar air system begins to deliver heat.

System and component design

This step by step procedure is described for the reference solar air system and is related to the nomograms described in Nomograms below.

Step 1: Define collector type and collecting area

WHICH COLLECTOR TYPE: THERMOSYPHON COLLECTOR OR FAN-DRIVEN?

If large surfaces are available a thermosyphon collector can be a good solution (low cost – low efficiency); if the collector area is limited the best solution is a fan-driven system to increase efficiency. The correction factors given in Chapter I.2 allow the use of nomograms when the collector is not south-oriented or not vertical.

COLLECTION AREA

Using the standard configuration, the area needed for collection can be determined by means of the graphs in Figure II.2.5; data needed are: building heating demand and climate. The area for variations of a thermosyphon and fan-driven collector can be estimated (see Nomograms below).

Step 2: Design for low pressure drop and high flow rate

Calculate the total air resistance: the collector-inlet, the collector-outlet and ceiling-outlet. The resistance is mainly related to the ratio A_{inlet}/A_{coll} and the spatial

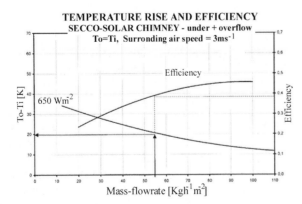

Figure II.2.7. Energy efficiency of a reference thermosyphon collector (Secco Solar Chimney)

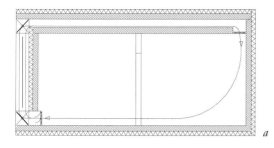
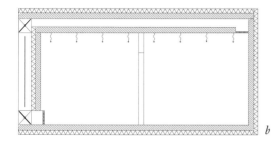
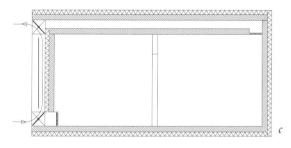
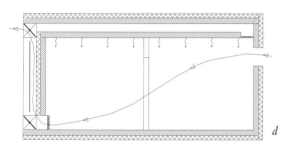

Figure II.2.8. Operation modes of the reference system: (a) winter sunny days, (b) winter night, (c) summer day – vented collector, (d) summer day – building ventilation

Table II.2.2. Air flow according to damper configuration

	Resistance	Description	A_{inlet}/A_{coll}	Peak air flow m^3/m^2_{coll}
R1	high	one compact damper	2.5%	20–30
R2	middle	extended damper	3%	30–40
R3	low	distributed damper	4%	50–60

- For designs with many spatial constraints use openings as in R1.
- For design where the south facade is two times wider than panel height, use openings as in R2 or R3 but distribute them in the center of each module of 2–2.5 m of panel.

distribution of return openings; these two variables are quantified from measured cases: see Table II.2.2.

Air return dampers are critical elements of the loop because their placement is often constrained by the interior design. The result may be less return openings and therefore higher resistance to air flow.

RULES OF THUMB
- Ducts in the ceiling, inlet openings and return air dampers have to maintain approximately the horizontal cross-section of the collector, or about 4% of the panel collecting surface. Thereby the pressure drop will be less.
- For general design the middle resistance, R2, is a good choice.

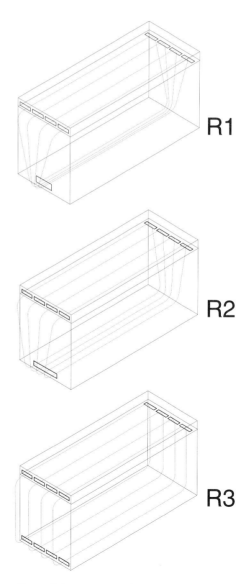

Figure II.2.9. Schematic view of the three different resistance codes R1, R2, R3

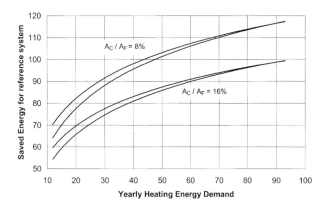
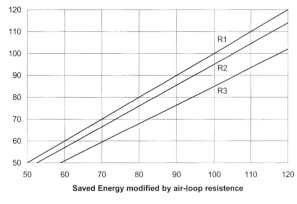
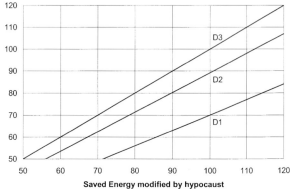

Figure II.2.10. Design in sunny mild climate

Step 3: Selection of hypocaust
Three types of hypocaust are suggested and studied: D1, D2, and D3 (see Appendix G). The first parameter that increases saved energy is the thickness of the insulation of a ceiling-hypocaust (increasing from D1 to D3); the number of ducts is not a design variable because many constraints fix this number. As a rule of thumb the hypocaust type and building insulation level can be related as follows:

- for high insulation buildings use type D3,
- for medium insulation buildings use type D2,
- for low insulation buildings use type D1.

Figures II.2.10–13 give the correction factors.

Step 4: System saved energy correction factors
Figures II.2.10–13 show correction factors for calculation of saved energy in a design case with respect to the standard configuration taking into account increasing resistances (from R3 to R1) and decreasing hypocaust insulation levels (from D3 to D1).

Step 5: Ducts, fans and controls
If a fan is coupled to a collector air loop laid out for thermosyphoning the pressure drop can be expected to be lower than 5 Pa at 50 m^3/(h m$^2_{coll}$). Then cross-flow (see Chapter V.1) fans can be used. The choice should maximize the impeller width but with as few fans and separated panels as possible. If a fan is used with reduced collector and hypocaust sections to improve efficiency, the design rules of System 4 can be applied.

Using Trnsair

Trnsair can be used for unusual design variations not included in this handbook and for climates very different from those given in nomograms or out of range for the ratio A_{coll}/A_{floor} or for yearly heating demand. Similarly, non-residential applications require computer modelling.

CONSTRUCTION (SYSTEM INTEGRATION)

General observations

The internal layout should maximize continuity of the storage system and have few interruptions by vertical functional elements (stairs, lifts). The architect and engineer must work together, otherwise the design of the building may force the fit of the system, compromising both performance and the design.

Subsystems integration

The collector is part of the building envelope and is operated under atmospheric pressure with specific technological

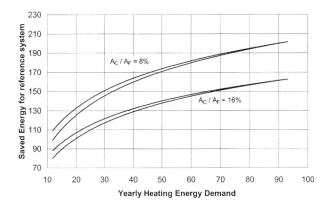
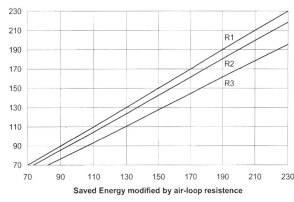
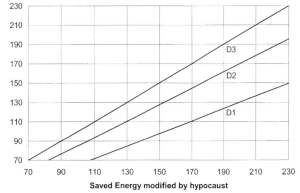

Figure II.2.11. Design in sunny cold climate

requirements (air tightness, waterproof) specified by international regulations on curtain walls.

It is convenient to use continuous facades based on a 'module' equal to 60 cm, also the module for the most common prefabricated reinforced concrete floor, which can serve as heat storage. The air panel in the south wall can be a prefabricated air panel assembled or mounted on site over the wall or a prefabricated 'solar air wall', mounted as a traditional opaque curtain wall.

The second option is much more convenient with regard to design integration.

Distribution/storage
The storage/distribution (see Chapter IV.6 on hypocausts) is a modified traditional prefabricated concrete ceiling which distributes heat to the adjacent space and provides thermal storage. The section of the hollow cores-ducts has to be similar to the horizontal section of the thermosyphon collector (4% of solar panel surface). Their length should not exceed 8–10 m.

The design requires a high degree of building integration because the air from the collector enters directly in the ducts of the ceiling (without any manifold). This requires a 'connection' which often conflicts with other building requirements (i.e. load-bearing).

Furthermore the layout of storage/distribution components must 'obey' the thermal zoning of inner spaces (north/south, heated/unheated) and structural constraints (i.e. staircases interrupt the ceiling).

Controls
There are two basic types of controls. Daily control is operated by:

- a back draught plastic sheet damper, lifted by hot air buoyancy force. It is located in the return inlet at the bottom side of the collector;
- automatic dampers driven by a preferred temperature sensor.

Seasonal controls:

- are manually actuated through access panels at the bottom and top of the air panel.

Advanced control logic for best efficiency
System 2 can be controlled by two means: the temperature difference between the collector absorber and room air and the temperature difference between the outgoing air from hypocaust openings and room air. The first type gives high dynamic stability and good energy performance, and the second gives poor stability and less efficiency.

Poor performance of the second type of controllers is mainly due to the small difference that the controller must act upon. This difference is often lower than anomalies introduced by under-ceiling stratified hot air coming from random internal gains (lighting, cooking) and from direct solar gain on sunny days. The effect of each internal

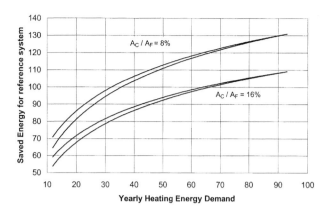

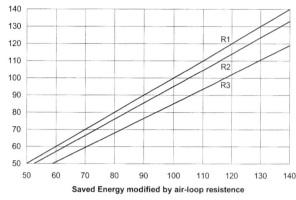

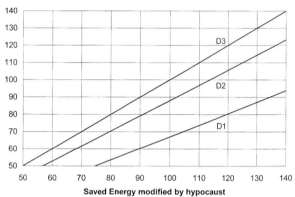

Figure II.2.12. Design in sunny temperate climate

gain can be short in time but there may be many different gains generated in a day, so the average 'noise signal' on ceiling air temperature sensors can be higher than the 'true signal', causing inappropriate control actions.

Controllers driven by hypocaust temperature differences can work correctly but this implies that a sensor for outgoing air temperatures is put inside the channels and far (more than two metres) from their output. The set point will be near to 29°C and hysteresis of the controller will be 3 K. For controllers driven by the collector temperature difference, the sensor should be at half the height of the panel. The set point, for air flow switch on, can be near to 27°C and the hysteresis can be 7–10 K. In this case the air temperature outgoing from the ceiling is about 3–4 K over roomtop stratified air and good heat exchange is possible without any error in the control action. This means higher than average energy is supplied from the panel to the room with no oscillation of the control system.

ENERGY PERFORMANCE AND ECONOMICS

Starting from the cost of a traditional energy system and the additional cost of System 2 and the required energy performance, the energy saved in a season by each square metre of collector = YSE/A_{coll} (see Figure II.2.14). Note that the additional collector cost is given by the sum of system construction and the present value of future

YSE/A_{coll} = Additional Collector Cost/m² / (Traditional Heating System Unitary Energy Cost) x (Payback Time)

Figure II.2.14. Equation for energy performance

running costs to repair and to maintain efficiency during the whole useful life of the air system.

For example, assume that the following data are known:

- unitary additional collector cost = €150/m²,
- traditional heating system unitary energy cost = €0.06/kWh,
- payback time = 25 years.

So we can calculate YSE/A_{coll} = 100 kWh/m². This value gives the lower energy performance of System 2, necessary to pay the investment cost during system life time. On the other hand, if the additional cost is greater (or lower) than €150/m² then payback time is longer (or shorter) than 25 years.

NOMOGRAMS

Nomograms (Figures II.2.10–13) are provided for residential buildings, in four climate types defined and dis-

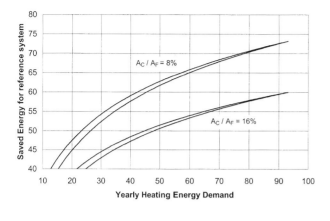
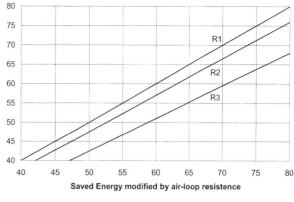
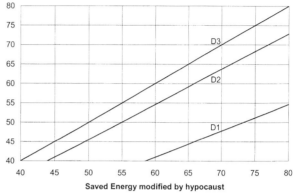

Figure II.2.13. Design in cloudy temperate climate

cussed above. Use of the nomograms is shown by this example.

1. Set your energy goal by an economical analysis tool. Assume that, calculated according to your national regulations on thermal energy saving:

- target yearly saved energy per m² collector surface is YSE/A_{coll} = 90 kWh/m²;
- climate in your design is sunny temperate;
- all building's properties are known;
- hypocaust is D2 type (middle insulation);
- air loop resistance is given by R2 (middle resistance);
- building thermal mass is high;
- you have enough collectors on the south facade;
- yearly net heating demand per m² floor surface is YHD/A_{floor} = 75 kWh/m²;

2. Choose the nomogram according to climate: in this example look at Figure II.2.12.
3. Put YSE/A_{coll} = 90 kWh/m² on the vertical scale of the lower right sector and draw a horizontal line to cross D2 line. From this point draw a vertical line to cross R2 line in the upper right sector. Then draw an horizontal line toward upper left sector to cross the left scale.
4. Put YHD/A_{floor} = 75 kWh/m² on the horizontal scale of the upper left sector and draw a vertical line to cross the horizontal one previously drawn.
5. Read (also by interpolation) the ratio A_{coll}/A_{floor} at the cross point defined by the previous procedure (bear in mind that only the upper line, corresponding to high mass, is relevant for you).
6. You read approximately 16%, so you can calculate $A_{coll} = A_{floor} \times 16 / 100$; this is the right collector area to meet your goal in your building and climate.
7. If your south facade is not as wide as needed you can put a fraction of the calculated collector area on the roof, using a fan-driven roof-integrated collector.

ACKNOWLEDGEMENTS

Work on this chapter was supported by ENEA. The authors thank especially M T Porfiri and M Citterio for their efforts in tuning up TRNSYS types for System 2.

BIBLIOGRAPHY

Barra O A and Pugliese Caratelli E (1979) A theoretical study of laminar free convection in 1-D Solar inducted flow, *Solar Energy*, 23, 211–215.

Barra O A and Franceschi L (1985) *The Barra Thermosyphoning Air System: Residential and Agriculture Applications both in Italy and in the UK*, UK-ISES Conference C40, London, April 1985.

Barra O A, Artese G, Costantini T, Costanzo G, Franceschi L and Carratelli E P (1980) *The Barra-Costantini Solar Passive System: Experimental performances*, in: *Building*

Energy Management, ed by E De Oliveira Fernandes, Pergamon Press, Oxford and NewYork.

Barra O A (1981) *La Conversione Fototermica dell' Energia Solare*, Etas Libri, Milano.

Commission of European Community (1987) *Project Monitor*, Marostica, 14 December.

Contadini G, De Giorgi G, Fumagalli S, Rizzi G, Ferrario A, Rossi G and Silvestrini G (1989) *Monitoring of a solar chimney and different ceiling storage/distribution systems in an office building located in northern Italy*, Proceedings of the 14th Passive Solar Conference, Denver, Colorado, June 19–23, 1989.

Hastings S R (editor) (1998) *Solar Air System – Built Examples*, James & James (Science Publishers) Ltd, London.

Hastings S R and Rostvik H N (editors) (1998), *Solar Air System – Product Catologue*, James & James (Science Publishers) Ltd, London.

Norton B, Hobday R A and Deal C R (1989) *The Development of a Passive Air-heating Solar-energy Collector for the Re-cladding of Buildings*, Proceedings of Applied Energy Research Conference, Swansea, Wales, Sept. 1989.

Romanazzo M, Rossi G and Scudo G (1988) *Performance of Chimney System Integrated to Residential Building*, European Conference on Architecture, 6–10 April 1987, Monaco, Kluwer Academic Publisher, Dordrecht, 727–732.

Scudo G, Fumagalli G and Rizzi G (1990) *Innovative Solar Components for Office Buildings*, in: T C Steemers and W Palz (editors) (1989), 1989, 2nd European Conference on Architecture, Paris, France, December 4–8 1989, Kluwer Academic Publishers, 260–262.

Chapter authors: Giancarlo Rossi and Gianni Scudo

II.3 System 3: Double envelope systems

INTRODUCTION

System description and principles

In a double envelope solar air system, solar heated air is circulated through cavities in the building envelope, surrounding the building with a layer of solar heated air which will reduce the auxiliary heat demand of the building. The indoor comfort is improved because inner surfaces of the external walls are warmer.

System variations and characteristics

Three main system variations of the double envelope system are described below. The variations relate to the nature of the air loop around the building and the type of collector used.

Active loop with solar air collectors and double envelope storage walls
The system is built as a shell outside a traditional building envelope, which makes it very convenient for retrofit applications and for new buildings, see Figure II.3.2. The system is well suited for multistorey buildings. If combined with external insulation, building materials can be doubly exploited. The collectors may be roof- or wall-integrated.

ADVANTAGES
- high degree of integration possible, even in retrofits;
- totally silent during the night since the system runs only during the daytime;
- low consumption of electricity for fans since they run only during the daytime.

Active loop with sunspace collector
In this variation a sunspace or a conservatory is used as a solar collector. The solar heated air is fan-forced through cavities in the building envelope, see Figure II.3.3. The heat is stored, either in the cavity walls if these have sufficient thermal mass, or in a separate heat storage located in the loop. The stored heat is released to the building by radiation to the room. If the heat is stored in the cavity walls, there is no need for night-time circulation and the heat is released passively to the room.

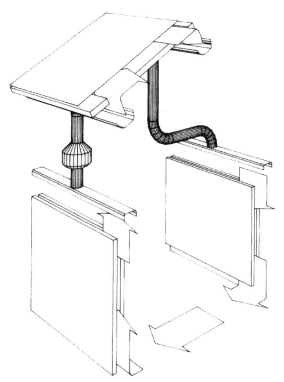

Figure II.3.1. System perspective drawing

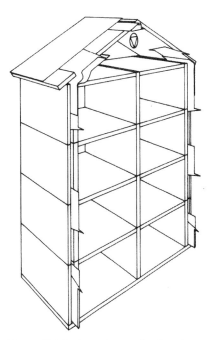

Figure II.3.2. Active loop with solar air collectors and double envelope storage walls

ADVANTAGES
- effective even at low temperatures;
- sunspace can be used for living.

LIMITATIONS
- sunspace is an inefficient 'collector' compared to flat plate air collectors;
- reduced daylight to living spaces if facades are covered with windows.

Multistorey loop with sunspace collector
The system can be described as a number of small sunspace systems placed on top of each other. Each of the systems functions independently, see Figure II.3.4. The systems can be run either with passive loops or fan-powered in active loops (more convenient in colder climates). Heat is stored in the thermal mass of walls and intermediate floors and discharged passively through radiation.

Suitable applications

In Gunnarshaug[1], Booth[2], Svenson and Vejen[3] and Nordström[4] practical applications and experiences are illustrated with built examples.

Apartments
A double envelope solar system can be installed when renovating or building a new apartment building. If the renovation includes adding external wall insulation, the marginal extra costs can be low.

Detached houses
If large, preferably unbroken wall areas exist, a double envelope system can be considered. A roof can also be included in the double envelope. Existing, uninsulated cavity walls can be transformed to double envelope walls.

Offices and schools
A double envelope system should not be built in large office and school buildings with overheating problems. Offices and schools in cold climates with large exposed surface areas and a space heating demand are good candidates for double envelope installations. A greenhouse used as solar collector will also add extra qualities to these buildings.

Industry and warehouses
These buildings often have a need for high ventilation rates and are more suited for simple, open solar air systems for preheating of ventilation air. If the building is massive and has envelope areas which need external insulation, a double envelope system can be considered.

ENERGY PERFORMANCE

This section starts with a description of the design of the reference solar system for the energy performance analysis.

A first, short presentation of the annual energy performance for double envelope systems under various climate conditions is given below. Calculated results concerning space heating are presented for a reference system which is a residential building with envelope design, thermal properties and utilization, according to descriptions in Appendix H (common building assumptions for all solar systems in this handbook).

In Energy performance of system variations below some examples of calculated results are presented for a residential building with external walls designed in other

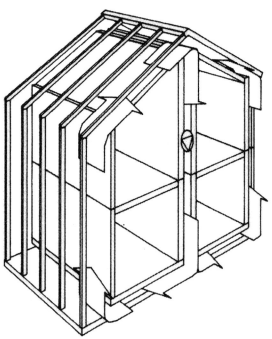

Figure II.3.3. Active loop with sunspace collector

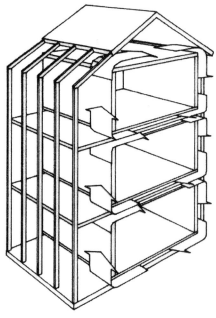

Figure II.3.4. Multistorey loop with sunspace collector

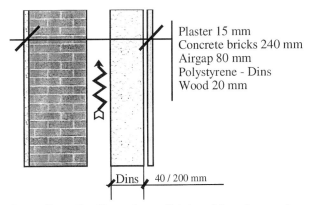

Figure II.3.5. Double envelope wall design of the reference solar system (indoors – left)

ways. The nomograms and tables give a more complete and systematic presentation of the energy performance for various wall designs as well as climate conditions.

All the results are obtained through use of Trnsair for solar air heating systems (see Appendix G).

Reference solar system

The principal design of the reference solar system is illustrated in Figure II.3.2 which shows a multistorey active loop with roof-integrated solar collectors combined with double envelope storage walls. This design has been applied in demonstration apartment blocks in Denmark and Sweden.[3–6] Here the air is supplied to and returned from the envelope gaps through ducts and manifolds integrated in the external walls. The airflow pattern is parallel in all walls.

Figure II.3.1 illustrates the system design and operation which was used to simulate and evaluate the energy performance for double envelope systems presented below. Here the air from the solar collector outlet is supplied to the north wall cavity and distributed further on in ducts to the south wall gap and back to the collector inlet. Consequently the airflow connects both walls in serial, which is why no manifold arrangements for upward air distribution are needed in the lower parts of the walls.

The double envelope wall design of the reference solar system is shown in Figure II.3.5. It is a heavy, insulated wall with the solar heated airflow of the gap in direct contact with a brickwork of hollow concrete blocks for efficient heat transfer, transport and storage. The highly efficient thermal insulation layer (expanded polystyrene) outside the air gap lowers the heat loss from the brickwork when its temperature is elevated for heat storage. The 40 mm layer of polystyrene represents a low insulation level and the 200 mm layer a high insulation level (see Appendix H for details). The wall design is applied in cloudy/temperate, sunny/temperate or sunny/cold climates.

Four climate types were selected for the analysis (Chapter I.2) and two levels of thermal insulation in the building were included. For the climate types cloudy/temperate, sunny/temperate and sunny/cold the low and high insulation level in the external walls is illustrated in Figure II.3.5 (building design, thermal properties and utilization – see Appendix H). The U-value of the wall is 0.58 W/m² K for the low insulation level and 0.17 W/m² K for the high insulation level. A reduction to 50% of the high insulation level has been selected for the climate type sunny/mild. In this case the U-value of the walls is 0.31 W/m² K for the high insulation level. Since this climate is considerably milder than the other three climates it has been assumed reasonable to use envelope insulation levels (walls, roof and ground) which result in total transmission plus air infiltration heat losses for the high insulation level in the same range as for the other climates. Also windows with higher transmission losses were selected for the sunny/mild climate compared to the other climates (see below).

Energy performance of reference solar system

The overall simulation results for the reference double envelope system (System Type 3) applied in residential buildings for space heating are shown in Figures II.3.6 and II.3.7. The figures present the simulation results for saved energy as a function of the specific collector area per m² floor for low insulation and high insulation level in the building (low and high insulation level – see Appendix H). 'Saved energy' is defined as the difference in net heat load (auxiliary heating demand) without and with the solar system in operation. This difference in relation to the net heat load without solar system operation defines 'saved energy fraction'.

For Figure II.3.7 a high insulation level corresponds to 50% of the levels given in Appendix H.

Energy performance of system variations

This section concentrates on variations of the external wall design Thermal properties of wall materials as heat capacity and thermal conductivity as well as the position of layers with high heat storage capacity are factors of certain and considerable importance in this context.

Four variations of the double envelope wall design are shown in Figure II.3.8. In all variations the thickness of the insulation layer (expanded polystyrene) has been made equivalent to the wall in the reference solar system. Indicated dimensions of the polystyrene layer represent low and high insulation levels for wall designs applied in cloudy/temperate, sunny/temperate and sunny/cold climates.

In the first wall variation (upper left in Figure II.3.8) the hollow concrete brickwork of the reference solar system is replaced by a homogenous concrete layer on one side of the air gap. The concrete has the same thickness as the concrete brickwork and represents a wall layer with 3.4 times higher thermal conductivity and 4 times higher volumetric heat capacity.

The second wall variation (upper right in Figure II.3.8) includes homogenous concrete layers on both sides of the air gap (same total thickness as in first

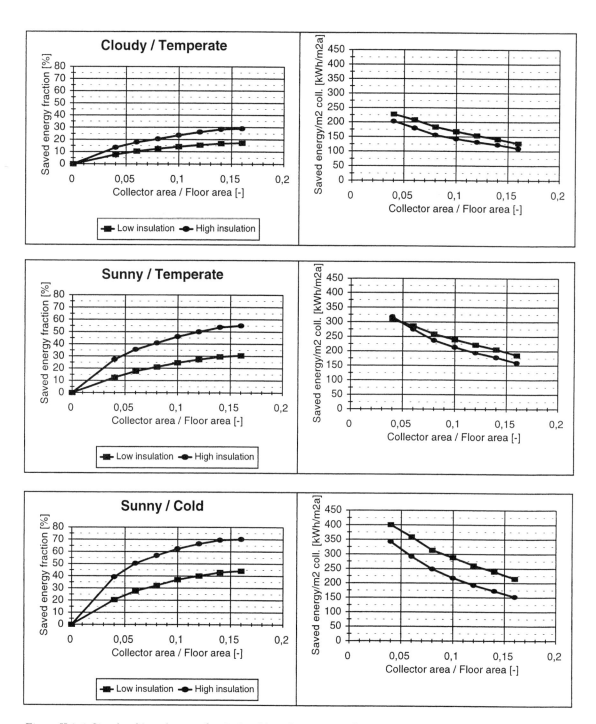

Figure II.3.6. Simulated 'saved energy fraction' and 'saved energy per m² collector' in cloudy/temperate (upper), sunny/temperate (middle) and sunny/cold (lower) climates

variation) in order to further increase the heat transfer to and transport in the storage material.

Instead of the heavy brickwork of the reference solar system the third wall variation (lower left in Figure II.3.8) includes a lighter brickwork type of LECA-blocks (Lightweight Expanded Clay Aggregates) on one side of the air gap (same thickness as in reference system). This represents a wall layer with 50% of the thermal conductivity and volumetric heat capacity for hollow concrete blocks.

The last wall variation (lower right in Figure II.3.8) includes a brickwork of hollow concrete blocks (same thickness as in reference system) which is separated from the air gap by a thermal insulation layer. This represents a wall type having a very low heat transport capacity from the air gap to the heavy material layer for heat storage.

The first and second wall variations as well as the reference wall type are presumed for new building projects, the third and fourth variations are primarily for retrofit projects. The reason for this is that these varia-

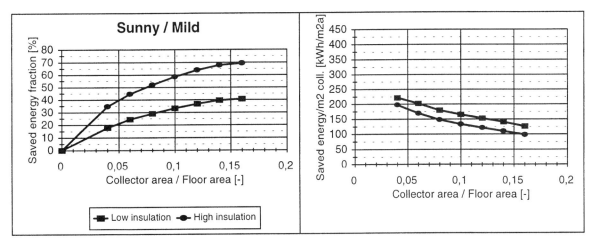

Figure II.3.7. Simulated 'saved energy fraction' and 'saved energy per m² collector' in sunny/mild climate

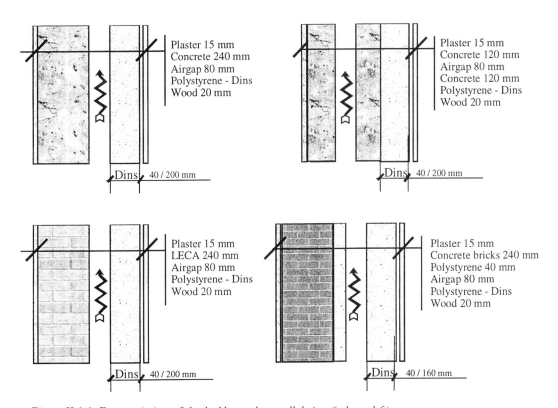

Figure II.3.8. Four variations of the double envelope wall design (indoors–left)

tions, even without the wall parts outside the air gap, have acceptably low U-values and could be found in residential building examples which are suitable for renovation measures.

Figure II.3.9 presents simulated saved energy fractions and saved energy as a function of the collector area per m² heated floor for low insulation and high insulation levels in the building. The presentation here is limited to a sunny/temperate climate in order to illustrate typical results and differences.

A conclusion from Figures II.3.9 and II.3.10 is that the first and second walls save about the same amount of energy. For the higher specific collector areas the saved energy fraction is about 5–10% higher than the saved fraction for the reference wall type. The saved energy for the other two wall variations is considerably lowered in relation to the saved energy for the reference case. The fourth wall variation has a particularly low energy performance as a result of the insulation between the thermal mass and the solar heated air gap. Consequently the second, third and fourth wall types have been excluded from the more complete presentations in Nomograms and worksheet below. However a short analysis of the third and fourth wall types

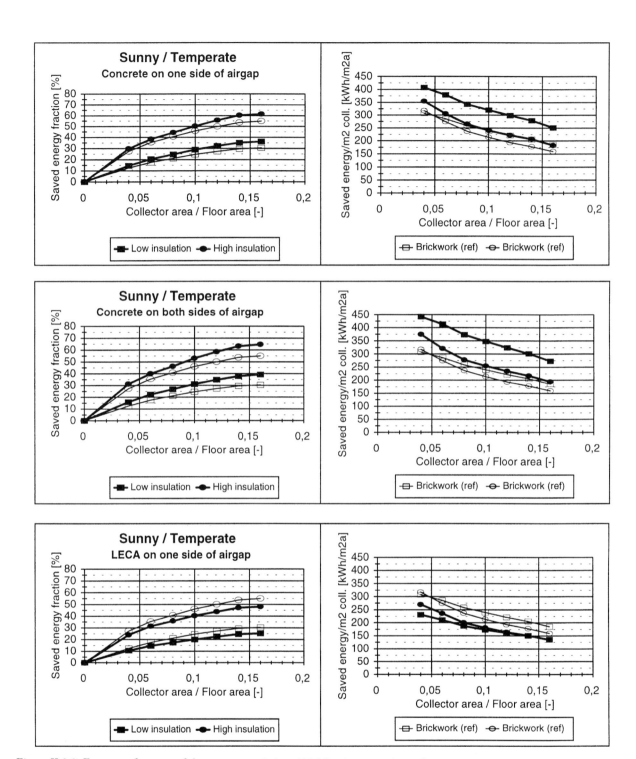

Figure II.3.9. Energy performance of three system variations (thick lines) compared to reference solar system (thin lines) for space heating of residential buildings in a sunny/temperate climate

concerning the total energy saving (before and after addition of the insulating wall parts outside the air gap and in final combination with a solar heating system) is included in Energy savings in building retrofit examples.

DESIGN

The design of the system must be linked to the design of the building itself. A double envelope system is in fact as much an architect's business as that of an engineer. The process must be carried out in close cooperation between these two disciplines.

System design

Step 1. Estimate building area, volume and heat load profile
Estimate the heat losses and space heating demand for the building. Calculations should show heat demand for space heating on a yearly and monthly basis.

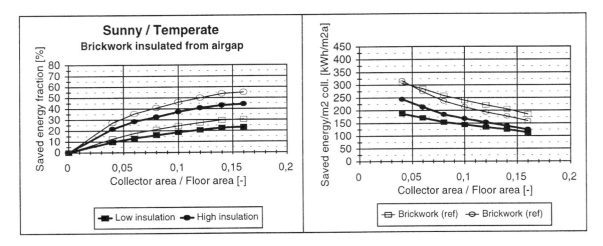

Figure II.3.10. Energy performance of one system variation (thick lines) compared to reference solar system (thin lines) for space heating of residential buildings in a sunny/temperate climate

Calculate building volume and floor area.

Step 2. Calculate solar collector system
Look for south-oriented areas which can be used as solar air collectors. If possible, use large, unbroken south-oriented roof or wall areas for solar collectors. Areas which are divided by windows and clerestories etc., are more complicated and expensive. It is also important to consider what is behind these areas inside the house. An empty space (like an attic) makes it easier to install ducting and fans than an apartment space.

Start the calculations with the largest possible collector area and then, later on, reduce the collector area in order to get an optimum system design. Consider also that manifolds and fittings will reduce the active collector area.

Read about flat plate solar collectors (Chapter IV.1) and select the collector type for the project. Try to find a type which can be well integrated in the building and allow you to doubly exploit existing construction to reduce costs.

Calculate the airflow and size of the air gap in the solar collector (the 'duct' cross-sectional area where the air is circulating).

As a general rule for standard and opaque building-integrated and site-built collectors, start with an air speed of 1–2 m/s. A solar air collector will perform best with an air speed of about 3 m/s (the thermal efficiency increases with increasing airflow rate). On the other hand, excess airflow will reduce the temperature and increase the air leakage, increase electrical consumption and will require larger fans.

Estimate the pressure drop and the efficiency of the solar collectors from the curves supplied from the manufacturer or found in the solar collector chapter.

If standard collectors are chosen, the manufacturer will provide recommendations and efficiency curves, from which airflow rates, air speed and pressure drop easily can be estimated.

In a *semi open system* where the system is open to the ambient air, the solar collector air channels will become dirty after some years and will need to be cleaned. This is very important, especially if the system is connected to the intake of fresh ventilation air. In this case, the solar collectors should be regarded as a conventional ventilation system with the same requirements for hygienic maintenance. Here filtration of incoming air is desirable.

In a *closed system* the air will be used only to transport heat from the collector to the air cavity in the wall, without mixing with the ventilation air. Since the air in the collector will have no contact with ambient air, it will stay clean. Maintenance will be minimal.

Step 3. Calculate double envelope system
Calculate the area and the width of the double envelope walls. For these walls, concentrate on large, unbroken parts of the external walls with as few window openings as possible. The possible connections to the double envelope from the collectors must also be taken into account.

Assume that the air cavity is 5 cm wide with thermal wall insulation on the outside and a wall with thermal mass between the air gap and the room.

From the collector airflow, the airflow and air speed in the double envelope air gap can now easily be calculated (see Figure II.3.11).

In order to distribute the solar heated air homogeneously in the air cavity, it may be necessary to install perforated manifolds. The pressure will drop and thus the air speed over the perforation should be increased in order to force the air to spread evenly. For a first estimation a recommended air speed is 5 m/s over the perforation. If, later on, the total system pressure drop is considered to be too high, the air speed over the perforation can be reduced until an optimal situation is reached. In order to reduce the air speed and pressure drop and to

---

envelope air speed (m/s) = airflow of collectors (m³/h) / horizontal section area of envelope air gap (m²) x 3600

Figure II.3.11. Equation for calculating airflow and air speed

maintain a good distribution of the air, the perforation holes should be smaller near the air inlet and larger with increasing distance from this point.

Step 4. Design distribution/duct system
The distribution/duct system connects the solar collectors to the double envelope walls, thus closing the system. The duct system is only used to transport air so heat loss to the surroundings should be minimized. Consequently the insulation requirement increases with the temperature of the air in the ducts.

The duct system should be kept as short and straight as possible to avoid pressure drops, noise and costs. Try to doubly exploit the building construction for the transportation of air to reduce costs and save space inside the building.

After designing and sizing the duct system, the air speed can be calculated from the airflow of the collector and the double envelope walls (the same volume of air will pass through these three parts of the system). The duct air speed should not exceed 5 m/s to avoid noise. If you find that the air speed is too high, it must be reduced, either by increasing the dimension of the ducts, or by reducing the total airflow of the system, starting with the solar collector.

In order to avoid high pressure drops and large, complicated systems, consider 'breaking up' a large system into several smaller systems. By doing this it is often possible to reduce the length of ducts and thus the costs. It is also easier to control the airflow and the distribution of air in a small system where the air only can 'take one way'.

Step 5. Calculate pressure drop and decide where to place the fans
Estimate the total pressure drop of the system, using the calculations of each system part. If the total pressure drop is unreasonably high, it should be reduced by redimensioning of the system components (wider air gap, larger/shorter ducts, reduction in the number of duct bends if possible).

It is a good idea to locate the fans in the loop, using the following procedure taking the pressure drops into account:

- Find the most leak-prone part of the system (the point where the zero-pressure of the system should be located).
- Locate the fan as far from this point as possible.
- Calculate the pressure drops from the fan to the weak point on both sides of the fan. If the pressure drops are equal on both sides, the impact of leakage over the weak point can be ignored.

Step 6. Choose fan and control equipment
If the fan is placed on the 'hot side' of the system (close to the solar collector), expect high temperatures in the airflow. In this case choose a fan with an external motor.

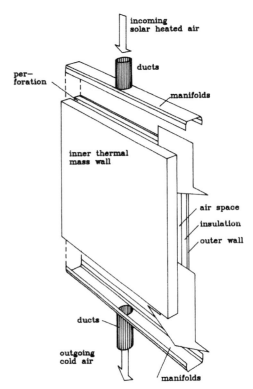

Figure II.3.12. The principle of a typical double envelope wall including duct arrangements

If the motor is placed in hot air (internal motor), the fan will halt due to overheating and the fan might have to be removed in order to reset the overheating switch. This can be avoided by placing the fan on the 'cold side' of the system (in front of the collector).

The advantage of fans with external motors is that they are independent of air temperatures and often more efficient. Fans with internal motors are often cheaper, easier to install and the motor heat is added to the air stream.

The fans are sized according to total pressure drop and the airflow of the system. Use the specific diagram for the fan from the manufacturer. Also consider the electric energy consumption.

In order to reduce the electricity consumption by the fans and maximize output of the system, choose a control strategy which is both efficient and reliable. A very advanced control system may be more effective when it functions, but on the other hand it is more complex and therefore sometimes less reliable. But, a too simple control system might result in bad system performance.

The control system should be easy to understand and run. Commercial service assistance usually is expensive and can quickly damage the budget. Examine the service provided by the manufacturer!

The location of sensors is critical. Since this kind of system is very much 'built in', access to sensors in the air channels after construction is nearly impossible. It might be a good idea to install a number of extra sensors in alternative spots. This will allow testing of alternative strategies at a low extra cost.

System 3: Double envelope systems

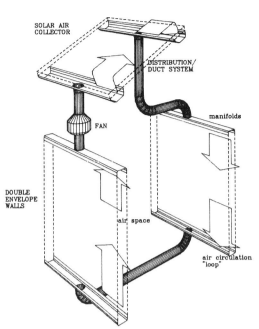

Figure II.3.13. Typical duct system layout

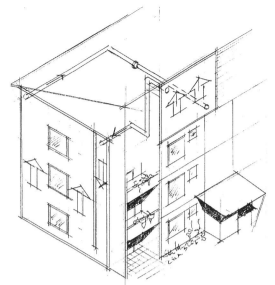

Figure II.3.14. Example of a system layout for a 3-storey multifamily solar retrofit building

Component design and specific design advice

Solar collector system

Any type of glazed, flat plate solar air collector can be used (see Chapter IV.1). The ducts must be kept short and the solar collector should be located close to the envelope loop. The possibilities for double exploiting are important for the system's budget. It is therefore important to coordinate the solar collector installations with the building or rebuilding plans in order to use materials and labour efficiently. It is sometimes possible to integrate the solar collectors in the roof or a south-facing wall, using the building materials for duct and absorber purposes. The solar collectors could either be prefabricated or site-built. If a site-built collector is chosen, much attention must be focused on air tightness and material durability. A collector can easily reach temperatures exceeding 100°C. Sealing and glazing materials must resist these temperatures for a long time. Plastic glazing materials, e.g polycarbonate, expand with heat. The fitting system must be designed for this expansion. Fittings are preferably made of EPDM rubber which is a soft and flexible material. A harder plastic fitting (e.g PVC) combined with aluminium or polycarbonate will make cracking sounds when the polycarbonate expands and shrinks in mornings and afternoons. This might be disturbing. It is important to get an even airflow through the collector. One way to achieve this is to divide a large collector area into a number of small sections. A perforation between the manifold and the air channels can even out the airflow.

Double envelope system

The double envelope is a critical part of the system. It serves as ducting, storage and distribution. Large envelope areas which are as unbroken by doors and windows as possible are desirable. The double envelope should also be in contact with the room spaces of the building which are planned to be heated.

It is important that envelope surfaces not covered by the double envelope (e.g. windows) have low heat losses. Otherwise these will lower the operative temperature and cause draughts.

If the envelope consists of heavy materials like concrete or masonry, solar heat can be stored in the double envelope and released passively to the building. If, on the other hand, the building envelope consists of light materials, heat storage must be arranged somewhere else within the loop, e.g. in the basement. In this case the air in the loop must be circulated by fans both day and night in order to distribute the heat. The double envelope must be insulated outside the cavity, otherwise the transmission losses will decrease the performance of the system. The width of the cavity should be designed taking the airflow and fan capacity into account. The airflow rate in the loop will influence the heat exchange to and from the thermal mass. This is explained more in detail in Chapter IV.6 on hypocaust storages. A double envelope system should be designed for an even airflow through all cavities. This can sometimes be achieved by dividing the envelope into a number of smaller sections. An even airflow can be achieved by placing the air inlet and outlet diagonally (on opposite sides of the air cavity). A perforated plate should be placed between the manifolds and the air cavity and the air speed should not exceed 5 m/s.

Finally, moisture, vapour barriers, air tightness and fire safety must comply with proper building construction methods as well as local building codes.

Duct system

The duct system transports solar heated air from the collector to the double envelope walls. In order to reduce

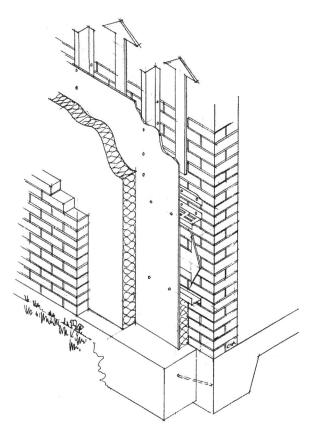

Figure II.3.15. Making a new double envelope cavity wall from an old (or new) brick wall

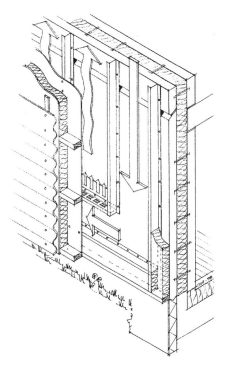

Figure II.3.16. Double envelope retrofit of an old concrete element wall; base section

the duct installations a split of the system into smaller systems might be considered. The duct system can be designed as a conventional air duct system. The ducts on the 'hot side' of the system (after the solar collector) should be insulated. If the fan is placed on the cold side of the system, before the solar collector, a fan with an internal motor can be used. The fan should be placed as far as possible from the least air-tight spot of the system to reduce air leakage (see Chapter V.1).

As in all ventilation systems, a solar air system can be noisy, but since a double envelope system is a 'closed' loop, it will not be in contact with the indoor space. This reduces the risk for low frequency noise caused by the fans. Fans and the duct system should have no or minor direct contact with the building structure. Since noise causes most problems during the night, it is an advantage if the fans do not have to run at that time. This is the case with passive heat discharge from thermal mass in direct connection to the room spaces.

Fans and dampers

Like all solar air systems, the double envelope system has a 'hot side' (after the collector) and a 'cold side' (before the collector). If located before the collector (on the cold side) a less expensive fan (i.e. with the engine placed in the airflow) can be used because it will not have to tolerate high temperatures. Since cold air is more dense than hot air, locating the fan on the cold side will be more efficient. Fans with the motor placed in the airflow often have a reset button for use if overheating and fan shutoff occurs. It is wise to assure that this button can be easily reached for resetting.

Dampers must be used, either for fire protection or to block cold 'back flow' in the loop during night time. It is sometimes possible to design the system in a way that will prevent back flow. If not, dampers should be installed, which close the loop when the fans are not running.

Control system

The simplest and probably most reliable control arrangement is a differential thermostat with one sensor in the top section of the solar collector and the other sensor in a central part of the double envelope. When the temperature in the solar collector exceeds the temperature in the double envelope by a few degrees the fans are switched on, otherwise not. In this way the fans will run only when there is solar heat to be gained. This will save electric energy and keep the system quiet during the night. This strategy requires thermal mass in the envelope which will allow passive discharge.

Fire protection

Fire safety is essential. In large buildings including several apartments, if the air channels cover more than one fire cell, precautions must be taken in order to prevent spreading of fire and smoke between cells. Fire detectors, which will switch off the fans in case of fire, should always be installed. Combustible materials should be avoided in the cavities of the solar collector and the double envelope. Local fire regulations and building codes for ventilation systems (see Chapter V.4) must be followed.

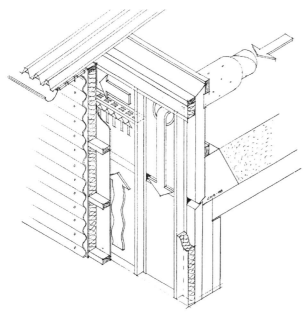

Figure II.3.17. Double envelope retrofit of an old concrete element wall; top section

CONSTRUCTION

Integration
Possibilities for integration lie hidden in all projects. For a double envelope house, focus on the building materials and the structure of the external walls. Almost always there are ways to doubly exploit building materials for solar purposes through small changes in the construction. The case study book produced within IEA-Task 19 presents different built examples.[4]

Double envelope
The building envelope must integrate the double envelope, the solar collector and the air circulation loop. Building materials must withstand the temperatures of the solar heated air and provide sufficient air tightness. During construction take precautions to keep the air cavity free of dirt. Once the wall is built, it cannot be reached for cleaning and repairing. If the manifold perforations are covered with rubbish from building materials during construction, this will decrease the function of the system during its entire lifetime. Such failures have to be prevented by careful site supervision.

Solar collectors
A simple site-built solar air collector is adequate. It can effectively be integrated in the roof if trapezoid corrugated sheet metal is used as roofing material. The sheet metal then can serve both as absorber and duct system for the solar air collectors.

New cavity walls
By adding a brick wall to a new or an existing brick or concrete wall, a new double envelope cavity wall can be created. Since a brick wall is very heavy, a new foundation must be added outside and fixed to the old wall. Remember that the outside brick wall must be allowed to have minor movements. The manifolds are arranged inside the cavity by using perforated sheet metal. When masonry work is performed, it is important that some kind of cover is placed over the perforation to prevent pieces of mortar and dirt from plugging up the perforation.

Concrete element walls
During the 1960s and 1970s, a large number of apartment houses were built using concrete elements. These houses suffer from bad insulation and severe cold bridges in the element joints. By turning these element walls into double envelope solar walls, the cold bridges can transfer solar heat to the rooms. It is important that the chemical composition of the sealant between the elements is considered; if it, for instance, contains PCBs (poisonous), it must be properly sealed to prevent leakage of toxic substances.

NOMOGRAMS AND WORKSHEET

This section presents a simple way to determine the energy performance of a double envelope system. The nomograms and tables were generated by Trnsair for solar air heating systems (see Appendix H) and are presented in Figures II.3.18 to II.3.25.

Also included are presentations concerning fan energy as well as correction factors for the type of solar collector and the collector airflow.

Finally, this section presents a short analysis of the energy savings owing to installation of a solar heating system compared to savings owing to additional wall insulation for double envelope wall types which are suitable for building retrofit projects.

Assumptions

The following assumptions have been made for cloudy/temperate, sunny/temperate and sunny/cold climates:

- system design and operation according to Figure II.3.1, solar heat first supplied to the north wall gap and further distributed to the south wall gap;
- two-floor, residential building, which is a part of the end of a rowhouse-block;
- external, double envelope walls according to Figures II.3.5 and II.3.8 facing south and north (with windows). External wall facing east with same design and insulation level but without air gap for heated air and windows. Internal, uninsulated wall facing west to adjacent zone with equal indoor climate (envelope design, materials etc.: see Appendix H);
- double-glass windows for low insulation level and low-energy windows (low-emission glass coating) for high insulation level (see Appendix H);
- roof-mounted, commercial and high performance flat plate solar collectors (selective absorber and single-glass cover, see Chapter IV.1);

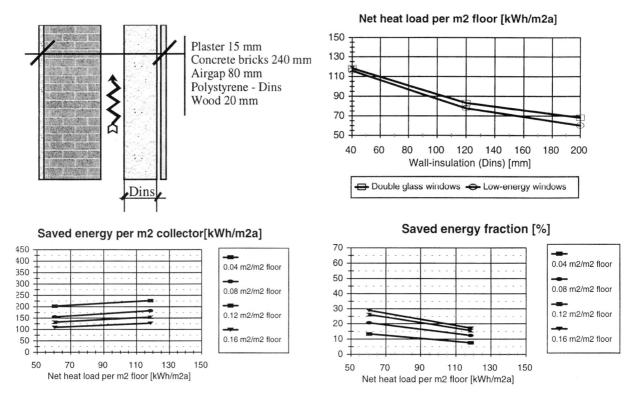

Figure II.3.18. 'Saved energy' in a cloudy/temperate climate. Concrete masonry on one side of air gap

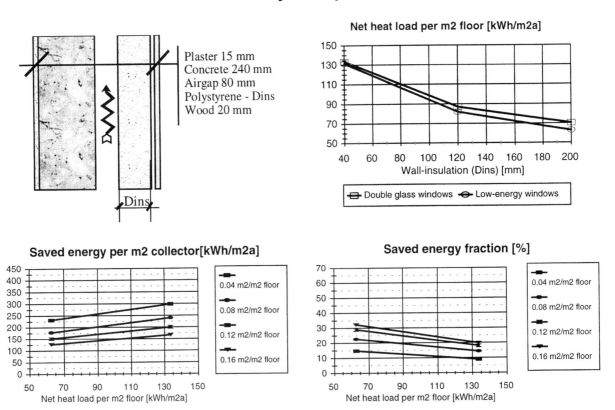

Figure II.3.19. 'Saved energy' in a cloudy/temperate climate. Homogenous concrete on one side of air gap

System 3: Double envelope systems

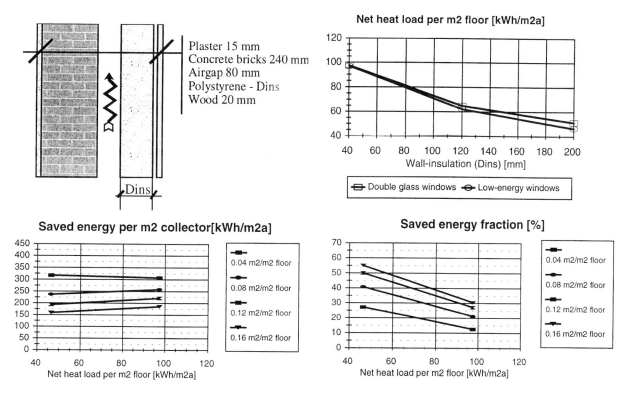

Figure II.3.20. 'Saved energy' in a sunny/temperate climate. Concrete masonry on one side of air gap

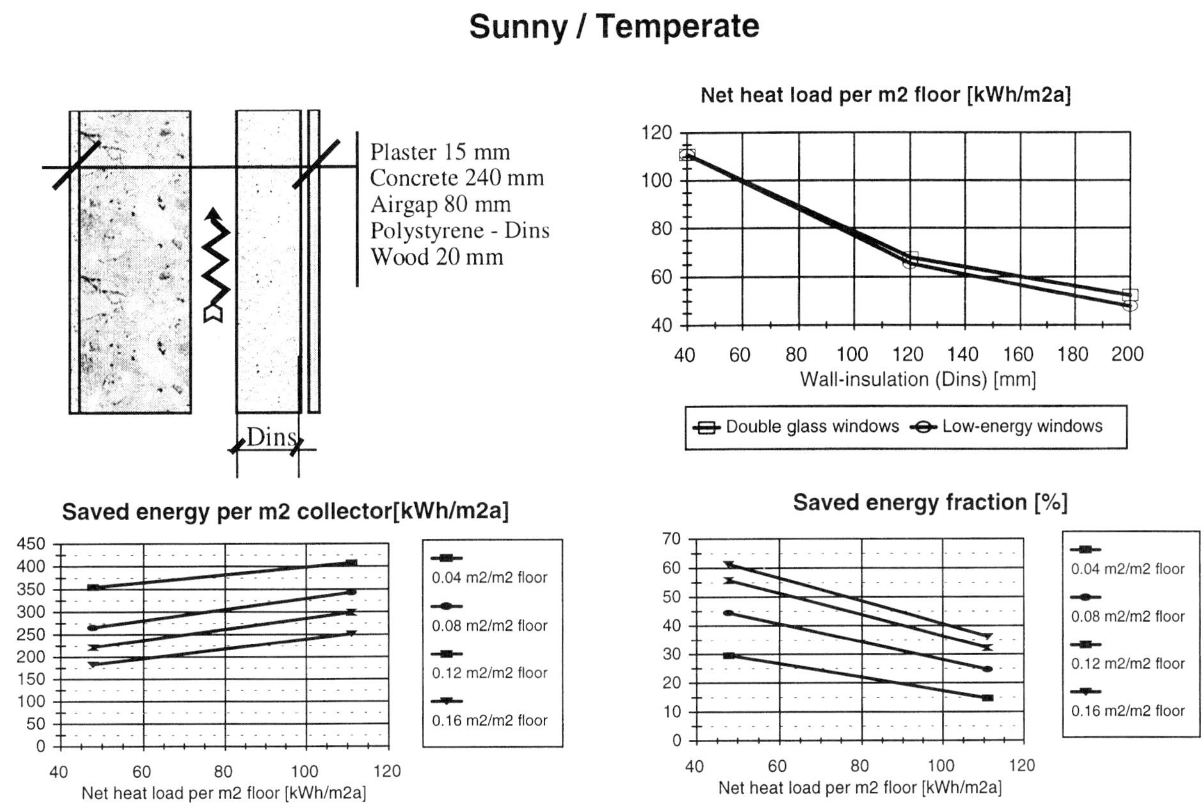

Figure II.3.21. 'Saved energy' in a sunny/temperate climate. Homogenous concrete on one side of air gap

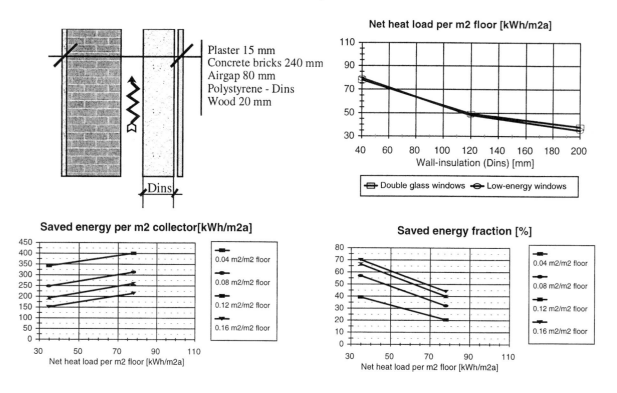

Figure II.3.22. 'Saved energy' in a sunny/cold climate. Concrete masonry on one side of air gap

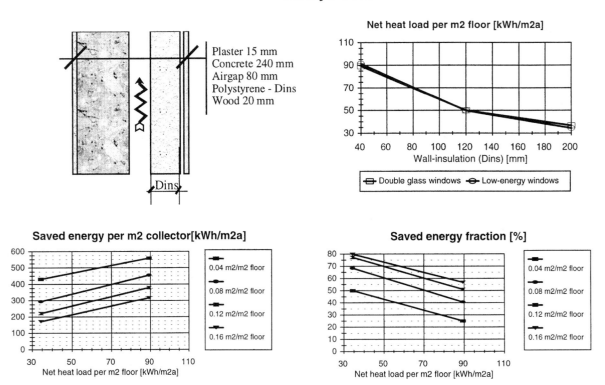

Figure II.3.23. 'Saved energy' in a sunny/cold climate. Homogenous concrete on one side of air gap

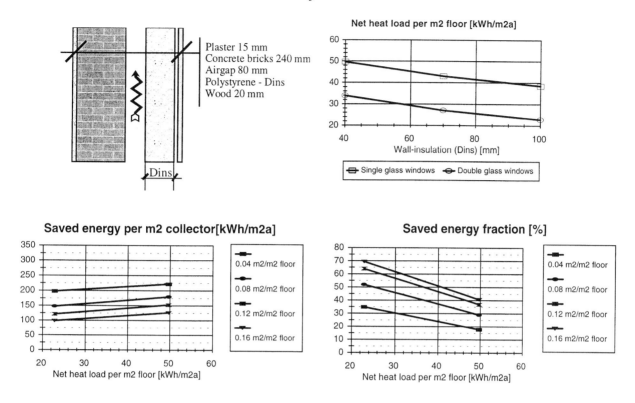

Figure II.3.24. 'Saved energy' in a sunny/mild climate. Concrete masonry on one side of air gap

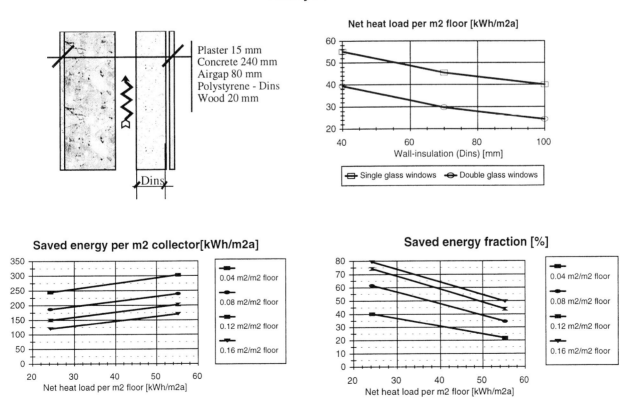

Figure II.3.25. 'Saved energy' in a sunny/mild climate. Homogenous concrete on one side of air gap

- solar system in operation September–May;
- collector tilt: 45°, collector azimuth: south;
- airflow rate through solar collectors: 50 m^3/m$^2_{coll}$ h
- fan-powered air circulation: 80 % of electric energy transferred as heat to the airflow is included in the energy savings;
- control strategy: fan turned on at collector temperature 3 K above north envelope temperature and off at 1 K difference. No fan operation if the indoor temperature exceeds 23 °C;
- heat losses from ductwork not considered, assumed of low importance for the total system performance.

For a sunny/mild climate assume:

- reduction to 50% of the envelope insulation level applied for the other three climate types for the high insulation level (same low insulation level);
- single-glass windows for low insulation level and double-glass windows for high insulation level (see Appendix H).

Nomograms

Two external wall designs of residential construction were analyzed. Figures II.3.18 to II.3.25 present simulation results for the saved energy as a function of the annual net heat load per m^2 heated floor (auxiliary heating demand without the solar system in operation) for specific collector areas between 0.04 and 0.16 m^2 per m^2 heated floor. The net heat load ranges between lower and higher values corresponding to high and low insulation level in the building envelope. The external wall design (concrete masonry on one side of air gap) and corresponding net heat load is illustrated in the top of the figure. For quick guidance and input in the nomograms

Table II.3.1. Simulated saved energy for residential buildings with external walls of homogenous concrete on both sides of the air gap (see Figure II.3.8, upper right)

Insulation level	Parameters	Units	Cloudy/temperate	Sunny/temperate	Sunny/cold	Sunny/mild
Low	Net heat load per m^2 floor	(kWh/m^2 a)	134	111	90	55
	Saved energy per m^2 collector					
	0.04 m^2/m^2 floor	(kWh/m^2 a)	325	443	611	328
	0.16 m^2/m^2 floor	(kWh/m^2 a)	183	273	334	185
	Saved energy fraction					
	0.04 m^2/m^2 floor	(%)	10	16	27	23
	0.16 m^2/m^2 floor	(%)	22	39	60	53
High	Net heat load per m^2 floor	(kWh/m^2 a)	63	48	35	25
	Saved energy per m^2 collector					
	0.04 m^2/m^2 floor	(kWh/m^2 a)	241	375	455	259
	0.16 m^2/m^2 floor	(kWh/m^2 a)	137	194	180	126
	Saved energy fraction					
	0.04 m^2/m^2 floor	(%)	15	31	52	42
	0.16 m^2/m^2 floor	(%)	35	65	83	82

Table II.3.2. Simulated saved energy of residential buildings with external walls of LECA masonry on one side of the air gap (see Figure II.3.8, lower left)

Insulation level	Parameters	Units	Cloudy/temperate	Sunny/temperate	Sunny/cold	Sunny/mild
Low	Net heat load per m^2 floor	(kWh/m^2 a)	106	86	69	45
	Saved energy per m^2 collector					
	0.04 m^2/m^2 floor	(kWh/m^2 a)	169	231	268	158
	0.16 m^2/m^2 floor	(kWh/m^2 a)	95	136	151	90
	Saved energy fraction					
	0.04 m^2/m^2 floor	(%)	6	11	16	14
	0.16 m^2/m^2 floor	(%)	14	25	35	32
High	Net heat load per m^2 floor	(kWh/m^2 a)	58	45	35	21
	Saved energy per m^2 collector					
	0.04 m^2/m^2 floor	(kWh/m^2 a)	177	270	280	148
	0.16 m^2/m^2 floor	(kWh/m^2 a)	95	135	131	77
	Saved energy fraction					
	0.04 m^2/m^2 floor	(%)	12	24	32	28
	0.16 m^2/m^2 floor	(%)	26	48	61	58

Table II.3.3. Simulated saved energy of residential buildings with walls of hollow heat load and the specific collector area for space heating of residential buildings under various climate conditions. The external wall includes a brickwork of hollow concrete blocks separated from the air gap by a thermal insulation layer (see Figure II.3.8, lower right)

Insulation level	Parameters	Units	Cloudy/temperate	Sunny/temperate	Sunny/cold	Sunny/mild
Low	Net heat load per m² floor	(kWh/m² a)	97	78	61	41
	Saved energy per m² collector					
	0.04 m²/m² floor	(kWh/m² a)	137	188	213	124
	0.16 m²/m² floor	(kWh/m² a)	80	113	120	75
	Saved energy fraction					
	0.04 m²/m² floor	(%)	6	10	14	12
	0.16 m²/m² floor	(%)	13	23	31	29
High	Net heat load per m² floor	(kWh/m² a)	59	45	35	21
	Saved energy per m² collector					
	0.04 m²/m² floor	(kWh/m² a)	160	245	250	125
	0.16 m²/m² floor	(kWh/m² a)	87	125	119	65
	Saved energy fraction					
	0.04 m²/m² floor	(%)	11	22	29	24
	0.16 m²/m² floor	(%)	24	44	55	51

Table II.3.4. Annual values for the heat dissipated from the fan motor to the collector airflow in various climates (fan electricity demand: 10 W/m² collector, 80% heat dissipation). The values are given per m² collector

Specific collector area (m²/m² floor)	Cloudy/temperate (kWh/m² a)	Sunny/temperate (kWh/m² a)	Sunny/cold (kWh/m² a)	Sunny/mild (kWh/m² a)
0.04	10.5	13.5	12.1	9.9
0.16	8.3	10.8	9.4	8.0

Table II.3.5. Correction factors for recalculating the energy savings by commercial collectors (selective absorber, single cover) to savings for the site-built collector type (black absorber, single cover) under various climate conditions

Specific collector area (m²/m² floor)	Cloudy/temperate (–)	Sunny/temperate (–)	Sunny/cold (–)	Sunny/mild (–)
0.04	0.70	0.75	0.75	0.75
0.16	0.70	0.75	0.80	0.80

Table II.3.6. Correction factors for recalculation of the energy savings for the collector airflow 50 m³/m²$_{coll}$ h to savings for other airflows under various climate conditions

Specific collector airflow (m³/m² h)	Specific collector area (m²/m² floor)	Cloudy/temperate (–)	Sunny/temperate (–)	Sunny/cold (–)	Sunny/mild (–)
25	0.04	0.9	0.9	0.9	0.9
25	0.16	1	1	1	1
100	0.04	0.95	1	1	1
100	0.16	0.95	1	1	1

the net heat load as a function of the wall insulation thickness and window type can be found from the wall design illustration.

'Saved energy' is defined as the difference in net heat load (auxiliary heating demand) without and with the solar system in operation. This difference in relation to the net heat load without solar system operation defines the 'saved energy fraction'.

For the remaining three external wall designs, according to Figure II.3.8, shorter presentations of the saved energy versus the annual net heat load and the specific collector area are given in Tables II.3.1 to II.3.3. It is relevant to use linear interpolation between table values for saved energy if alternative values are selected for the specific collector area. As can be seen from Figures II.3.18 to II.3.25 linear interpolation with respect to insulation thickness is less relevant in the tables below.

Fan energy consumption and dissipation

The energy savings presented in Figures II.3.18 to II.3.25 and Tables II.3.1 to II.3.3 include heat from the fan to the collector airflow.

Table II.3.7. Energy savings from solar heating compared to savings from additional wall insulation. The wall is LECA-blocks on one side of the air gap (lower left in Figure II.3.8) and the additional insulation is 40 mm of expanded polystyrene. Specific collector areas (left column) are expressed in m^2 collectors per m^2 heated floor

Saving measure	Units	Cloudy/temperate	Sunny/temperate	Sunny/cold	Sunny/mild
Additional insulation	(kWh/m^2 a)	33	30	27	13
Solar heating 0.04 m^2/m^2 floor	(kWh/m^2 a) (%)*	7 21	9 31	11 39	6 49
Solar heating 0.10 m^2/m^2 floor	(kWh/m^2 a) (%)*	12 38	17 58	20 71	12 91
Solar heating 0.16 m^2/m^2 floor	(kWh/m^2 a) (%)*	15 47	22 72	24 88	15 113

* Related to saving for additional insulation

Table II.3.8. Energy savings from solar heating compared to savings from additional wall insulation. The wall is hollow concrete blocks separated from the air gap by 40 mm of expanded polystyrene (lower right in Figure II.3.8) and the additional insulation is a similar layer. Specific collector areas (left column) are expressed in m^2 collectors per m^2 heated floor

Saving measure	Units	Cloudy/temperate	Sunny/temperate	Sunny/cold	Sunny/mild
Additional insulation	(kWh/m^2 a)	19	17	15	7
Solar heating 0.04 m^2/m^2 floor	(kWh/m^2 a) (%)*	6 30	8 44	9 57	5 71
Solar heating 0.10 m^2/m^2 floor	(kWh/m^2 a) (%)*	10 55	15 85	16 103	10 136
Solar heating 0.16 m^2/m^2 floor	(kWh/m^2 a) (%)*	13 69	18 107	19 128	12 171

* Related to saving for additional insulation

Concerning fan energy the following assumptions have been made:

- specific electricity demand: 10 W/m^2 collector;
- 80% of electric energy transferred as heat to the airflow (fan motor inside ducts, minor heat losses from motor not supplied to the airflow).

Table II.3.4 presents calculated, annual values for the energy dissipated from the fan to the collector airflow under various climate conditions. The annual electric energy demand for the fan is a factor 1.25 times the table values. The annual operation time for the fan expressed in hours is a factor 125 times the table values. If alternative values are selected for the specific electricity demand or the percentage of heat dissipation, linear corrections of the table values are relevant in order to calculate energy dissipation and demand.

Correction factors for collector type and airflow

The energy savings presented in Figures II.3.18 to II.3.25 and Tables II.3.1 to II.3.3 are calculated for roof-mounted, commercial and high performance flat plate solar collectors with selective absorber and single-glass cover.

A simple alternative which could be less expensive is roof-mounted, site-built flat plate collectors with a black absorber (non-selective) and single-glass or single-plastic cover. This type of collector has a reduced thermal performance compared to the commercial type (see Chapter IV.1).

Table II.3.5 presents correction factors for recalculation of the energy savings for the commercial collector type to savings for the site-built type under various climate conditions. Linear interpolation between table values for specific collector area is relevant.

The energy savings presented in Figures II.3.18 to II.3.25 and Tables II.3.1 to II.3.3 are calculated for an airflow rate through the solar collectors of 50 m^3/m^2_{coll} h.

Table II.3.6 presents correction factors for recalculation of the energy savings for the airflow 50 m^3/m^2_{coll} h to savings for other airflows under various climate conditions. Linear interpolation between table values for specific collector area and airflow is relevant.

Energy savings in building retrofit examples

Two variations of the double envelope wall design were presented above. These are suitable for retrofit projects (see Figure II.3.8). For the two wall designs a short analysis is presented below concerning the energy savings due to additional wall insulation compared to savings due to installation of a solar heating system.

The first wall design (lower left in Figure II.3.8) includes LECA-blocks (Lightweight Expanded Clay Aggregates) on one side of the air gap. Energy savings from solar heating for this wall are presented in Table II.3.2. Table II.3.7 presents these energy savings related to floor area for a specific collector area of 0.04, 0.10 and 0.16 m^2_{coll}/m^2_{floor}. Table II.3.7 also presents the energy saving related to floor area specifically for additional wall insulation comprising of a 40 mm layer of expanded

System 3: Double envelope systems

Worksheet

Step	To be calculated	Calculations
1	building net heat load	Figure II.3.21 (upper): **68** kWh/m²$_{floor}$ a
2	saved energy per m² collector (high performance collectors)	Figure II.3.21 (middle): **265** kWh/m²$_{coll}$ a
3	saved energy per m² collector (site-built collectors)	correction from Table II.3.5: **0.75 × 265** = 199 kWh/m²$_{coll}$ a
4	saved energy fraction (high performance collectors)	Figure II.3.21 (lower): **44** %
5	saved energy fraction (site-built collectors)	correction from Table II.3.5: **0.75 × 44** = 33 %
6	saved energy per m² floor area	saved energy fraction (5) × net heat load (1): **0.33 × 68** = 22 kWh/m²$_{floor}$ a
7	dissipated fan energy (heat) per m² collector	Table II.3.4: 12.2 kWh/m²$_{coll}$ a
8	dissipated fan energy (heat) per m² floor area	fan energy (7) × specific collector area: **12.2 × 0.10** = 1.2 kWh/m²$_{floor}$ a
9	dissipated fan energy related to saved energy	(8) / (6): **1.2 / 22** = **0.055**
10	fan electric energy demand per m² collector	(7) × **1.25**: **12.2 × 1.25** = **15.3** kWh/m²$_{coll}$ a
11	fan electric energy demand per m² floor area	fan electric energy (10) × specific collector area: **15.3 × 0.10** = **1.5** kWh/m²$_{floor}$ a
12	annual fan operation time	fan electric energy (10) × 1000 / fan specific electricity demand: **1.5 × 1000 / 10 = 1500** h/a

Blank worksheet

Step	To be calculated	Calculations
1	building net heat load	Figures II.3.18 to II.3.25 (upper) or Tables II.3.1 to II.3.3: _ kWh/m²$_{floor}$ a
2	saved energy per m² collector (high performance collectors)	Figures II.3.18 to II.3.25 (middle) or Tables II.3.1 to II.3.3: _ kWh/m²$_{coll}$ a
3	saved energy per m² collector (site-built collectors)	correction from Table II.3.5: _ × _ = _ kWh/m²$_{coll}$ a
4	saved energy fraction (high performance collectors)	Figures II.3.18 to II.3.25 (lower) or Tables II.3.1 to II.3.3
5	saved energy fraction (site-built collectors)	correction from Table II.3.5: _ × _ = _ %
6	saved energy per m² floor area	saved energy fraction (5) or (4) × net heat load (1): _ × _ = _ kWh/m²$_{floor}$ a
7	dissipated fan energy (heat) per m² collector	Table II.3.4: _ kWh/m²$_{coll}$ a
8	dissipated fan energy (heat) per m² floor area	fan energy (7) × specific collector area: _ × _ = _ kWh/m²$_{floor}$ a
9	dissipated fan energy related to saved energy	(8) / (6): _ / _ = _ -
10	fan electric energy demand per m² collector	(7) × 1.25: _ × 1.25 = _ kWh/m²$_{coll}$ a
11	fan electric energy demand per m² floor area	fan electric energy (10) × specific collector area: _ × _ = _ kWh/m²$_{floor}$ a
12	annual fan operation time	fan electric energy (10) × 1000 / fan specific electricity demand: _ × _ 1000 / _ =h/a

polystyrene which represents the low insulation level. Table values are given which express energy savings owing to solar heating in percent of savings owing to additional insulation.

The second wall design (lower right in Figure II.3.8) includes a brickwork of hollow concrete blocks separated from the air gap by 40 mm of expanded polystyrene. The additional wall insulation outside the air gap is also 40 mm of expanded polystyrene. Energy savings due to solar heating for this wall are presented in Table II.3.3. Table II.3.8 presents energy savings for this wall design in a format equal to Table II.3.7.

It can be concluded that the energy savings from solar heating are on the same level as savings from additional wall insulation for sunny climates and specific collector areas of 0.10–0.16 m²$_{coll}$/m²$_{floor}$.

Worksheet

The following example illustrates the practical use of the diagrams and the tables in this section. After the example an incompleted worksheet is provided for the reader's own calculations.

Assumptions:
- new residential building with external walls including 240 mm of homogenous concrete, air gap for solar heat, 120 mm of expanded polystyrene and 20 mm of

wood cladding (upper left in Figure II.3.8) and double-glass windows;
- sunny/temperate climate;
- site-built solar collectors with black absorber (non-selective) and single-glass cover;
- specific collector area: 0.10 m^2/m^2 floor;
- collector tilt: 45°, collector azimuth: south;
- specific collector airflow: 50 m^3/m$^2_{coll}$ h;
- fan specific electricity demand: 10 W/m$^2_{coll}$;
- 80% of electric energy transferred as heat to the airflow.

Tasks
- calculate annual values for 'saved energy per m^2 collector' and 'saved energy fraction';
- estimate the fan energy dissipated as heat to the collector airflow, the electric energy demand and the annual operation time for the fan.

REFERENCES

1. Gunnarshaug J (1989) *Experience with Norwegian Double-envelope Houses with an Active Solar Heating System*. IEA Status Seminar on Solar Air Systems, March 1–3, 1988, Utrecht, The Netherlands. NOVEM, The Netherlands.
2. Booth D (1980) *The Double Shell Solar House*. The Community Builders, Canterbury, New Hampshire, USA.
3. Svendsen B and Vejen N K (1993) *Project Solar House*. Danacon A/S, Consulting Engineers and Thermal Insulation Laboratory, Technical University of Denmark, Report no. 258.
4. Nordström C (1998) A solar air apartment block (Gothenburg, Sweden). In: Hastings R (ed) *Solar Air Systems – Built Examples*. London, James & James (Science Publishers) Ltd, 113–117.
6. Gustén J and Jagemar L (1992) *Hybrid Solar House in Gothenburg, Sweden, Measurement and Evaluation* (in Swedish). Department for Building Services Engineering, Chalmers University of Technology, SE-412 96 Gothenburg, Sweden. Internal Report no. I37:1992.
5. Nordström C (1985) *Hybrid Solar System in Connection to Renovation of a Multifamily House in Järnbrott – Gothenburg, Sweden* (in Swedish). The Swedish Council for Building Research, Stockholm.

Chapter authors: Christer Nordström and Torbjörn Jilar

II.4 System 4: Closed collection loop with radiant discharge system

INTRODUCTION

System description

In this system an air collector is connected to building integrated heat storage. The air is circulated in a closed loop by natural or fan-forced convection through the collector to the storage and back to the collector. The room-facing surface of the storage discharges heat by radiation and convection to the heated space. A simple temperature difference control between the collector and the storage operates the fan. Reverse air circulation during the night is avoided by a damper located below the collector.

This type of system has been successfully applied in residential buildings throughout central and northern Europe as well as in North America. If well designed it may cover a substantial fraction of the annual heating load (Figure II.4.5).

System variations

Depending on the type of collector and storage and on the way the storage is connected to the collector several variations of the system can be imagined. Flat-plate air collectors or venetian blind window air collectors are both suitable and the heat from the collector may be delivered to either a rockbed, a hypocaust or a PCM storage which then radiates heat into the room. Some variations are described below. The single collection loop system will be used as the reference system for design advice and calculations throughout this chapter.

Single loop systems with radiant discharge
(FIGURE II.4.2) CHARACTERISTICS

- proven solution for residential buildings;
- collectors may be roof-mounted or facade-integrated;
- heat storage in hypocaust floors or walls (murocausts) uses the building structure in an efficient way;
- thermocirculation in the air loop is possible.

(FIGURE II.4.3) CHARACTERISTICS

- proven solution for residential buildings;
- venetian blind air window collectors are part of the building envelope and offer both collector and daylighting control functions;
- heat storage in rockbeds is used in many countries;
- fan-forced air circulation is recommended if rockbeds are to be used.

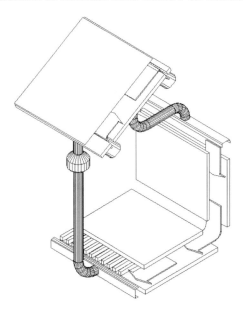

Figure II.4.1. Principle of air collector connected to building-integrated heat storage

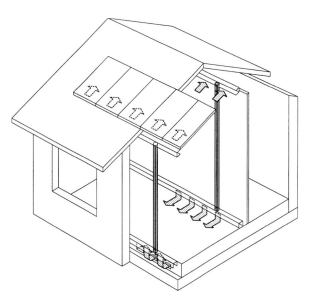

Figure II.4.2. Roof-mounted air collector connected to a hypocaust storage

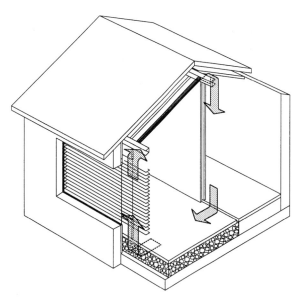

Figure II.4.3. Venetian blind air window collector connected to a rockbed storage

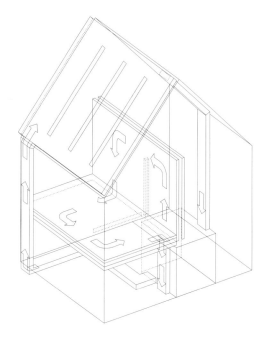

Figure II.4.4. System with roof-mounted collector, rockbed storage in the cellar and natural discharge to hypocaust and murocaust (after S Larsen)

Two loop systems with radiant discharge

For better control and distribution of heat throughout the building a separate discharge loop may be used (see Figure II.4.4). For this purpose a first storage element, preferably a well-insulated rockbed storage, is solar heated to a relatively high temperature level. Air is circulated by natural convection in a closed loop from this first storage to a second storage unit which will remain at a lower temperature level and discharges itself by radiant heat transfer. A hypocaust and/or murocaust will preferably be used as low temperature storage. This combination of two storage elements at different temperature levels is elaborate and requires much experience. Several successful applications exist in Vorarlberg (Austria).

(FIGURE II.4.4) CHARACTERISTICS

- distribution to every part of the building possible;
- sharing of hypocaust and duct system with auxiliary heating system for the distribution of heat;
- simple inexpensive controls and mechanical equipment;
- very low consumption of electric power;
- sensitive to obtaining correct pressure drop in the hypocausts.

Advantages

- using the collector as part of the building envelope lowers the extra costs for the solar system;
- building integrated storage will facilitate heat distribution;
- sharing of hypocaust and duct system with auxiliary heating system will lower extra costs for the solar system;
- simple inexpensive controls.

Limitations

- window air collectors tend to overheat the adjacent rooms;
- rockbed storage is bulky;
- high pressure drop and relatively high consumption of electric energy compared to saved energy.

Suitable applications

By building usage

The closed collection loop system with radiant discharge is applicable to a wide variation of building types. It is

Table II.4.1. Suitable applications by building usage

Building usage	Air collector type		Storage type	
	Flat-plate	Window collector	Rockbed	Hypocaust
Housing	Suitable	Suitable	Suitable	Suitable
Office buildings	Suitable	Overheating risks	Not recommended	Suitable
Schools	Suitable	Overheating risks	Suitable	Suitable
Sports halls	Suitable	Glare risks	Suitable	Suitable
Warehouse	Suitable	Not recommended	Suitable	Suitable

recommended when the occupancy internal gains are low (i.e. housing and small office buildings). If window air collectors are used, these can also provide glare and solar control.

Suitable climates

Most continental European and North American climates are suitable.

ENERGY PERFORMANCE

Reference system

The *reference system* for residential buildings (see Figure II.4.2) is defined as follows:

- roof-mounted commercially available flat-plate collectors;
- collector tilt 45°, azimuth 0° south;
- hypocaust storage;
- fan-powered air circulation;
- radiant discharge.

Energy performance of reference system

The graphs in Figure II.4.5 allow a quick estimate of the energy performance of the reference system for residential and office buildings in four different climate types. The graphs cover different insulation and mass levels of the building, two types of collectors and three different hypocaust types (see Design for details).

The energy performance of the reference system is expressed as *saved energy per m^2 of collector* per heating season. The heating season has been set from September until May for all climates for better comparison. The shape of the curves shows the typical non-useable solar heat, which occurs with increasing collector areas. The system performs differently in the various climate zones and has its highest values in a sunny/cold climate. Due to the larger heat demand in residential buildings compared to office buildings, the closed collection loop system is most economical in residential buildings. The results have been obtained using Trnsair to simulate the parameter variations.

Energy performance of system variations

The performance of the system variations described in the Introduction above are compared in Table II.4.2. The various technology factors allow a rough estimate of the energy performance of a system, which differs in one or more components from the reference solar system. A factor of 1.0 for all components is assigned to the reference system.

Example: A site-built flat-plate collector integrated into the facade and coupled to a rockbed storage will

Table II.4.2. Technology factors for components

Technology		Factor
Collector	Commercial flat-plate collector	1.0
	Site-built flat-plate collector	0.8
	Window air collector	0.5
Building integration	Roof-mounted 45 degrees	1.0
	Wall/facade	0.8
Storage	Hypocaust storage	1.0
	Rockbed storage	0.9
Air circulation	Fan-forced circulation	1.0
	Natural circulation	0.5

suffer a decrease in energy performance compared to the reference system. The values in Table II.4.2 are based on estimates and experience from built examples in central Europe. The following general observations can be made:

- The thermal performance of window air collectors is lower compared to the performance of flat-plate collectors;
- Collectors on the south wall perform less well than steep roof-mounted collectors;
- Rockbeds require fan-forced circulation;
- Systems with natural circulation are estimated to yield 50% less than systems with fan forced circulation.

The rules of thumb in Table II.4.3 apply for residential buildings in a cloudy/sunny temperate climate.

Table II.4.3. Predicted energy performance for different system variations in a cloudy/sunny temperate climate

Collector type used in system	System performance kWh/m^2a (heating season)
Window air collector	80–150
Site-built wall collector	100–190
Site-built roof-mounted collector	120–240
Commercial flat-plate roof-mounted collector	150–300 (reference system)

Energy performance of built examples

Many investigators have obtained system performance in kWh/m^2a (heating season) from in situ measurements. Because these data are related to the thermal quality of the building envelope they serve only as general guidance. Energy performance from measured data is usually somewhat lower than predicted by computer simulations because of construction weakness and 'non-standard' behaviour of the inhabitants.

DESIGN

System and component design

The objective here is to optimize components of the solar system in order to achieve a cost-effective system performing well. Both thermal and hydraulic aspects are discussed. In this section the step by step design procedure for the

Residential buildings

Sunny/temperate

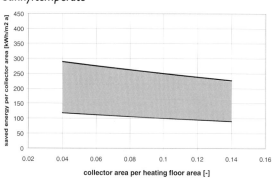

Cloudy/temperate

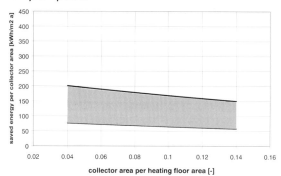

Sunny/cold

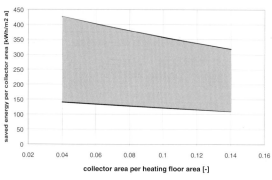

Sunny/mild

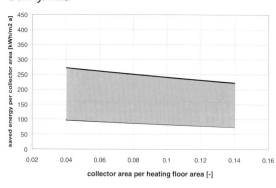

Office Buildings

Sunny/temperate

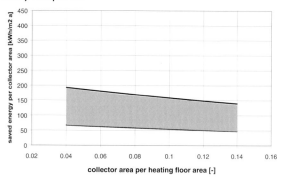

Cloudy/temperate

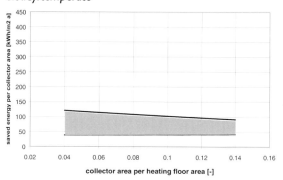

Sunny/cold

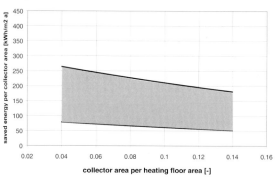

Definition of axis

Y-axis: Saved energy per collector area and heating season

X-axis: Ratio of collector area per heated floor area

Figure II.4.5. Energy performance of reference system expressed as saved energy

Table II.4.4. Energy performance of selected systems in residential buildings

Collector type	Storage type	Location	Climate	Solar system output kWh/m² collector
Commercial flat-plate, roof-mounted	Hypocaust	Düsseldorf (D)	Cloudy temperate	278
Site-built flat-plate, roof-mounted	Rockbed/Hypocaust	Vorarlberg (A)	Cloudy temperate	115
Window air collector	Hypocaust	Swiss middle land	Cloudy temperate	140

reference system is described. Information is also given for component variations. The step by step procedure is closely related to the nomograms for the reference system. This section also describes the use of the nomograms.

Step 1 Choose collector type and size the collector field

Depending on the available building surfaces, roof-mounted or facade-mounted collectors will be selected. Which collector type will best fit the needs: flat-plate or window collectors? Is a combination possible? For detailed description and more information about the performance of air collectors see Chapters IV.1 and IV.2.

The efficiency of the air collector will be best when the air flow rate is larger than 50 m³/h m² collector. A maximum temperature rise of 20K for window air collectors and 30K for flat-plate collectors will then result. For the window air collectors the air flow has to be in contact with both sides of the venetian blind inside the cavity.

Collector orientation and tilt
The collector tilt and orientation influence the system gains. Table II.4.5 gives correction factors obtained from Trnsair simulations for the standard collector in a well-insulated residential building and for cloudy temperate climate.

Collectors facing west have a smaller useable output due to reduced heat needs of the building in the afternoon. The Trnsair simulations do not allow rotation of the building. However, this is only a small additional uncertainty in the correction factors. In addition, the Trnsair model favours the use of internal and direct passive gains when calculating the contributions to cover the net heat load.

Area of collector field
The necessary collector area depends on the collector type, flow rate through the collector and orientation and tilt. The ratio of collector area per m² of heated floor area is likely to range between 0.05 and 0.2. The collector area for a desired energy saving can be estimated from Figure II.4.5. If other than commercial flat-plate collectors are to be used, the technology factors given in *Energy performance of system variations* have to be considered. When using window air collectors the collector area per m² of heated floor area is larger than for flat-plate collectors and ranges between 0.1 and 0.3.

Step 2 Select appropriate air flow rate

In forced circulation, air flow rates are typically 40 to 70 m³/h m² for all types of collectors considered in this chapter. The nomograms allow interpolation within this range. Air velocities in the flat-plate collector are between 2 to 5 m/s, for the window air collector approximately 0.5 m/s. Flow rates exceeding approximately 80 m³/h m² cause undesired pressure drop, while flow rates less than approximately 40 m³/h m² lead to very high collector temperatures and therefore high thermal losses. Simulation results also show a decrease of the system efficiency for larger flow rates.

Air flow by natural circulation is very slow and consequently lower saved fractions result. Typical flow rates for a two-story high collector are 15 to 35 m³/h m². The nomograms have been developed for forced circulation.

The air flow rate is always expressed per m² collector area. If the collectors are arranged in series, each collector will be subjected to the total air flow rate.

Step 3 Select appropriate storage type and capacity

Three hypocaust geometries have been considered for the nomograms. Type A1 (see Chapter IV.6) may be characterized by a high convective transfer in closely spaced air channels. This yields the best results for the simulations for the nomograms. Type A3 has a larger channel diameter and therefore a smaller convective heat exchange coefficient. The distance between the air channels is about twice as large compared to both other types. It has the lowest efficiency. Type A9 is intermediate and has also been selected for the reference system. The whole structure may easily reach 50 cm of thickness. The hypocaust storage may be replaced by a rockbed storage, if the rules of thumb in Table II.4.6 are accepted.

The values in Table II.4.6 come from experience with built projects. The density for rockbeds was taken as 1600 kg/m³ and the specific heat 0.84 kJ/kgK. Because approximately 30% of a rockbed is air gaps, the heat

Table II.4.5. Correction factor as a function of collector orientation and tilt

Orientation	Tilt	Correction factor
0° (S)	30°	0.9
0° (S)	45°	1.0
0° (S)	60°	1.0
+30° (SSW)	45°	0.9
-30° (SSE)	45°	1.0
+60° (SW)	45°	0.7
-60° (SE)	45°	0.85

Table II.4.6. Storage capacity for closed loop systems with radiant discharge

Storage type	Volume per collector area m^3/m^2	Heat capacity per collector area kWh/m^2K
Hypocaust/murocaust	0.5–1.0	0.2–0.4
Rockbed	0.3–0.7	0.1–0.25

Table II.4.7. Pressure drop of components in a closed loop system

Component	Pressure drop
Flat-plate collector	3–5 Pa/m^2 of collector
Window air collector	0.5–3 Pa/m^2 of collector
Duct system	0.5–3 Pa/m length of duct
Hypocaust storage	0.5–5 Pa/m length of storage
Rockbed storage	15–40 Pa/m length of storage

capacity is less than for a hypocaust of the same size. On the other hand the better heat exchange in the rockbed compensates for the lower heat capacity. The specific storage volume and consequently the heat capacity is larger (by a factor of two) for closed loop systems compared to open loop systems (Chapter II.5). See also Chapters IV.6 and IV.7 for hypocausts and rockbeds.

Step 4 Read system output and saved energy

Use the nomograms (Figure II.4.8–21) to determine the system output as well as the saved energy. Several iterations may be necessary to obtain the optimal configuration. The nomogram is entered with the net heat load per m^2 of floor area and the fraction of collector area per floor area.

For well-insulated buildings with heat recovery from ventilation air a smaller fraction of collector area to floor area has to be used for optimum system gains.

Step 5 Size ducts, fans, controls and auxiliary heating system

Ducts
Layout of the duct system connecting the various components in the air system is standard HVAC engineering. The following rules should be observed:

- Keep ducts as short and straight as possible to reduce pressure drop, heat loss and costs.
- Insulate ducts with at least 50 mm of insulation with a conductivity of 0.04 W/mK in non-heated spaces. For duct diameters of 0.35 m approximately 0.8 WmK of heat will be lost per meter of duct run.
- Air speed in the ducts should not exceed 3–5 m/s in order to avoid noise.
- Use dampers to avoid night losses through reverse air circulation.

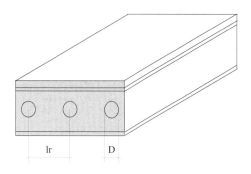

Figure II.4.6. Hypocaust geometries for closed loop system nomograms

Calculate pressure drop of the system (see Chapter V.1). Table II.4.7 gives values from built experience.
Example for the pressure drop of an entire system:

15 m^2 of window air collector:	15 Pa
20 m of ducts:	30 Pa
10 m of hypocaust:	<u>25 Pa</u>
	70 Pa

Fans
Proceed as follows to determine where to place fans, the sizing of fan power and consumption of electrical energy.

Select a low pressure drop fan from Chapter V.1. Radial duct fans can move an air volume of 150 m^3/h to 1000 m^3/h at pressure drops between 50 Pa and 350 Pa.

If the fan is to be placed on the hot air side of the collector there are two options: either install a fan with an external motor, or use a fan with a motor which can tolerate the 100–120°C air stream. Note that the fan efficiency drops by approximately 10% because of lower air density if the fan is placed on the hot air side of the collector.

A rule of thumb for the fan power is 5–15 W/m^2 of collector or 3–8 kWh/m^2a assuming 700 running hours per year. In general fans with two speeds will be sufficient. Variable speed control will be used in connection with PV-driven fans. Fans will be traditionally on the hot side of the system. It will be necessary to include dampers in order to prevent any unwanted backwards air flow during hours when the fan is not running.

Controls
Use a thermostat with an adjustable on-off differential with one sensor in the collector and another sensor near the cold side of the storage. Storing heat will be initiated if the temperature in the collector is approximately 6K above the average storage temperature for hypocausts or above the temperature of the cold side of the rockbed. The system will turn off if the temperature difference drops below approximately 3K. The on-off temperature difference will influence the running time of the fan and therefore the system output. Built examples show that the operation time of fans will range between 600 and 1000 hours per heating season in a cloudy/sunny temperate climate.

Auxiliary heating system
Choose a back-up heating system which matches the remaining heat demand and which offers a large flexibility if components must be replaced. Sometimes the solar air system can be combined with an auxiliary air heating system.

Using Trnsair

Trnsair offers the possibility to perform design calculations for a closed loop collection system with radiant discharge, as discussed in this chapter. Only a limited number of components and variations are possible in the analysis. Rockbeds can be modelled in connection with System 4. However, window air collectors are not available in the input.

CONSTRUCTION

System related integration

A critical design step is fixing the layout of ducts between the collector and storage for the given building design. Manifolds collect and distribute air from or to the collector and also to and from the storage. The design of the manifolds is demanding and often the cause of system performance not meeting expectations.

Component related integration

Collector

In closed loop systems the collector will be part of the building envelope. Since the room heating demand is largest during winter steep slopes or vertical surfaces in facades are preferred. The collector field is usually a large fraction of the roof or facade, in contrast to the small collector field for DHW supply.

Flat-plate collectors are commercially available and benefit from the experience with water collectors. Window collectors are site-built usually, though commercially available products are used in Scandinavian countries ('air flow' or 'exhaust air windows').

Ductwork

The ductwork includes manifolds to collect and distribute air to or from the collector and the storage. In order to distribute air to all hypocaust channels large manifolds are needed. Plan the installation of the fan so that it can be easily replaced. Experience has shown that the fan power is often too small for the system as built.

Ductwork, dampers and fans are standard ventilation components and request common HVAC-engineering. Noise protection such as acoustic dampers in the ducts should not be neglected.

Storage

Hypocaust storage offers large surface areas for radiant heat discharge and provides comfort. Large surface area rockbeds have been successful in one-story buildings but are not recommended for multi-story buildings because of construction difficulties.

The surface area of the hypocaust may in many cases be larger than the values in the design recommendations, just because the floor area is given. The amount of direct solar gains through the windows will compete with the energy gained from the air system, thus influencing the level of insulation of the storage.

The thermal storage is part of the building structure. The storage will have to be well insulated when in contact with the ground. Room-facing areas will be insulated at different levels to control the heat release. The insulation level will be higher in rooms with direct solar radiation.

NOMOGRAMS

Assumptions

The nomograms have been produced for the reference system and a heating season from September to May. They have been calculated for standard and high performance flat-plate collectors at a tilt of 45 degrees. Set points were 3 K to start running the fan and 1 K to turn off the system. The running time of the fan was 1100 hours. The dissipated fan heat is about 7% of the collected energy.

Example of nomogram use

Starting from a net heat load of 71 kWh/m^{2}a, a specific collector area of 0.04, an air flow rate slightly above 40 m^3/m^{2}h, and a highly efficient hypocaust a system of 280 kWh/m^{2}a is obtained from the nomogram. This result is in the range of the measured value of 278 kWh/m^{2}a.

BIBLIOGRAPHY

Hastings S R (ed) (1994) *IEA Task 11: Passive Solar Commercial and Institutional Buildings. A Sourcebook of Examples and Design Insights*, John Wiley, New York.

Hastings S R (ed) (1998) *Solar Air System - Built Examples*, James & James (Science Publishers) Ltd, London.

Kornher S (1984) The Complete Handbook of Solar Air Heating Systems, Rodale Press, Emmans, Pa.

Kurer T, Filleux C, Lang R and Gasser H (1982) *Research Project SOLAR TRAP; Final Report* (in German), Zurich.

Labhard E, Binz A and Zanoni T (1995) *Erneuerbare Energien und Architektur (Fragestellungen im Entwurfsprozess - Ein Leitfaden)*. EDMZ Nr. 724.215D, Bundesamt für Konjunkturfragen Bern (also available in Italian).

Short W and Kutscher C F (1984) *Analysis of an Active Charge/Passive Discharge Solar Space Conditioning System Employing Air Collectors and Hollow-core Concrete Block Floor*. Report SERI. Available from the National Technical Information Service, Springfield, VA, USA 22161.

Chapter authors: Charles Filleux and Peter Elste

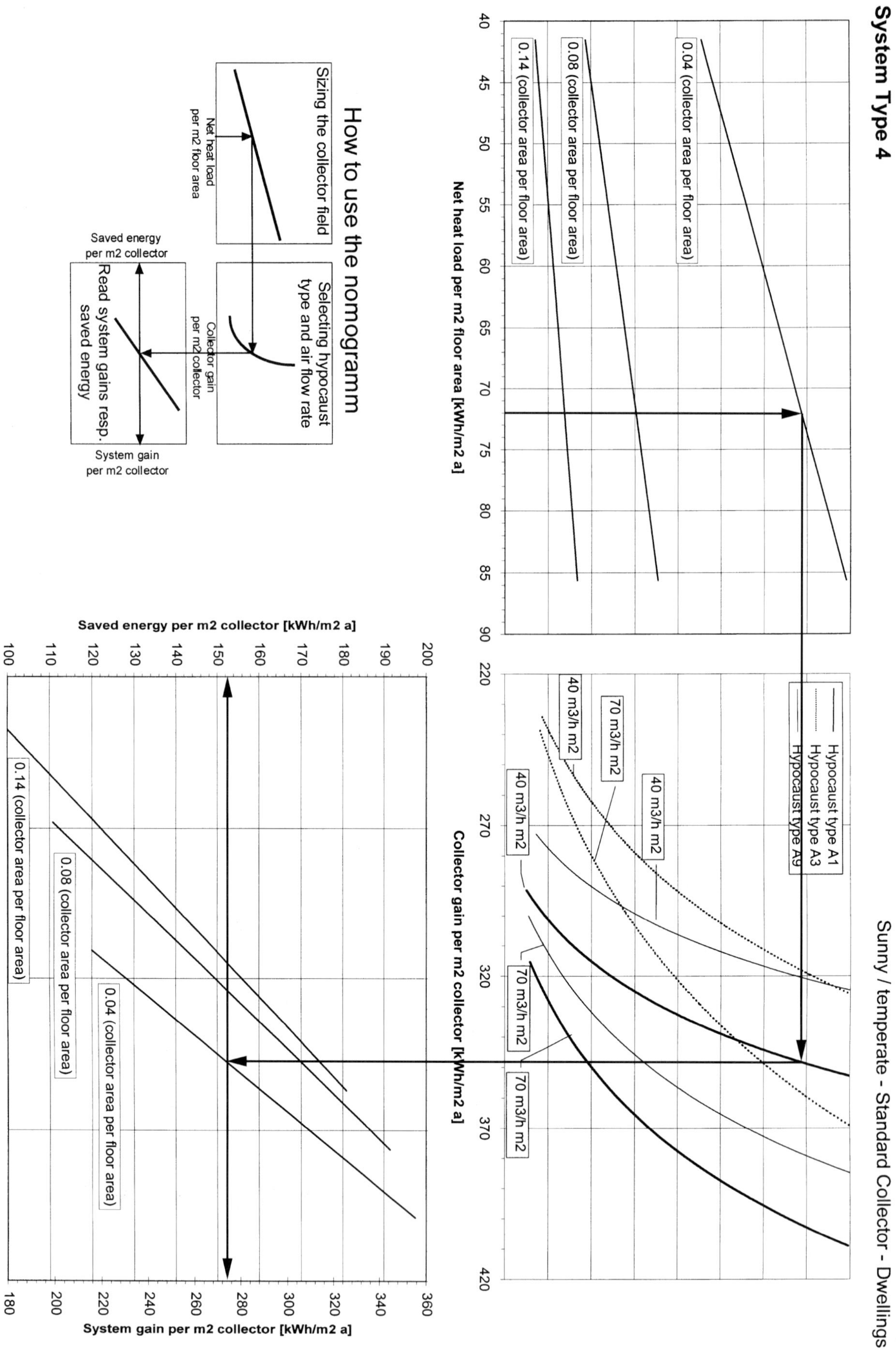

System Type 4

Figure II.4.7. Example of nomogram use

Sunny / temperate - Standard Collector - Dwellings

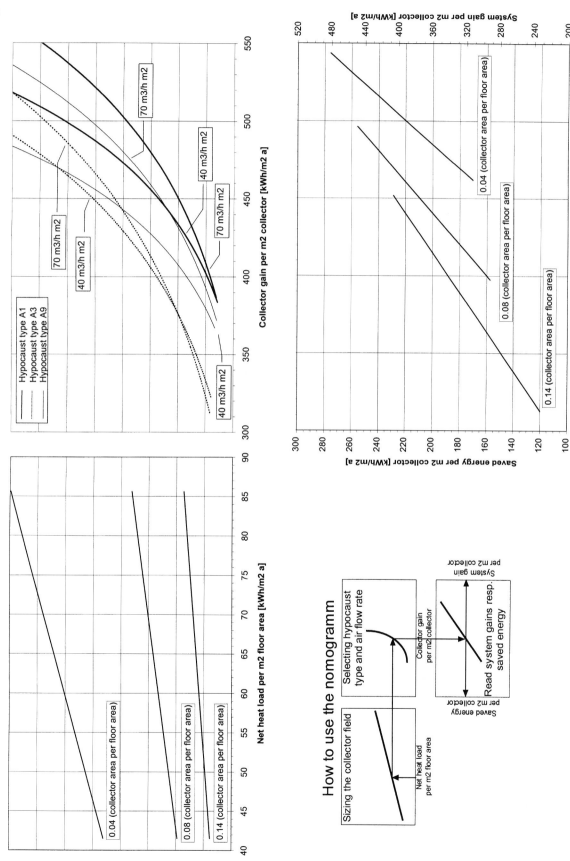

Figure II.4.8. Nomogram for sunny/temperate, high performance collector, dwellings

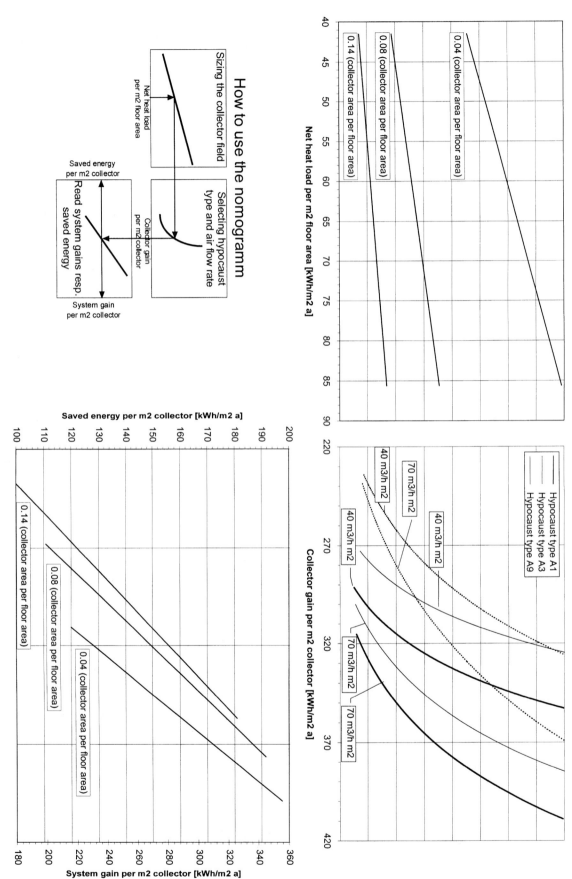

Figure II.4.9. Nomogram for sunny/temperate standard collector, dwellings

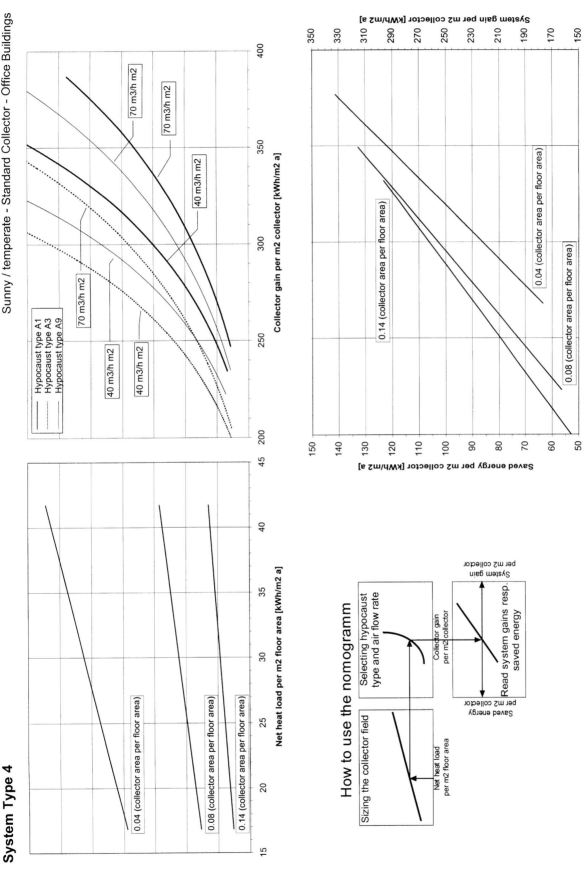

Figure II.4.10. Nomogram for sunny/temperate, standard collector, office buildings

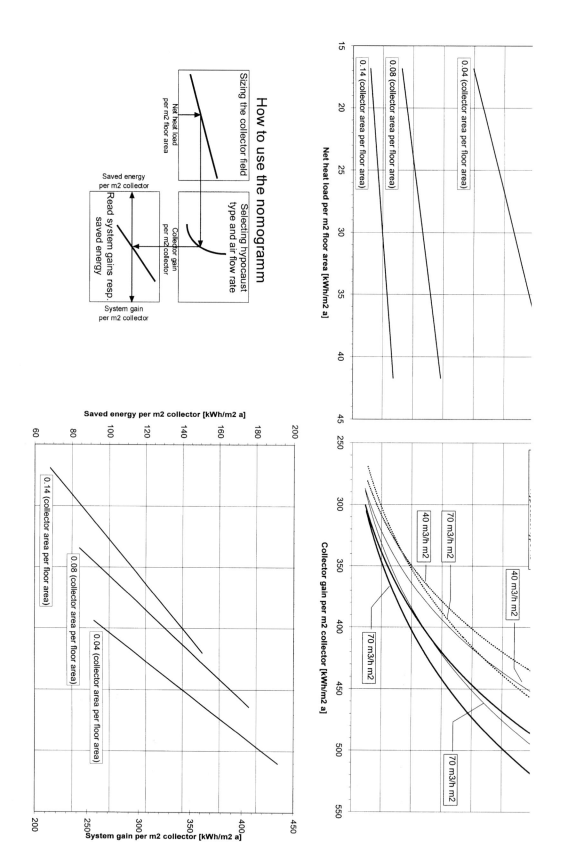

Figure II.4.11. Nomogram for sunny/temperate, high performance collector, office buildings

System 4: Closed collection loop with radiant discharge system

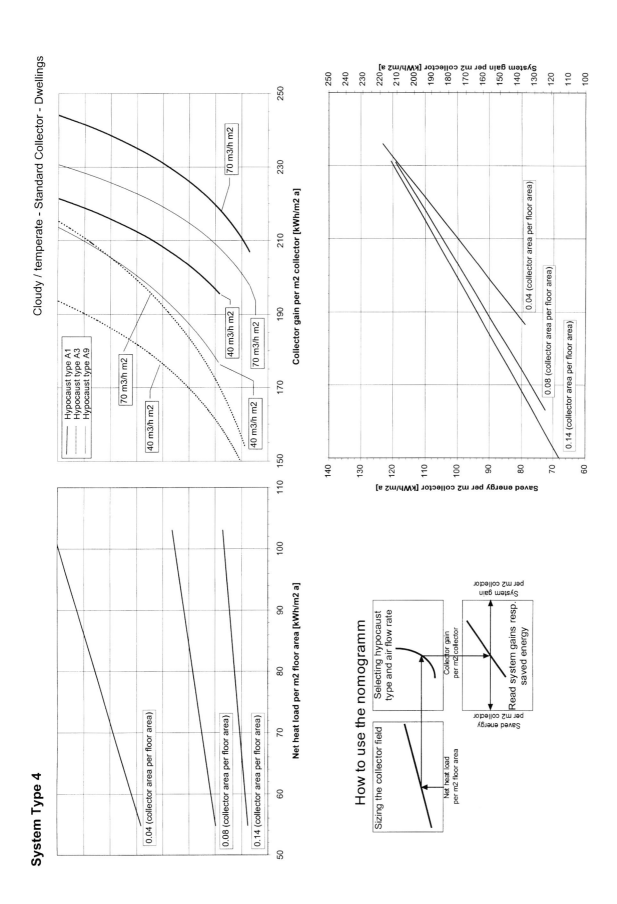

Figure II.4.12. Nomogram for cloudy/temperate, standard collector, dwellings

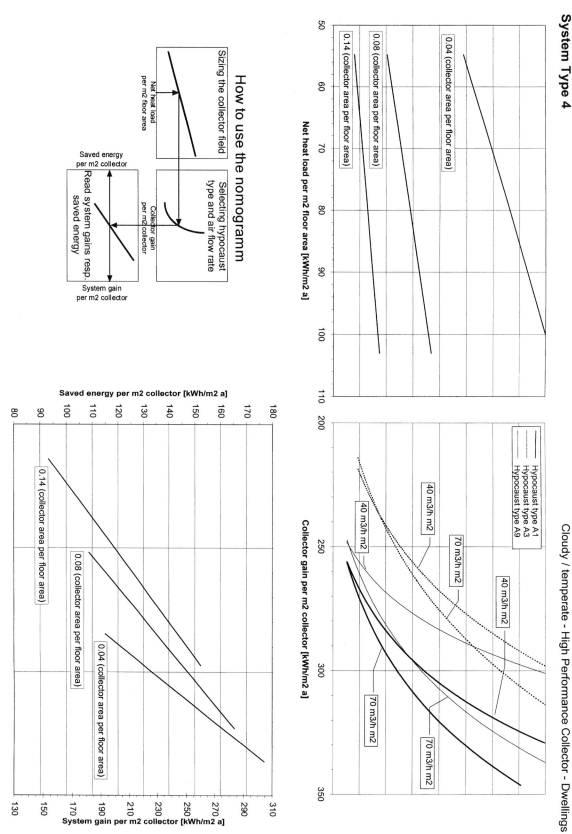

Figure II.4.13. Nomogram for cloudy/temperate, high performance collector, dwellings

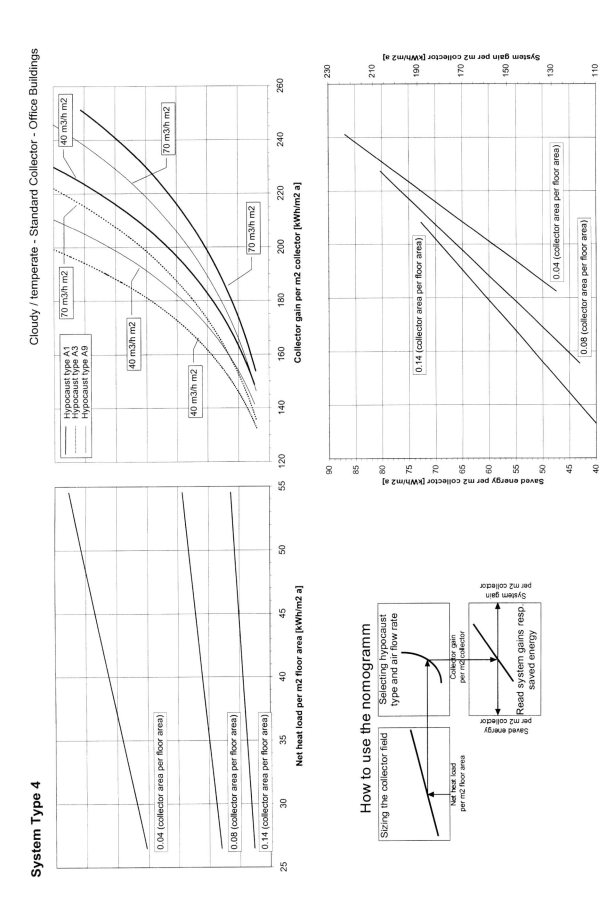

Figure II.4.14. Nomogram for cloudy/temperate, standard collector, office buildings

System Type 4

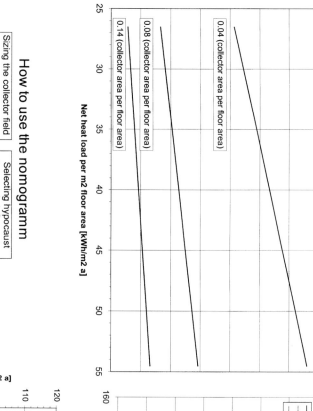
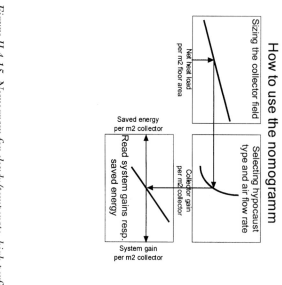
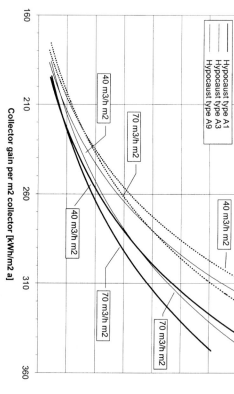
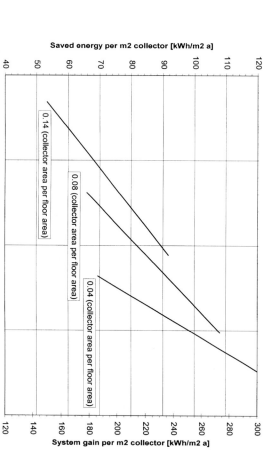

Figure II.4.15. Nomogram for cloudy/temperate, high performance collector, office buildings

System 4: Closed collection loop with radiant discharge system

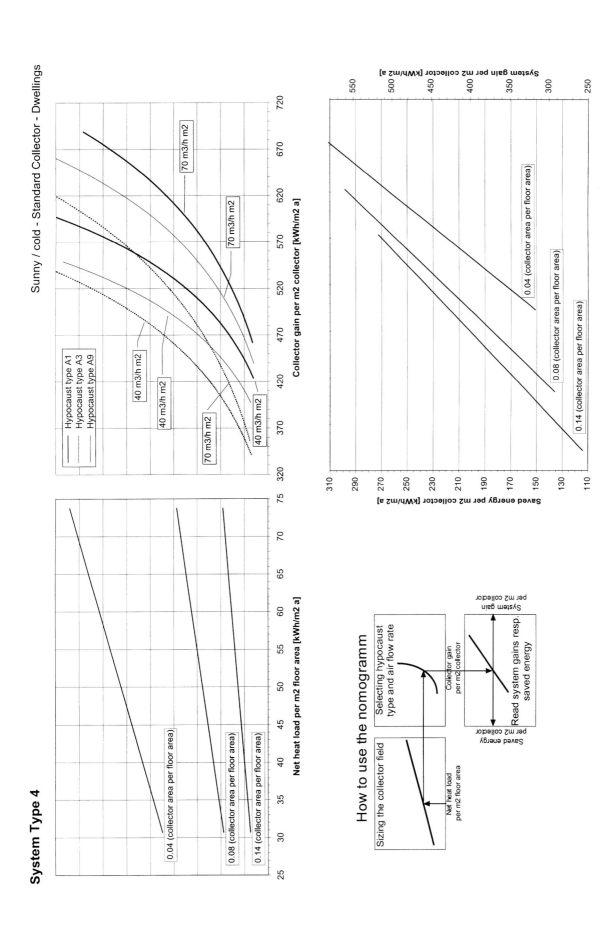

Figure II.4.16. Nomogram for sunny/cold, standard collector, dwellings

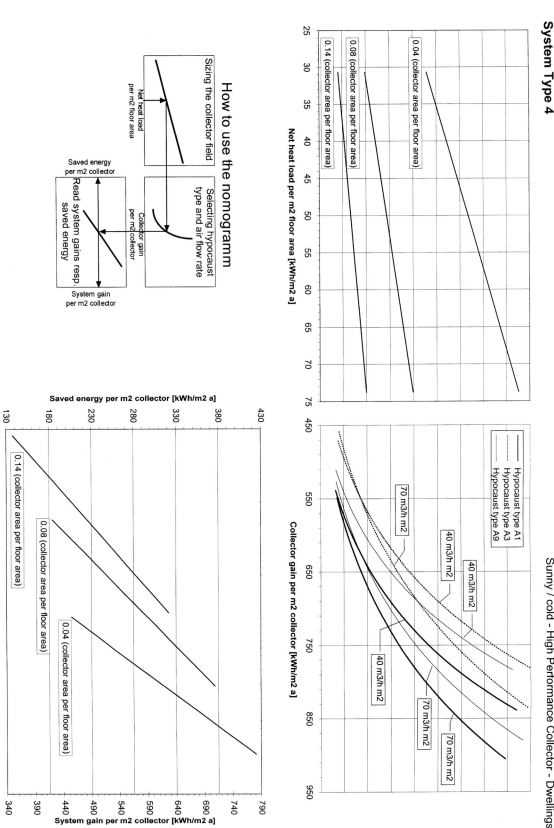

Figure II.4.17. *Nomogram for sunny/cold, high performance collector, dwellings*

System 4: Closed collection loop with radiant discharge system

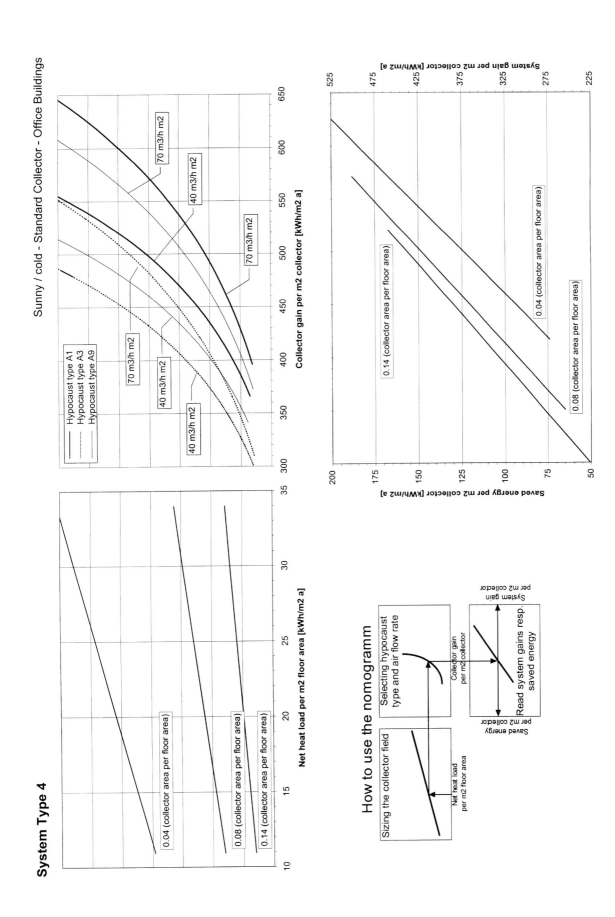

Figure II.4.18. *Nomogram for sunny/cold, standard collector, office buildings*

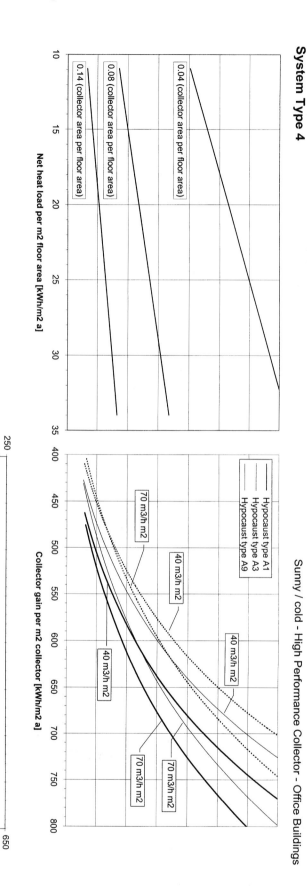

Figure II.4.19. Nomogram for sunny/cold, high performance collector, office buildings

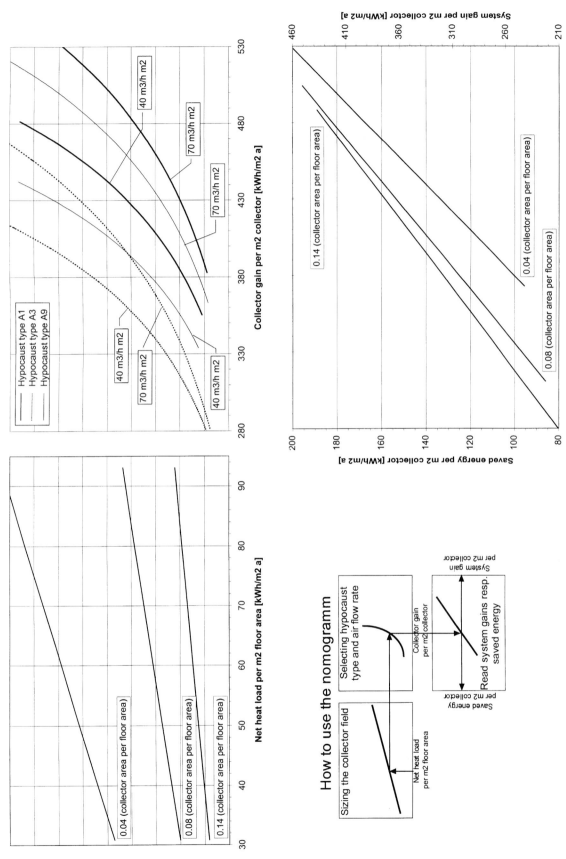

Figure II.4.20. Nomogram for sunny/mild, standard collector, dwellings

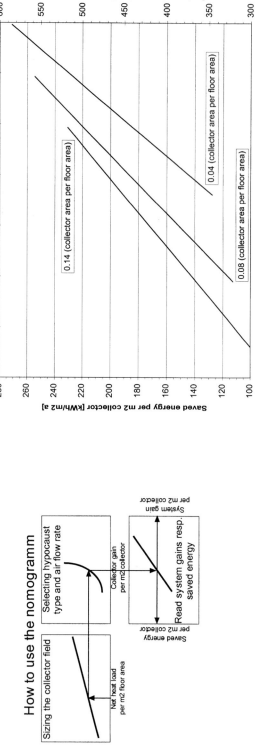

Figure II.4.21. Nomogram for sunny/mild, high performance collector, dwellings

II.5 System 5: Closed collection loop with open discharge loop

INTRODUCTION

This system provides comfort, even in rooms with high internal and solar gains and small losses, because it allows controlled discharging of stored solar energy to the heated room. This increases the efficiency of the solar system and reduces the risk of overheating.

System description

System 5 has a closed loop between the collector and hypocaust or rock bed storage and a temperature-controlled fan. The storage heats room air circulating in an open loop. Figure II.5.1 shows the first variation of this system, the system with a rockbed. To avoid loss of heat from the storage to the ambient at night, the storage has to be equipped with dampers. The system consists of the following main components:

- air collector;
- air distribution (should be integrated in building components if possible);
- heat storage;
- fans (charge, discharge);
- control systems (charge, discharge);
- dampers.

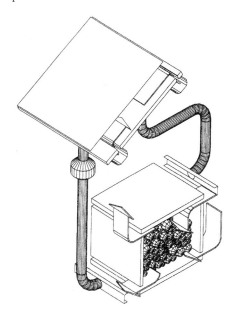

Figure II.5.1. Principle of the first implementation of system 5

System variations and characteristics

Many variations of this system have been built. Besides the typical air collectors, there exist system configurations where parts of the roof are glazed, thus forming a so-called attic space collector. In other applications the air is circulated behind transparent insulation to improve efficiency, sunspaces may be used as sun traps and sometimes simply a second glass facade is installed in front of external walls. Common to these variations is the goal to reduce costs by using existing facade elements. Similarly hollow concrete floors, cored block masonry, hollow filler block floors or gravel filling at the floor slab are the storage building elements most frequently used. The discharging is done either via the same ducting network used to charge the storage, or via a separate ducting network or through the surface of the storage.

Three typical variations are:

- roof-integrated collectors using interior building components;
- system integrated in external wall;
- attic space collector with rock bed.

Roof-integrated collectors using interior building components

Air collectors are installed on the south-facing roof of the building or the projecting sunspace which supply the heat to the building's massive interior walls and ceilings. Discharge is controlled by an insulated facing shell. Figure II.5.2 shows the operating mode of an actual system. The usable solar gains in a Central European climate were measured to be between 80–100 kWh per m² collector area during the heating period. The air flows through the exterior loop, passes through the roof-mounted air collector and releases the energy to the storage wall. The controlled discharge of the wall is done by the air circulating in the interior loop which thus heats the room.

Characteristics
ADVANTAGES

- existing building components are used;
- combination with heating and ventilation systems possible;
- storage losses used as gains for the room;

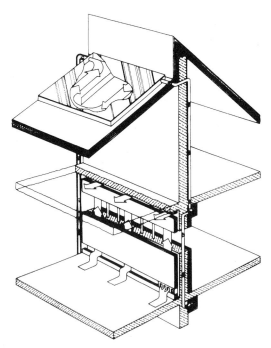

Figure II.5.2. Diagram of a roof-integrated solar system

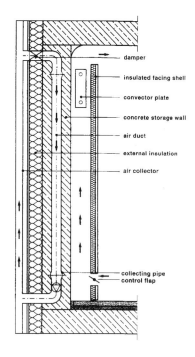

Figure II.5.3. Diagram of a hybrid solar energy gaining system integrated in the external wall

- thermal storage mass available;
- small extra costs;
- simple to install in new houses.

DISADVANTAGES

- no furniture can be placed at interior walls;
- increased installation costs because of facing shell;
- only small usable gains.

System integrated in external wall
The external wall of a building is constructed as collector and storage system. Facade-mounted collector systems, ventilated transparent insulation systems or glazing units installed in front of the external wall serve as solar collectors, whereas the static external wall system situated behind is used as storage. To avoid uncontrolled discharging, an insulated facing shell is installed in front of the internal side of the wall. A room thermostat opens a slotted grille, allowing room air to pass over the storage surface. The discharging process can be accelerated by a fan or, more efficiently, by a convective radiator installed at the end of the hybrid wall. Figure II.5.3 shows the operating mode of this system. The air flows through the exterior loop, passes through the wall-mounted air collector and releases the energy to the storage wall. The controlled discharge of the wall is done by the air circulating in the inner loop which thus heats the room. The usable solar gains in a Central European climate were measured to be between 50–100 kWh per m² collector area during the heating period.

Characteristics
ADVANTAGES

- integration of existing building units;
- can be combined with heating systems;
- transmission heat losses of external wall are reduced.

DISADVANTAGES

- collector angle is not optimal;
- acoustic weakening of external wall;
- storage volume is limited;
- increased storage losses through external wall.

Attic space collector with rockbed
The roof space of a building is rearranged as a collector. If the solar heat is not used directly for heating purposes, it is stored intermediately in rockbed storage situated near the foundation. Figure II.5.4 shows the operating mode of this system. The usable solar gains in a Central European climate were measured to be between 10–50 kWh per m² collector area during the heating period. Somewhat better results were obtained by investigations carried out in Colorado.

Characteristics
ADVANTAGES

- existing building units are used;
- very inexpensive system;
- simple to install.

System 5: Closed collection loop with open discharge loop

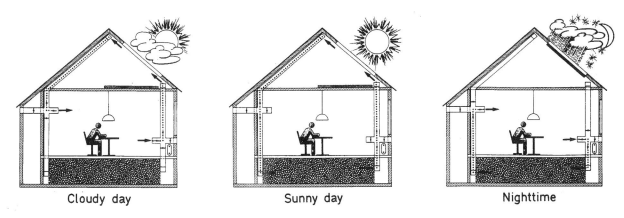

Figure II.5.4. Different operating states of the solar trap system in the heating period: (cloudy day) direct space heating gained in the roof space; (sunny day) storing the surplus energy gained in the roof space by loading the rockbed storage; (night-time) space heating by using the warm rockbed storage

Table II.5.1. Test results obtained from system 5 collectors constructed in Germany (D) and Denmark (DK)

Collector system	Storage system	Location	Climate heating period Degree-day Kd	Horiz. global radiation kWh/m²	Heating contrib. of solar system kWh/m²coll.	Cost US$/m²coll.
Air collector (attic integrated) 12 m²	8 m³ rock bed storage 177 m² hollow core floor	Zirndorf (D)	3866	586	92	–
Air collector (attic integrated) 26 m²	28 m³ rock bed storage	Darmstadt (D)	3083	606	81	–
Air collector (attic integrated) 17 m²	12 m² hollow core wall	Berlin (D)	4502	546	93	221
Air collector (wall integrated) 9 m²	30 m² hollow core floor	Berlin (D)	3202	459	98	–
Air collector (wall integrated) 12 m²	36 m² hollow core floor	Zaberfeld (D)	4260	650	81	657
Air collector (wall integrated) 7 m²	7 m² hollow core wall	Berlin (D)	4226	544	90	631
Air collector (wall integrated) 15 m²	100 m² hollow core wall /-floor	Fulda (D)	–	–	60	286
Air collector (wall integrated) 7 m²	9 m² hollow core wall	München (D)	–	–	127	414
Glazed attic space 10 m²	1.5 m³ rock bed storage	Essen (D)	3545	548	24	–
Glazed attic space 44 m²	2.7 m³ rock bed storage	Tastrup (DK)	–	–	46	–
Sunspace 27 m²	25 m³ rock bed storage	Rastede (D)	4035	497	12	–
Sunspace 59 m²	10 m³ rock bed storage	Landstuhl (D)	4208	581	–	–

Disadvantages

- only small solar gains;
- high storage losses;
- roof space cannot be used for other purposes (high temperature);
- odour problems may occur.

Comparison of various measured systems

Table II.5.1 lists tested solar systems, the amounts of heat they supply per m² collector and the climates in which the systems operate. In Figure II.5.5 the results are shown by system types.

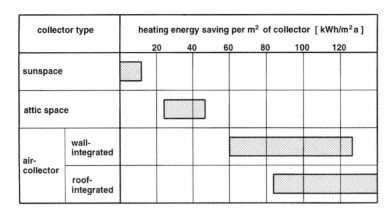

Figure II.5.5. Comparison of the measured heating energy savings per m² collector area gained by hybrid systems constructed in Germany and Denmark

The results made evident that the largest gains were achieved by the roof-integrated systems, as wall-integrated collectors have up to 20% less output. The collector output from roof spaces or sunspaces as collectors was negligibly small. Experience gained in practice shows that the collectors should be constructed such that air at high temperatures (>50°C) is fed to the storage. It is vital that the storage is incorporated in the heated area of the building. Otherwise the system's energy balance is dominated by the storage losses. The external wall provides inefficient storage, but internal building components are highly efficient storage units. The storages should be discharged in a counterflow circulation to ensure that the highest possible temperatures can be released to the room from the storage. If the storage is discharged via a parallel flow circulation the usable temperature level of the storage is reduced.

Characteristics, advantages and limitations

In existing buildings lacking channels through which air can flow, a solution may be to install an airtight suspended ceiling into which the heated air from the collector can be released. Rockbeds are recommended as storage only in buildings occupied by one user. (single-family housing, small offices), because of the transmission of odours and other hygiene problems that arise when they are used in multi-user buildings. Rockbeds are recommended as storages only in buildings occupied by one user (single family housing, small offices) because of the transfer of odours and hygienic problems occuring when being used in multi-user buildings.

By using controlled discharge, compared to passive discharge as in system 4, the risk of overheating the rooms is reduced. In order to further reduce overheating it is recommended to construct the storage elements with a limited thermal conductivity ($\lambda \approx 0.8$–1.2 W/mK). Thus a sufficient time lapse is attained between the energy input into the building component and the energy release from the component's surface area to the room.

Suitable applications

This system is suitable for many building types. The highest efficiency of the system is obtained in buildings if there is a heating demand at night, but buildings used only during daytime (i.e. office buildings or school buildings) can also benefit from the solar gains during non-occupancy when the heat is set back or shut off. The storage can ensure that the building does not cool down too much. Thereby the heating power required to reheat the building in early morning hours can be reduced. During night set back the storage cools down more than in continuously heated buildings so more solar energy can be stored the next day.

The systems are recommended for buildings with large south-oriented collection surfaces. A storage surface area about 1 to 1.5 times larger than the collection surface area is sufficient. For normal thicknesses of building components of 15–25 cm this is equivalent to a storage/collector ratio of 0.1–0.2 m³ per m² collector area. The efficiency of the system can be increased if it is combined with a ventilation system as in system 1. As the outdoor air is cooler than the indoor air, the storage can be cooled down even further which thus increases the collector efficiency.

ENERGY PERFORMANCE

Reference solar system

The reference system comprises the collector, hollow-core or rock bed heat storage unit, circulating fan and auxiliary heater. Also necessary are connecting ducts, motorized dampers and a controller. Numerous arrangements of ducts, motorized dampers and additional fans have been used.

Reference for dwellings and office buildings:

- flat–plate collectors, a vertical wall, south-orientated hypocaust storage;
- fan powered air circulation, 40 m³/m²h;
- discharge volume flow rate 4 m³/m²;
- solar generated heat is delivered 60 % to the north and 40 % to the south side of the building.

Energy performance of reference system

The graphs in Figure II.5.6 allow the energy performance of the collection system to be quickly estimated. Two sets of graphs have been developed: one for

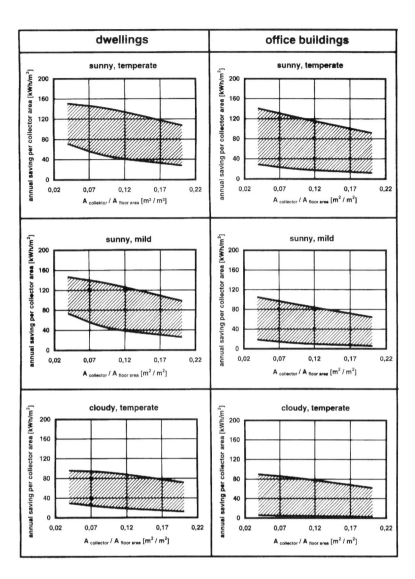

Figure II.5.6. Range of annual saving of heating energy per m² collector area due to system 5 as a function of the ratio collector area/floor area

dwellings and one for office buildings, both for three different climates.

As for other solar air systems the yearly energy performance for a specific climate depends on the size of the collector field, the insulation level of the building and the free heat. The energy performance is expressed as the saved energy in Figure II.5.6. The data points were obtained using Trnsair.

Energy performance of system variations

The system variations appear to perform very similarly. Table II.5.2 compares saved energy relative for the system variations. It can be seen that there are no differences between rock bed and hollow-core storage systems. An

Table II.5.2. Comparison of saved heating energy with different system 5 configurations

	Roof-mounted flat-plate	Flat-plate collector on vertical wall
Hypocaust storage	120 %	100 %
Rock bed storage	120 %	100 %

optimally orientated collector on the roof collects about 20 % more solar gains during the heating period than the wall-integrated system.

DESIGN

System design

Solar air heating design options

In some systems, a by-pass with motorized damper permits use of the auxiliary heater without air passage through the rock bed when little or no heat is in storage. Fan power is thus reduced.

Another variation provides only one fan for all modes of air circulation. An air handling unit contains the fan and two pairs of motorized dampers, one pair directing solar heated air to the fan, either from collector or from storage, and the other pair directing hot air from the fan either to the storage unit or to the auxiliary heater and the rooms. Back-draft dampers are located in the cold air duct, and a by-pass duct with manual dampers provides water heating only. The elimination of the second fan saves power, but the damper system is more complex.

A third variation, used in the experimental solar air heating system at Colorado State University also used only one fan, but the ducting arrangement requires six motor-driven dampers for the three heating operating modes. A motorized back-draft damper and a manual summer by-pass damper are also provided. The complexity of this duct pattern and numerous motorized dampers are disadvantages.

It should be noted that the standard configuration and all variations involve the collector fan handling hot air from the collector, and not the cool air supply to the collector. Hot air leaking out through many connections in and between collectors would result in poor performance if the collector were even slightly pressurized from a fan on the air supply side. On the discharge side, however, moderate leakage of outdoor air into the hot air stream is not an appreciable problem.

The above variations are representative of numerous design options for solar air heating. An additional variation is in the control strategy. Most systems employ temperature sensors in duct locations near collector entrance and exit and in the cold portion of the heat storage. When sufficient positive temperature differences are sensed, the collector fan is powered, damper motors are actuated and collection occurs. Air is directed to rooms from storage when a room thermostat senses the need for space heating. In most systems, air from storage passes through the auxiliary heater, and, if not at sufficient temperature, supplementary heat from fuel or electricity is added. This requirement is usually sensed by a two-stage room thermostat, auxiliary heating being initiated if room temperature decreases one or two degrees below the upper setting.

Another possible auxiliary control, not generally used, is immediate supply of backup heat if the temperature of solar storage or air from storage is less than a preset level of about 25°C. Room temperature variation is reduced by this strategy. A by-pass duct and damper which provide cool air directly to the auxiliary rather than via the heat storage unit can also minimize room temperature fluctuation when storage temperature is below a preset low limit.

An innovative system design and operation, developed in theory but not confirmed by construction and testing, is the dual fan system operated at varying collector flow rate to deliver constant temperature air to the storage. Substantial improvement for the constant flow-varying constant temperature operation has been predicted: improved system efficiency, lower fan energy use, much improved storage temperature stratification and heat retrieval and greater comfort in the heated space. Powering the fan by photovoltaic cells may also simplify system control.

A two-fan system achieves the most effective air flow rate in each circuit: *collector to storage* and *storage to rooms*. A single motor-operated damper is required in the hot duct from the collector. One self-opening and self-closing 'back-draft' damper is required to prevent reverse air flow at the bottom of the collector. The collector fan runs when air is being heated in the collector, and the second fan, frequently part of an auxiliary heater, delivers air to the heated space from heat storage.

Component design and specific design advice

Efficient performance of the solar system requires not only that the individual components be well designed and correctly installed, but that the components are compatible with each other. Oversized heat storage relative to collector area, undersized collector fan capacity, inadequate air ducts, poorly sealed dampers, inaccurate and/or improperly located temperature sensors, all can contribute to poor operation of otherwise good equipment. Effective integration of properly sized components and their operation under conditions and settings to be expected are essential.

The rules of thumb below can be used for an initial global dimensioning of the system.

Collector
A roof that is not essentially flat, but sloping to the south, usually at a steep angle, is essential for the collector tilt. Options include supporting the collector a few centimeters above the roof and at the same slope; setting the collector into the roof so that the glazing serves as a roofing surface; using the collector as a sunshade over patio and porch areas; and, in high latitudes, vertical collectors on south walls. The collector location should minimize the length of duct runs. A collector area of one fifth to one fourth of the floor area can be used as a starting point.

Rock bed storage
A starting point is for each m² of collector to provide 0.15 m³ of rock with a density of 1600 kg/m³ and specific heat of 0.84 kJ/kgK. A recommended rock depth of 1.5 m minimum and 2 m maximum is based on a minimum pressure loss of 40 Pa. A rock bin cube or cylinder about two meters high will contain about ten tons of small stones. In buildings with a basement or other reinforced concrete floor, the rock bin can be built on the floor near or against a wall or corner: insulation of the floor and walls is critical. The minimum U-value is 0.2 W/m²K. Access for duct openings near the bottom and top of the bin must be considered. If there is not suitable space within the building, the bin may be built against an outside wall and provided with suitable thermal insulation and rainwater protection. The optimum place for a rock bin storage is within the heated volume of a building since storage losses stay within the building.

Hypocaust element
Start with 1.5 m² of hollow-core wall or roof element per m² of collector area. To avoid uncontrolled discharge an insulated facing shell is required. Provide easily accessible cleaning opening. Take precautions to protect connections of the ducting system during construction. A utility room in most buildings can usually accommodate the heat storage unit as well as fan, duct, and damper

equipment, the auxiliary heater, and a hot water exchanger and hot water tank.

Air handling unit flow rate
The optimal flow rate is 40 m³/h per m² of collector area.

Ductwork for the system
Maximum static pressure loss is 1 Pa per m of duct, and maximum velocity for collector manifolds is 3.5 m/s. Rectangular or circular metal ducts are usually employed, with rigid internal insulation in the hot air ducts. Flexible external insulation is another option. Ducts should be sized so that air velocity and pressure drop are in well-accepted ranges. An air velocity of about 3 m/s is practical. At that velocity, a pressure loss of about 0.7 Pa per meter of 25 cm diameter duct results.

Auxiliary heating system
Because heat discharge from this solar system is controlled there are practically no restrictions concerning the installed heating system. It only has to be controllable by room.

Do's and don'ts

- For cost-effectiveness, determine collector area and storage volume by a design procedure, e.g., f-chart, Suncode or Trnsair.
- To assure comfort, provide auxiliary heating to fully cover the maximum heating demand (the design load).
- Employ a complete system design, which has been well proven in numerous practical installations and avoid speculative concepts.
- Design the system with the fan between the collector outlet and storage inlet, to provide negative pressure in the collector.
- To avoid excessive collector and storage temperatures provide a hot air by-pass.
- To avoid cooling down of the storage by natural convection during the night, install controlled flaps that close the collector's in- and outlet when there is no solar radiation.

CONSTRUCTION

Collectors are detailed in Chapter IV.1, hypocausts in Chapter IV.6.

Checklist for installation and commissioning

- Employ persons with experience in installing forced warm air heating systems.
- Provide air tight air duct connections between collectors and throughout the solar/storage air handling system.
- Avoid fracturing and chipping the rocks when filling the rock bed and spread rocks uniformly to avoid rolling and size segregation.
- Adequately insulate all heated surfaces.
- Thoroughly test all modes of operation after installation is completed.

NOMOGRAMS

Two sets of nomograms have been developed: one set for the dwellings and one set for office buildings. As the results for rock beds and hypocaust elements are similar the diagrams are valid for both types of storage systems.

Assumptions

- Heated air is delivered to both the south and north side of the building.
- A fan of 100 W is assumed to be included in all the system variations. 20 % of the power to the fan is transferred to the air as heat.
- 50 % of the heat loss from the ductwork is considered as potentially useful - i.e. the duct work is mainly located inside the thermal envelope of the building.
- No heat recovery unit in the ventilation system is considered.
- Only solar collectors with small heat capacity are considered – i.e. not solar walls where e.g. a concrete wall acts as absorber.

Diagrams

The diagrams in Figures II.5.7 to II.5.12 are developed for a standard collector type.

Example of nomogram use

A multi-family residential building in Munich is characterized by the following data:

54 kWh/m²a heating demand (Q_{aux})
77 m² floor area
6.6 m² collector area (Standard collector)
0° orientation (south)
85° slope of collector
9 m² hypocaust system
25 m³/m²h specific volume flow for charging
140 m³/h specific volume flow for discharging
reference climate for Munich (cloudy temperate)
127 kWh/m²a saved energy per collector area

With Figure II.5.11 the result can be achieved as follows. In the top left graph with Q_{aux} = 54 kWh/m²a and V_{hypo} = 25 m³/m²h you can calculate a saved heating demand of 64 kWh/m²a. This saved heating demand has to be corrected with the collector areas and the floor area ratio of about 0.09 m²/m². The correction factor is 1.03. Finally a correction has to be done with the specific discharge volume flow of 140 m³/h with 9 m² hypocaust. The correction factor is 1.01. Thus the saved heating energy comes to 64 × 1.03 × 1.01 = 66 kWh/m²a the total comes

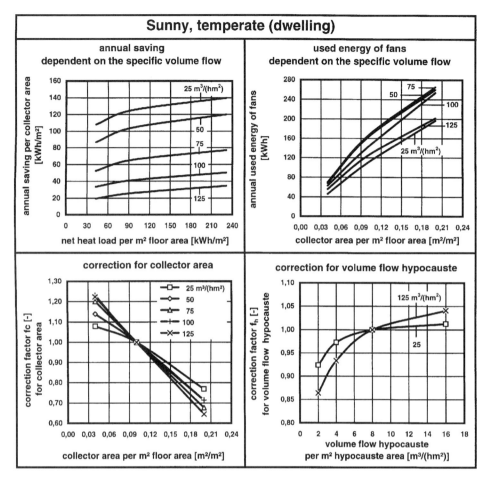

Figure II.5.7. Set of diagrams for determination of the performance of system 5 for dwellings in sunny/temperate climate

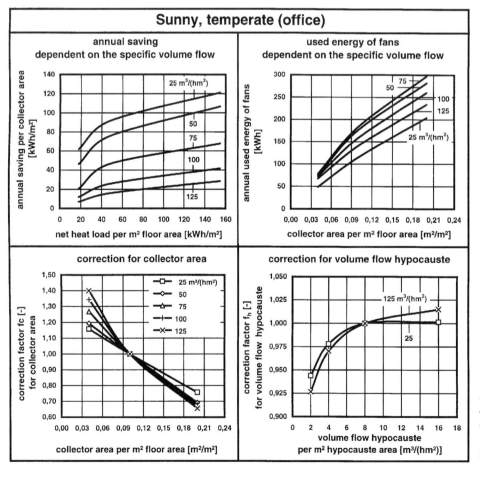

Figure II.5.8. Set of diagrams for determination of the performance of system 5 for offices in sunny/temperate climate

System 5: Closed collection loop with open discharge loop

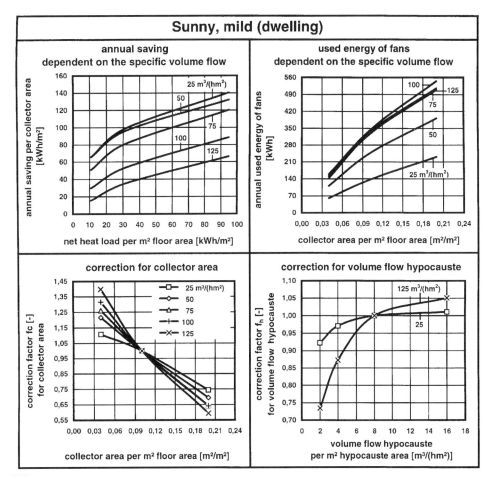

Figure II.5.9. Set of diagrams for determination of the performance of system 5 for dwellings in sunny/mild climate

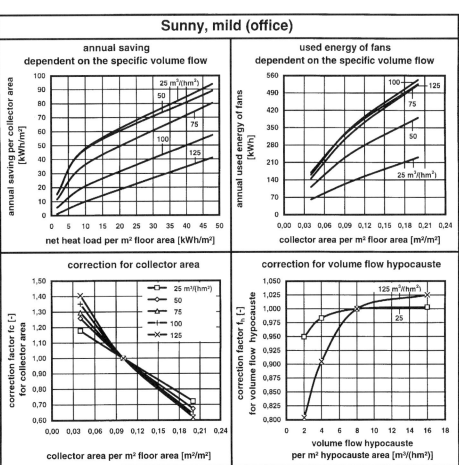

Figure II.5.10. Set of diagrams for determination of the performance of system 5 for offices in sunny/mild climate

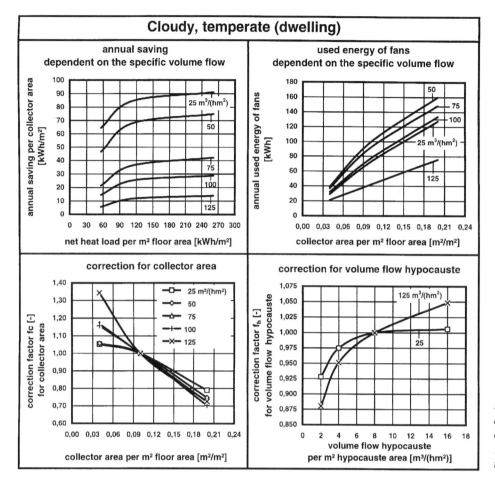

Figure II.5.11. Set of diagrams for determination of the performance of system 5 for dwellings in cloudy/temperate climate

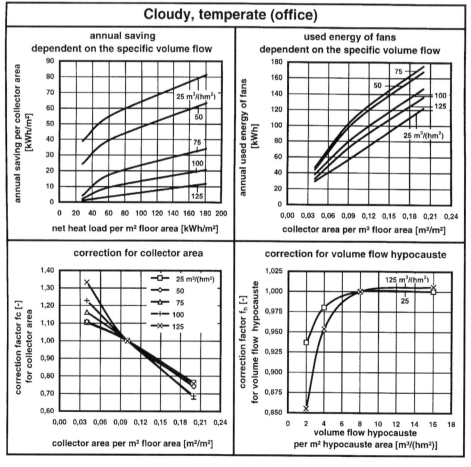

Figure II.5.12. Set of diagrams for determination of the performance of system 5 for offices in cloudy/temperate climate

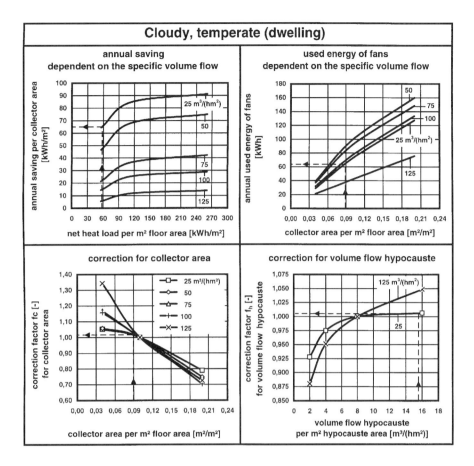

Figure II.5.13. Use of diagrams for the Munich case study

to about 450 kWh/a. The energy used by the fan over the same period is about 95 kWh/a. The result is shown in Figure II.5.13.

BIBLIOGRAPHY

1. de Boer J and Erhorn H (1999) *Solarspeicherhaus Koch in Fulda*. Report WB 108/1999, Fraunhofer-Institut für Bauphysik, Stuttgart.
2. Erhorn H, Gertis K, Rath J and Wagner J (1987) *Ein einfaches Hybridsystem zur Heizenergieeinsparung mittels Kiesspeicher und verglastem Dachraum*. Report WB 8/87, Fraunhofer-Institut für Bauphysik, Stuttgart.
3. Jensen S O (1987) *Roof Space Collector*. Report No. 87-15, 12/87, Technical University of Denmark, Thermal Insulation Laboratory, Copenhagen.
4. Oswald D, Erhorn H, Kuhl W, Reiß J and Steinborn F (1988) *Passive Solarenergienutzung in bewohnten Eigenheimen. Meßergebnisse und energetische Analyse für das Solarhaus in Rastede*. Report WB 34/88, Fraunhofer-Institut für Bauphysik, Stuttgart
5. Oswald D and Wichtler A (1998) *Bauteilintegrierte Heiz- und Haustechnik für zukünftige Baukonzepte (Hybride Heizsysteme). Projektstufe 3/2 – Umsetzung am Einfamilienhaus Zaberfeld*. Report GB 141/1998, Fraunhofer-Institut für Bauphysik, Stuttgart.
6. Reiß J and Erhorn H (1997) *Solare Hybridsysteme in einer Reihenhaus-Wohnanlage am Weinmeisterhornweg in Berlin*. Report WB 88/1997, Fraunhofer-Institut für Bauphysik, Stuttgart.
7. Reiß J and Erhorn H (1997) *Solare Hybridsysteme in einer Reihenhaus-Wohnanlage an der Wannseebahn in Berlin*. Report WB 91/1997, Fraunhofer-Institut für Bauphysik, Stuttgart.
8. Reiß J, Erhorn H, Kuhl W, Oswald D and Steinborn F (1988) *Passive Solarenergienutzung in bewohnten Eigenheimen. Meßergebnisse und energetische Analyse für das Solarhaus in Zirndorf*. Report WB 37/88, Fraunhofer-Institut für Bauphysik, Stuttgart.
9. Reiß J, Erhorn H, Kuhl W, Oswald D and Steinborn F (1988) *Passive Solarenergienutzung in bewohnten Eigenheimen. Meßergebnisse und energetische Analyse für das Solarhaus in Darmstadt*. Report WB 38/88, Fraunhofer-Institut für Bauphysik, Stuttgart.
10. Reiß J, Erhorn H and Rief M (1999) *Energetische Bewertung ausgeführter, optimierter solarer Hybridsysteme in der Wohnanlage in München-Sendling*. Report WB 88/1997, Fraunhofer-Institut für Bauphysik, Stuttgart.
11. Reiß J, Erhorn H and Stricker R (1991) *Passive und hybride Solarenergienutzung im Mehrfamilienwohnhausbau. Meßergebnisse und energetische Analyse des deutschen IEA-TASK VIII-Gebäudes in Berlin*. Report WB 64/91, Fraunhofer-Institut für Bauphysik, Stuttgart.
12. Reiß J, Erhorn H, Kuhl W, Oswald D and Steinborn F (1987) *Passive Solarenergienutzung in bewohnten Eigenheimen. Meßergebnisse und energetische Analyse für das Solarhaus in Landstuhl*. Report WB 18/87, Fraunhofer-Institut für Bauphysik, Stuttgart.

Chapter author: Hans Erhorn

II.6 System 6: Closed collection loop with heat exchange to water

INTRODUCTION

The closed loop solar air system has advantages over liquid systems. It runs no risks of leakage, boiling or freezing. It might also be chosen for its economy or for architectural reasons. Solar air heated water can provide space heating, domestic hot water heating or even be used for industrial applications. Apart from the collector the system consists of standard HVAC components.

System description/principles

In this system air-heated air from the collector is circulated to and from an air to water heat exchanger in a closed loop. This loop resembles the traditional active solar liquid system where a liquid (a water/glycol mixture) is used as the heat transfer medium, and is straightforward to construct. In addition to the air collector (see Chapter IV.1) these standard HVAC components are used:

- duct system
- heat exchanger
- frost damper
- fan
- control system.

The system border is the heat exchanger pipe connections. See Figure II.6.1 for a schematic diagram of an example of a closed collection loop system.

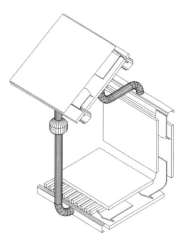

Figure II.6.1. Schematic diagram of a typical System 6 configuration

In Figure II.6.1 air is heated in the solar collector. The heat is transformed via a heat exchanger to a water tank. The diagram presents only one possible way in which the secondary side may be constructed. The actual configuration depends on the application and may be with or without storage. This chapter focuses on the primary side, its configuration variations and design.

System variations

Variations in this system relate to the choice of components and the layout of the closed loop, the connection of collectors, the location of the fan in relation to the collector and heat exchanger and the control system used.

This system corresponds in principle to domestic hot water (DHW) heating in summer mode operation of the other systems described in this handbook (see Chapter III.1).

Characteristics, advantages and limitations

Characteristics:

In order to deliver useful heat via a heat exchanger this system must operate at higher temperatures than the other systems described in this book. Therefore the components must be more efficient. The duct system, high efficiency collector, heat exchanger, flow rate, fan, control system and insulation levels must be optimized.

ADVANTAGES:

- heat distribution can be built as a traditional hydronic heating system with radiators or floor heating, familiar to installers;
- no problems with freezing, boiling and leakage in the air collector;
- standard ventilation equipment is familiar to installers and improves chances for obtaining a good price on a public tender and ensuring a good installation with few, if any, problems.

DISADVANTAGES:

- bulkier systems than liquid systems (ducts and air to water heat exchanger);

- potential risk of a high electrical consumption by the fan;
- risk of freezing in heat exchanger;
- the overall efficiency of this type of system will be reduced due to the temperature drop over the heat exchanger, compared to a direct air heating system. The constraints are adequate areas and spaces for the collectors, the ducts, the heat exchanger and the fan. Because noise levels from the duct system can be considerable for System 6 the installations must be placed outside a sound insulation barrier.

Suitable applications

This system can be used for:

- hot water heating (dwellings, sports halls, etc.);
- space heating (individual houses, apartment blocks, schools, etc.);
- combined space and domestic hot water heating;
- industrial applications.

ENERGY PERFORMANCE

Closed loop air systems will partly cover constant or varying load patterns. As with other solar systems the yearly performance depends on the load profile and the climate. A large number of Trnsair dynamic performance simulations have been performed with the three predefined dwelling insulation and mass levels and the four climates identified for this work (see Appendix H). The results of these simulations have been compiled into a set of nomograms and tables, which can be used to estimate the performance of a closed loop solar air system.

In spite of the fact that this particular system will obviously always be used for heating domestic water during the summer the yearly performance estimation nomograms are based on space heating performance only. This makes the treatment of this system consistent with the other systems within this book. To estimate the performance of a combined heating and domestic hot water system the performance estimations from the nomograms developed for a (pure) space heating system (in this chapter) must be added to those for a summer mode operation as a DHW heating system (Chapter III.1).

As this system may be used for other applications than space heating the load profile of these applications will probably be quite different than the profile for a dwelling. Therefore the nomograms only roughly indicate the performance of such applications, and a detailed simulation will need to be performed.

Reference system

The curves presented are all based on Trnsair simulations of a reference System 6. The simulations have been designed to show the influence of parameter variations of the main system variables, collector area, heat exchanger effectiveness and storage size. All other parameters are constant or have been varied as a function of the main parameters. The system simulated is shown in Figure II.6.1. The list of assumptions for the reference system is presented below in Assumptions.

Reference system performance

For a quick indication of what performance can be expected for a closed loop system in various climate types, see Table II.6.1. For more detailed information the reader is referred to the nomograms. Table II.6.1 shows the calculated solar system performance for dwellings in the different climates considered (specific collector areas between 0.04 and 0.08 m^2/m^2). To these saved energy figures the summer performance of a domestic hot water system should be added (see Chapter III.1).

Energy performance of built examples

Hastings (1998) presents three examples of this system type and their performance. The energy performance of these systems is presented in Table II.6.2.

The Stavanger Squash Centre illustrates what is well known from liquid-based active solar systems, that a high system performance can be achieved for applications designed to have a relatively small solar fraction of a constant load profile.

DESIGN

The design procedure addresses both the energy performance and economic efficiency of the system and the technical performance of the individual components. This includes the detailed design of the individual components and their interaction in the system, e.g. sizing of heat exchanger in relation to collector output, duct insulation thickness, etc.

System design

The system design covers the overall sizing of the system and the main components based on the energy performance nomograms. The starting point is the load profile.

Step 1: Load profile
Establish the yearly net heat load per m^2 heated floor area as input to the curve, kWh/m^2a. If the required temperatures of the heating system are different from what was

Table II.6.1. System performances for dwellings in different climates

Climate	Building mass & insulation levels	Net solar system saved energy, kWh/m^2
Cloudy/temperate	Medium, light	120–130
Sunny/temperate	Medium, light	210–220
Sunny/mild	Medium, light	165–200
Cloudy/temperate	High, high	70–80
Sunny/temperate	High, high	140–160
Sunny/mild	None	260–290
Sunny/cold	Medium, light	350–400

used for the curve calculations or the load profile is in other ways different Trnsair must be used.

Step 2: Select the collector type (see Chapter IV.1) and area
Using the curve select a collector area and type to match a required system output, saved energy per m² of collector, kWh/m²a.

Step 3: Decide on air mass flow(s) (heat capacity flows)
The nomograms have been constructed using a certain air mass flow. If another air mass flow is chosen, the performance of the collector must be recalculated using Trnsair.

Step 4: Specify the heat exchanger (see Chapter V.2)
The curve shows the sensitivity to heat exchanger temperature effectiveness. Based on this and costs provided by the manufacturer select the optimum heat exchanger.

Step 5: Size the storage and determine its heat losses
The curve shows the overall system performance dependency of the storage size. Based on this and the cost of the storage the size is fixed.

Components design/specific design advice

When the overall sizing of the system has been completed the individual components and their interaction must be checked to ensure that the system performs well technically. The reader is referred to the component chapters of the handbook for the details on each component.

Collector
Select the collector according to the desired temperature range. For low temperature applications, e.g. preheating of domestic hot water, a standard collector design is adequate and for a high temperature application high performance collectors should be chosen. The collector efficiency factor increases with the air mass flow. The higher the air mass flow, the higher the efficiency will be, but the electricity consumption by the fan will be greater. The collector manufacturer should provide a curve showing the efficiency as a function of air mass flow rates. This information should also include the pressure drop through the collector at different flow rates. The collector should have a satisfactory efficiency for the air mass flow(s) which are reasonable considering pressure drops. It should be emphasized that even if the collector efficiency increases with increasing air mass flow rates usually an optimum can be found for the system efficiency. Collectors connected both in series and parallel must be considered and compared. Generally a total length of 4 m for collectors in series should not be exceeded.

Depending on the application and related temperature range, the performance of this solar air system type is generally quite sensitive to the collector efficiency, therefore consider at the outset high performance collectors, then optimize the cost/performance relation (see Chapter IV.1).

Heat exchanger
Calculate temperature rises on both sides of the selected heat exchanger for different combinations of primary and secondary side inlet temperatures and mass flow rates. Locate the heat exchanger in the heated part of the building if possible and minimize the duct length as heat losses from air ducts are considerably higher than from the water pipes.

The heat exchanger must be selected with respect to pressure drop, heat transfer efficiency and economy. A temperature efficiency of 0.9 is reasonable to achieve without too high pressure drops and extensive costs.

Frost damper
The frost damper, a critical component, can be a simple one-way damper to prevent air backflow to the heat exchanger. Safer protection can be obtained by using a gasket, motorized, impedance-protected damper which opens when the fan runs (power is on) and closes by a mechanical spring. Place the damper motor in the air stream to recover its electric energy consumption

Storage
Compare the loads estimated for the time period which the storage is intended to cover (i.e. a three-day load period). Based on this and an estimated solar input calculate the storage energy balance for this period. This gives an impression of the temperature levels in the storage and the rest of the system. Also storage geometry and heat loss need to be established. The nomograms are made assuming a certain storage heat loss coefficient (see Table II.6.2) and, if other insulation levels are chosen, a Trnsair simulation must be performed.

Ducts
Pressure drop, heat losses and costs must be minimized. A small duct diameter means low heat losses and cost saving but increased pressure drop. A series connection of the collectors results in a simpler and less expensive duct system, but can result in reduced efficiency and high pressure losses. Design for air speeds in the order of 6–8 m/s in the duct system if located outside the building insulation; if not, lower values must be used to avoid noise problems.

- Dimension the duct system and calculate air velocity/pressure drop/noise. Standard ventilation engineer-

Table II.6.2. Three building examples

	Net solar system contribution
Havrevangen (DK): Space and domestic hot water heating	240 kWh/m²A_c
Toftegaard (DK): Domestic hot water heating all year	175 kWh/m²A_c
Karl Highschool Sportshall (D): Summer hot water heating	283 kWh/m²A_c [1]
Stavanger Squash Center (N): Hot water heating	375 kWh/m²A_c [2]

[1] Whole year performance
[2] Actual performance was lower due to less hot water consumption than estimated for the design.

ing calculation tools and tables apply. The price of the air ducts increases sharply with the diameter, but fan energy rises with smaller diameters. Therefore the diameter has to be optimized, the optimum may be a stepwise increase in duct diameter for long ducts with several collectors in parallel.
- Minimize duct heat losses. Too often this is neglected, because it is generally not very important for ventilation systems which operate at temperatures close to the surrounding temperatures in a building. In this closed loop air system, however, the temperatures during operating hours will often be 40–60 K above ambient. Figure II.6.2 illustrates the heat losses from a duct system as a function of duct diameter and length.

For a particular system design yearly duct losses should be compared to the yearly output of the collector to make sure that only a reasonable percentage is lost (<5%). To obtain precise information about the duct air temperatures a Trnsair simulation must be performed. Use insulation corresponding to 100 mm mineral wool as a recommended minimum.

Fan
A high efficiency fan should be chosen. An overall efficacy of above 0.7 can be requested. In selecting the fan (see also Chapter V.1) consider:

- fan type: often a ventilator with backward-bent wings will be most efficient for this purpose. To obtain a high relationship between the collected solar energy (thermal) and the fan power (electrical) the use of a variable speed fan is sometimes advisable. For example, a two-speed fan can use one speed in the heating season and the other speed during summer DHW operation. A variable speed fan can be used to maintain constant collector outlet temperatures at different levels for respectively summer and winter operation.
- capacity: size the fan in relation to pressure drop and required air mass flow rate.
Calculate the fan power from the formula:

$$P = \rho \times \Delta p / \eta$$

where:
η is the fan efficiency
Δp the overall system pressure drop (Pa)
ρ the air volume flow rate, m³/s.

The graphs in Figure II.6.3 show, as an example, the fan power requirements as a function of pressure drop and air volume flow rate for a fan efficiency of 0.5, as derived from the above formula.

The location of the fan motor should generally be outside the hot air stream, because of the high air temperatures of this system. If guarantees can be achieved for the fan motor to withstand the design temperatures it is advantageous to place the motor in the air stream. A two-speed or variable speed fan may lower the electricity consumption if properly controlled.

Control system
An optimal control strategy and reliable control system are of crucial importance to any solar system, passive, hybrid or active. However, due to the higher electricity consumption of the fan (compared to that of the pump in a liquid-based system) an optimal control strategy is especially critical for the air-based system. Due to the high primary energy use of electricity production the net thermal output of the collector circuit should always be more than three times higher than the rated fan power when the system runs (See also Chapter V.3).

The control strategy for starting and stopping the fan (and the pump in the secondary circuit) must be defined based on the characteristics (operating temperatures, solar heating energy uses, system layout) of the system(s) to which the air collectors are delivering heat. The design of the control system addresses:

- set points
- time delays

SET-POINTS
Select the correct set-points to assure that the fans are only running when the useful collected solar energy output is significantly higher than the fan electricity power.

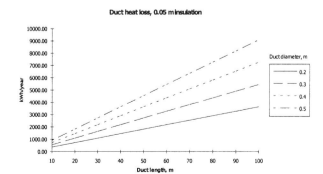

Figure II.6.2. Duct heat losses as a function of duct length and diameter

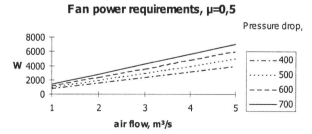

Figure II.6.3. Fan power requirement as a function of air mass flow rate and pressure drop.
As part of the fan selection process the total yearly electricity consumption for the fan should be estimated from the estimated number of running hours per year and be compared to the total estimated solar system output: COP=Q_{sys}/Q_{fan}. A COP of 15 to 20 should be achieved

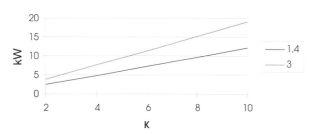

Figure II.6.4. Collected energy as a function of control temperature difference

A help procedure for choosing an appropriate stop delta T (ΔT) is illustrated by an example on Figure II.6.4.

Figure II.6.4 shows the collected energy as a function of the difference between collector inlet and outlet temperatures for two different fan speeds (and corresponding mass flow rates), one consuming 1.4 kW and the other 3 kW of electric power. The collected energy (power) is a product of the temperature difference and the mass flow rate. It is seen that at a ΔT of 6 K the collected energy by the 3 kW fan is 11–12 kW. The thermal output should equal at least three times the fan power to account for the difference in primary energy use between the production of electricity and heat. Therefore 6 K is an appropriate ΔT for the control system.

In type 6 systems the difference between the storage temperature and the collector outlet air temperature will generally be used to control the fan. Depending on the heat exchanger temperature efficiency (which is decisive for the temperature raise over the heat exchanger, see Chapter V.2) and a specific system curve made similar to Figure II.6.4 an adequate temperature difference to control the system can be found.

TIME DELAYS

The thermal capacity of air in the air ducts and the ducts themselves influences the time it takes to heat up the complete circuit. To avoid the control system making the fan start and stop many times during start-up it is necessary to allow some time to heat up the air in the ducts, the ducts themselves and the heat exchanger to operating temperature. As a rule of thumb it might take 2–5 minutes to heat up the duct system. The correct time delay should be calculated as part of the design process. If the control system does not allow for this delay the system fan will start and stop a number of times before proper operating temperatures are reached. The start of the pump on the secondary side can be delayed correspondingly. Introducing a higher start ΔT and using the hysteresis integrated in most controllers, an effect similar to the time delay can be achieved.

Do's and don'ts

- When using motorized dampers for freeze protection of the heat exchanger make sure that the control opens and closes the damper simultaneously with switching on and off of the fan.
- Do not exaggerate high collector performance by using high air mass flow rates which will lead to poor overall performance, because of excessive fan electricity consumption.
- Do not use fresh water in the heat exchanger if the water is hard. This will lead to mineral deposits within the heat exchanger, which will quickly lower the performance of the heat exchanger.
- Select the sensors for the control of the solar collection fan and the pump for the water circuit on the other side of the heat exchanger to be able to withstand the relatively high temperatures they will be exposed to (see Chapters IV.1 and V.3).

CONSTRUCTION

Building integration

Collector location
Collectors can be integrated in the roof, on top of the roof or in a facade. Also adequate space for the ducting must be reserved (a 90° bend of 60 cm diameter duct with 10 cm insulation requires considerable space).

It is important to ensure equal flow distribution in collectors when placed in parallel in a long array. The ducting must therefore include easy access dampers and balancing of the air mass flow distribution between the collectors in the array.

Heat exchanger
An equal distribution of the air flow is required to assure optimum performance of the heat exchanger. If the heat exchanger is placed in the duct system right after the fan there is a risk that the air flow will be a jet of air unevenly distributed over the surface of the heat exchanger. A location before the fan is generally recommended unless a long distance between the fan and the heat exchanger, bends in the air ducts or other measures assure an evenly distributed air flow. This also has a second advantage, that the fan will be on the cold side of the air stream, meaning that the requirement to withstand heat can be loosened up a little. A third advantage is that the colder the air, the higher the specific mass of the air. Thus the mass flow is higher at the same volume flow.

There are several good reasons to place the heat exchanger close to the collector.

- Ducts are more expensive than water pipes.
- Ducts take up more space than water pipes.
- Ducts have higher heat losses than water pipes.
- The pressure drop in the primary circuit should be minimized.

Freeze protection
The back-siphoning of cold air into the heat exchanger must be prevented by installing a motorized damper with a spring return function. A freeze protection thermostat might switch on the secondary circuit pump to heat the heat exchanger momentarily as a further precaution. If

there are special reasons glycol might be added to the water on the secondary side. However, placing the heat exchanger in a heated space is no guarantee. Cold back draughts from the collector might still cause freezing within the heat exchanger.

Air tightness
Components with leaks should be located in areas with a small pressure difference to the ambient. The pressure distribution in the system should be balanced by a careful selection of the location of dampers and fan(s).

Secondary side
The secondary side is constructed of conventional HVAC components and will not be discussed in detail. It should be noted, though, that this is part of a solar installation and care should be taken not to spoil the overall performance of the system by poor insulation, poor controls, etc.

Cost reduction hints
- Integrate the collector in the roof or wall.
- Optimize the tilt and orientation of the collector: this is a high performance system!
- Optimize the air mass flow throughout the system. Costs of ducts increase with the diameter. Money might be better spent on an effective heat exchanger.

NOMOGRAMS AND TABLES

Nomograms generated by Trnsair show the yearly energy performance of the solar system as a function of the load, the collector area and type and climates. As the performance also depends on the heat exchanger effectiveness (see Chapter V.2) the storage volume correction factors for these parameters are also presented as tables and nomograms.

In total the solar system performance, presented as the solar system output per square meter of solar collector, can be expressed as a function of all the above parameters:
$Q_{sys}/A_c = f (Q_{load}$, collector type, heat exchanger effect, storage volume, climate), (kWh/m²/year)

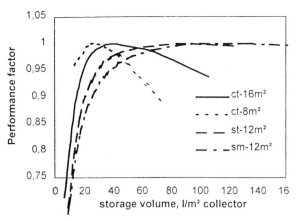

Figure II.6.6. Storage performance for different collector areas and climates

Assumptions

The Trnsair simulations have been performed for a reference System 6 as shown on Figure II.6.1 and defined by the parameters given in Table II.6.3. This table shows the system parameters used as the starting point for the calculation of the nomograms:

The specific collector area (collector area per floor area) has been varied (0.04, 0.08, 0.12, 0.16 and 0.2 m²/m²) and the collector circuit duct length, diameter and heat loss coefficient have been varied to correspond to these areas. The collectors are assumed to be connected in series.

The simulations have been performed for three pre-defined dwelling insulation and mass levels and the four climates identified for this work (see Appendix H).

Nomograms

Figures II.6.6 to II.6.14, and Tables II.6.4 to II.6.7 can be used to estimate a System 6 performance. Start with establishing the ratio of the load to the floor area Q_{load}/A_f and enter the curve on the x-axis at this point. Then for the appropriate climate and collector type find a suitable collector area/floor area ratio and read out the net solar contribution on the y-axis. Use Tables II.6.4 to II.6.7 to

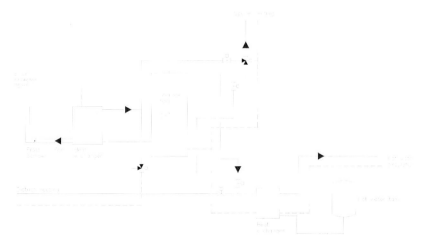

Figure II.6.5. Example of System 6 installation as a combined space heating and domestic hot water system

correct the obtained number for a potentially poorer heat exchanger efficiency and finally correct, if necessary, the system output in relation to storage volume from Figure II.6.6.

REFERENCES

Duffie J and Beckman W (1991) *Solar Engineering of Thermal Processes*, J Wiley, New York.

Hastings S R (editor) (1998) *Solar Air System – Built Examples*, James & James (Science Publishers) Ltd, London.

Jensen S Ø, Geneser K and Radisch N (1996), *Evaluering af tagrumssolfangersystemerne i Toftegård*, Herlev, Solar Energy Center, Denmark.

Mørck O C (1996) *Hybrid Solar Low-Energy Dwellings in Denmark*, EU-THERMIE Project no. 140/91-DK. Cenergia Energy Consultants, Ballerup, Denmark.

Chapter author: Ove C. Mørck

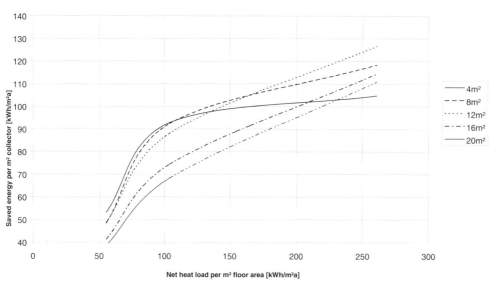

Figure II.6.7. System 6, climate: cloudy/temperate, standard collector

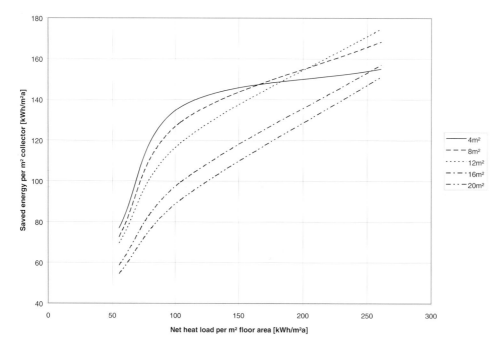

Figure II.6.8. System 6, climate: cloudy/temperate, high performance collector

Table II.6.3. Initial system parameters for the nomograms

System parameter	Parameter value
Air mass flow rate	20 m³/m²h
Heat exchanger efficiency	0.9
Storage volume	500 l
Storage tank heat loss	0.6 W/m²K
Fan power	100 W
Useful fraction of fan power	0.2
Heating system supply temperatures	Max: 42 °C
	min: 30 °C

Table II.6.4. Heat exchanger correction factors, cloudy/temperate climate

| Building type | Collector type | |
Insulation level/mass	standard	high performance
High/heavy	0.77	0.81
Medium/light	0.77	0.81
None/light	0.73	0.77

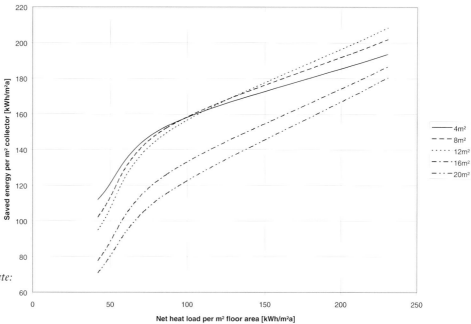

Figure II.6.9. System 6, climate: sunny/temperate, standard collector

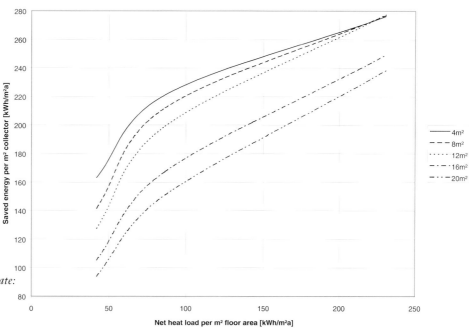

Figure II.6.10. System 6, climate: sunny/temperate, high performance collector

Table II.6.5. Heat exchanger correction factors, sunny/temperate climate

Building type insulation level/mass	Collector type standard	high performance
High/heavy	0.78	0.83
Medium/light	0.77	0.81
None/light	0.74	0.78

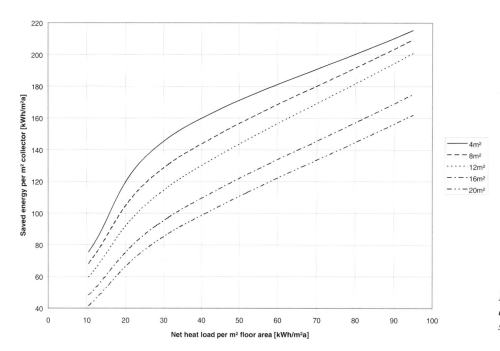

Figure II.6.11. System 6, climate: sunny/mild, standard collector

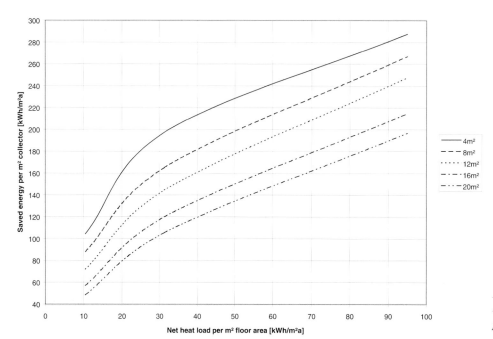

Figure II.6.12. System 6, climate: sunny/mild, high performance collector

Table II.6.6. Heat exchanger correction factors, sunny/mild climate

Building type insulation level/mass	Collector type	
	standard	high performance
High/heavy	0.80	0.85
Medium/light	0.79	0.83
None/light	0.77	0.81

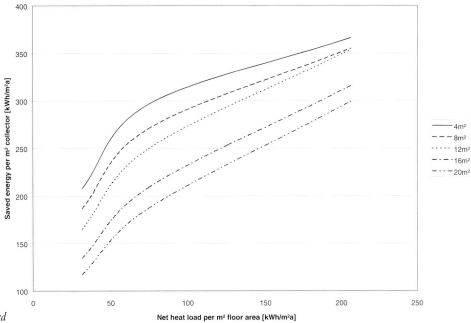

Figure II.6.13. System 6, climate: sunny/cold, standard collector

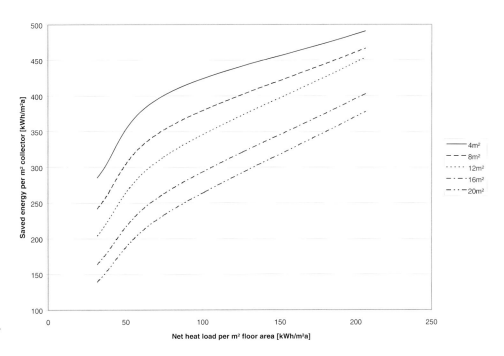

Figure II.6.14. System 6, climate: sunny/cold, high performance collector

Table II.6.7. Heat exchanger correction factors, sunny/cold climate

Building type insulation level/mass	Collector type	
	standard	high performance
High/heavy	0.80	0.84
Medium/light	0.77	0.82
None/light	0.75	0.79

III. Alternative Use of Systems

III.1 Domestic hot water heating

INTRODUCTION

In most climates a solar system designed for space heating will have excess heating capacity in summer. By installing an air to water heat exchanger and introducing a summer mode excess heat can be put to work heating domestic hot water (DHW). This simple, low cost addition to the system will greatly improve its economic return.

The basic system

Solar heated air passes through to an air to water heat exchanger located in a separate duct, bypassing the winter ducting. The heated water is circulated through a water to water heat exchanger immersed in the domestic hot water storage tank.

Figure III.1.1 illustrates a typical solar air DHW system. When the control system activates the fan, hot air heated in the collector is blown through the duct system to the heat exchanger, where it delivers its heat to the water. See also Table III.1.1.

System variations

Variations stem from combining this function with the six system types described in the preceding chapters. There are two main variations:

- a closed loop system (heats air in a closed loop);
- an open loop system (heats outside air).

Table III.1.1. Different collector types and their applicability to DHW systems

Collector type	Applicability
High performance flat plate	ideally suited
Standard flat plate	good application
Perforated collectors	possible for preheating
Window collectors	possible for preheating
Space collectors	poor
Combined air–liquid collectors	good application

The system can also be designed as a stand-alone system heating domestic hot water all year. Such systems should be designed as a closed loop system.

Advantages and limitations

- Energy and economic benefits of using solar energy otherwise wasted in summer.
- Seldom possible to optimize the system for DHW use, when it was designed for another purpose. However, the solar heating capacity in summer will likely exceed the DHW heating demand so maximal efficiency is not important.

DESIGN (INTEGRATION/SIZING)

Design options/design parameters

For the additional use of the system as a summer DHW system the following components and parameter specifications are important:

Figure III.1.1. Schematic diagram of a DHW closed loop solar air system

- collector
- heat exchanger
- storage size
- fan (if needed)
- air flow
- controls.

Collector

The summer performance of the system depends on the collector type, area and tilt. The type and area are likely to be fixed in relation to the design requirement of the main system. The tilt of the air solar collectors might either have been optimized for winter space heating or the tilt may be vertical in the case of facade integrated collectors. Thus the tilt will not be optimal for summer DHW heating. The influence of the tilt of the standard collector (up from horizontal) on the saved energy (Qsys) is presented in Figure III.1.2 for the open loop system.

Heat exchanger

If the collector area is oversized for the summer demand the heat exchanger does not have to be very effective but it should have a low pressure drop. If the pressure drop is low enough, it may be possible for the air to circulate through the exchanger by natural convection. At the other extreme is the case of a closed loop system with a relatively small collector. Such a configuration requires an efficient heat exchanger to get the most out of the system. Tables III.1.3 and III.1.4 show the effect of using different temperature efficiency for the heat exchanger (see also Chapter V.2).

Storage

The energy performance simulations show the sensitivity of system performance to storage size. The domestic hot water load is most critical. The simulations have been made based on a standard single family hot water consumption of 200 l/day. Results showed that a storage size of 300 l seems adequate for a DHW load of this size.

Fan

The fan already installed for winter operation might be used (perhaps at an alternative speed) for the summer mode operation, or the fan might be switched off and the

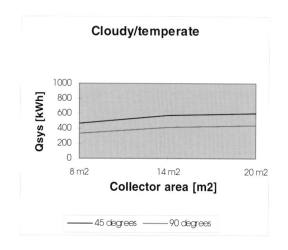
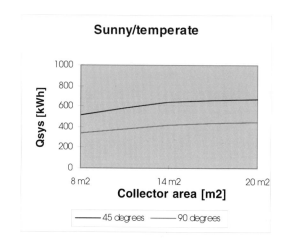
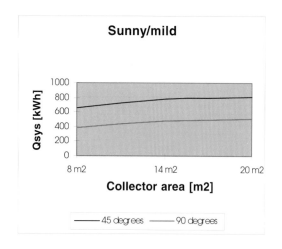
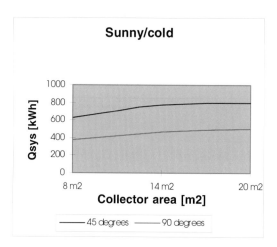

Figure III.1.2. Summer energy performance of an open loop system for four different climates, 45° and 90° slope as a function of collector area

system run solely by thermal circulation. In some situations it might be best to install a secondary smaller fan for the summer operation (see Chapter V.1).

Air flow
The optimum air mass flow depends on whether the system considered is an open or a closed loop system and the size of the air collector in relation to the load. Figures III.1.3 and III.1.4 illustrate the system performance sensitivity to the air volume flow for respectively a closed and an open loop system, where:

QDHWax = auxiliary energy consumption for DHW heating during summer
Qsys = saved energy.

It appears that for the closed loop system the sensitivity curve is very flat with an optimum at 60 m³/m²h: for the open loop system the optimum is 20 m³/m²h.

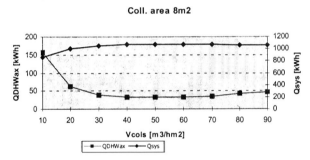

Figure III.1.3. Summer energy performance as function of air volume flow for a closed loop system with a high performance collector for DHW heating

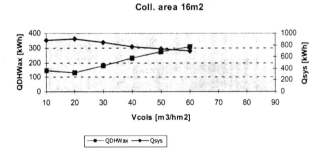

Figure III.1.4. Summer energy performance as function of air volume flow for an open loop system with a standard collector

ENERGY PERFORMANCE CURVES AND TABLES

Standard configuration

The curves and tables developed to estimate performance are based on two standard configurations: a closed loop and an open loop system. The simulations assume a fixed daily load of 200 l distributed over the day. Table III.1.2 shows the main system parameters for the two system types, their variations and the load profile used for the simulations. The bold numbers are values used in the simulations to generate the curves (Figure III.1.5).

The energy performance of the DHW system presented in this chapter refers to the summer mode operation of systems 1–6. The simulations have been performed for June, July and August. In practice the DHW system runs longer, possibly even the whole year.

System performance

The curves and tables presented in this section can be used to estimate the system performance of either a closed or an open loop system based on these parameters.

The energy performance of the two systems for four climate types is presented in eight diagrams (Figure III.1.5) showing the additional solar systems gains for operation from June through August.

Correction factors

Corrections for storage size
The simulations show that increasing storage from 200 to 500 l does not significantly improve performance, but a storage of only 100 l (half the daily load) reduced the performance by 20% to 30% in the case of a large collector area. Therefore for storage sizes between 1 and 2.5 times the daily load no correction to the system performance is needed. If the storage size equals only half the daily load, the performance read off the curves on Figure III.1.6 needs to be multiplied by a factor of approximately 3/4.

Corrections for heat exchanger effectiveness
Results presented in Figure III.1.5 have been obtained with a heat exchanger temperature effectiveness of 0.9. Repeating the simulations with a heat exchanger effec-

Table III.1.2. System parameters used for the simulations

System parameter	Closed loop system	Open loop system
Collector area (m²)	4–8–14–20	8–14–20
Heat exchanger effectiveness	0.6 0.9	0.6–0.9
Storage volume (l)	100–200–300–500	100–200–300–500
Air flow rate (m³/m²h)	60	20
Hot water temperature (°C)	50	50
Daily load (l)	200	200
Load profile (l) / time (h)	06/100; 12/30; 20/70	06/100; 12/30; 20/70

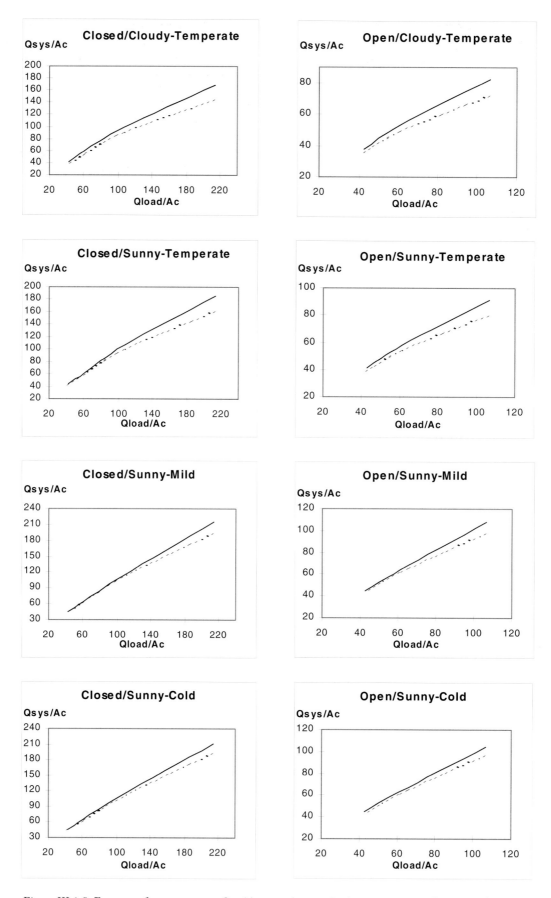

Figure III.1.5. Energy performance curves. Qsys/Ac = saved energy in three summer months per m² of collector [kWh/m²]. Upper, unbroken line: High performance collector; lower, dashed line: Standard collector

Table III.1.3. Correction factors for heat exchanger effectiveness (Open systems)

Climate \ Qload/Ac (kWh/m²)	107	61	43
Cloudy-Temperate	.83	.89	.93
Sunny-Temperate	.83	.89	.93
Sunny-Mild	.83	.95	.98
Sunny-Cold	.84	.95	.97

Table III.1.4. Correction factors for heat exchanger effectiveness (Closed systems)

Climate \ Qload/Ac	214	100
All climates	0.92	1.0

tiveness of 0.6 reduced system performance between 0 and 17% for the different climates and the two systems. The correction factors shown in Tables III.1.3 and III.1.4 can be used to estimate the performance of a system with a heat exchanger effectiveness of 0.6. For a heat exchanger effectiveness between 0.6 and 0.9 a linear interpolation may be used. From the tables it appears that open systems are more sensitive to heat exchanger effectiveness than closed systems. This can be explained by the fact that the open systems generally operate at lower collector temperatures than closed systems and therefore are more sensitive to the temperature difference needed across the heat exchanger.

Chapter author: Ove Mørck

III.2 Cooling applications and additional use of air systems

INTRODUCTION

This chapter discusses the possibility of using some components or parts of the solar air systems for cooling. The cooling application can be an additional benefit of already installed equipment during periods when the solar air system is not usually used for room heating. Such double use of the system improves the economy of a solar air system.

System description

Source of cool air and availability:

- outside air: when $T_{out} < T_{in}$, mainly nights;
- earth coupled channels: all the time;
- mechanical cooling: on request.

Cool air is used for:

- covering air demand for fresh air requirements;
- direct cooling: direct ventilation, night ventilation;
- indirect cooling: cooling of the hypocaust/murocaust storage ≥ time delay.

Modifications needed for the solar air system

The following feasibility study is based on dynamic simulations of selected cases. The Trnsair model has been modified to meet the assumptions as described below. The criteria for the design and operation of the systems were set as follows (only the items not included in the general definition of the building and solar air systems are listed here):

Fresh air ventilation

- fresh air supply 30m^3/(h person);
- supply temperature 20°C or higher in summer for cases without mechanical cooling or without pre-cooling of air in earth coupled channels.

Table III.2.1. Additional cooling application. Reference to solar air systems 1–6

Solar air system	Options for cooling	Modifications required
1: Ventilation air pre-heating	Use outside air for free cooling during night Use air pre-cooled in earth coupled channels during the day	No modification required (fan may not be powered by PV) Provide bypass of air collector Install earth coupled channel
2: Open loop	Use outside air for free cooling during night	Install possibility to open system to outside (to suck and discharge the outside air) during summer nights Optionally install fan to increase performance in thermosiphon systems
3: Double envelope	Use outside air for free cooling during night Use chiller-cooled air during day or/and night	Install possibility to open system to outside (to suck and discharge the outside air) during summer nights Install chiller with air/liquid heat exchanger to cool air of the closed loop
Closed loop, radiation discharge	Same as System 3	Same as System 3
Closed loop, forced discharge	Same as System 3	Same as System 3
Closed collection loop, heat exchange to water	Use outside air to cool water to a temperature of approx. 18–21°C Use cool water in floor heating for cooling	Install possibility to open system to outside (to suck and discharge the outside air) during summer nights. Use outside air to cool liquid circuit in air/liquid heat exchanger Install possibility to bypass domestic water storage tank. Use cooled liquid directly for floor (heating) cooling

Night ventilation

- same amount of supply air as fresh air ventilation;
- for cases with mechanical cooling or with earth coupled channels the supply air is cooled during the night if necessary.

Hypocaust cooling

- same amount of supply air as for solar air systems;
- in cases with mechanical or earth coupled channels the supply air is cooled during the night if necessary;
- when outside air temperature is low enough the hypocaust is operated as an open loop system;
- auxiliary cooling capacity is used as the first priority for the ventilation system and is applied for hypocaust cooling only when excess cooling capacity is available.

Earth coupled channel

- design air flow rate for the earth coupled channels is determined by the fresh air requirement. The desired supply temperature establishes the necessary length and depth of the channel. The number and diameter of channels are defined by the required air velocity of approx. 1.5 m/s.

Investigated system variation

Tables III.2.2 to III.2.4 show the system variations.

Characteristics, advantages, limitations

General advantages
The cooling application improves the system's economy by providing a double purpose. It is environmentally beneficial since it reduces or eliminates the use of mechanical cooling.

Advantages of particular subsystems
FREE COOLING
Flushing the building with outside air is a simple and effective cooling measure. In buildings with ventilation systems (like office buildings) the same system can be used at night when the outside temperature is below the inside temperature.

PRE-COOLING OF AIR IN EARTH COUPLED CHANNEL
When the outside air is not cool enough air can be cooled in an earth coupled channel. The cooling capacity of such a channel increases with the depth of the channel below the earth surface.

HYPOCAUST (CORE) COOLING
Hypocausts/murocausts allow cooling with very low temperature differences because of their large heat transfer area and storage capacity. They are therefore very suitable for cooling with outside night air.

Hypocausts/murocausts can be combined with:

- free cooling: considerably improves thermal comfort during the summer;
- mechanical cooling: reduces the design load that the chiller must meet.

However, the combination of hypocaust and earth coupled channels in cold or temperate climates is not necessary. Here, the use of outside air is sufficient. In sunny and mild climates cooling with the earth coupled channel and hypocaust is not alone sufficient to guarantee thermal comfort.

THERMAL COMFORT AND HYGIENE
The cooling of buildings (particularly offices) by large amounts of surface cooling rather than with high ventilation rates improves comfort. Pre-cooling of ventilation air in earth coupled channels improves the fresh air ventilation during periods of high outside temperature.

'BREAKING THE PEAK LOADS'
Cooling the core of the building mass not only improves summer comfort, it may even obviate the need for a chiller since the cooling load can be distributed over a much larger period of time. If this is the case the energy and investment savings are indeed very large.

LIMITATIONS
Thermal comfort
Do not admit air with temperatures lower than 19°C directly into the room.

Condensation
To avoid local condensation the air should not be below 18°C in any part of the system inside of the building.

Electrical energy for air transport
Large volume air flow rates may be required for high cooling loads. This results in high electrical consumption.

Climate conditions
Climate is a strong limiting factor. A thermal simulation is imperative to verify the feasibility using the local climate data, internal load and building envelope.

ENERGY AND COMFORT PERFORMANCE

Energy performance has been evaluated using Trnsys simulations modified for the cooling applications. A sunny/temperate and a sunny/mild climate were considered. Comparative performances are shown in Tables III.2.5 to III.2.7 for the system variations as described in Tables III.2.2 to III.2.4.

Table III.2.2. System variation: ventilation with outside air only

Schematics of system	Operation	
	Ventilation	Cooling
Figure III.2.1. Case A: Free cooling with outside air	For air quality during office hours	Without cooling (Reference Case)
		Night ventilation Case A
Figure III.2.2. Case B: Free cooling and hypocaust cooling with outside air		Night ventilation + Hypocaust cooling Case B

Table III.2.3. System variation: ventilation with outside air and with air cooled in earth coupled channel

Schematics of system	Operation	
	Ventilation	Cooling
Figure III.2.1. Case A: Free cooling with outside air	For air quality during office hours Cooled by earth coupled channels if required	Without cooling (Reference Case)
		Night ventilation with outside air Case A
Figure III.2.3. Case B: free cooling with outside air and with air cooled in earth coupled channel		Night ventilation with air cooled in earth coupled channels Case B

Table III.2.4. System variation: combination with mechanical cooling

Schematics of system	Operation	
	Ventilation	Cooling
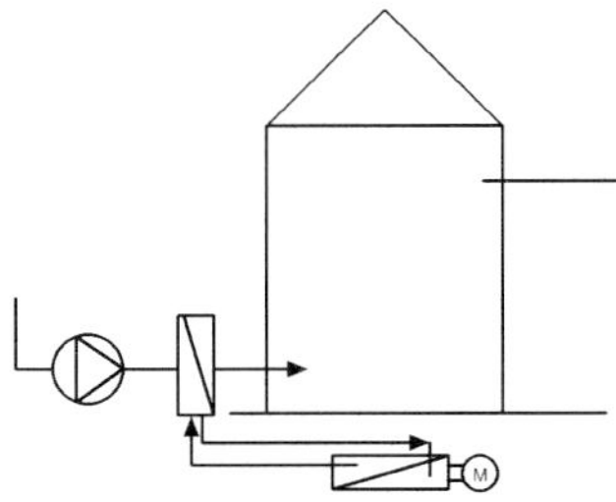 Figure III.2.4. Case A: night ventilation mechanically cooled if required	For air quality during office hours. Mechanically cooled if required	Peak cooling during office hours (Reference Case)
		Night ventilation mechanically cooled if required Case A
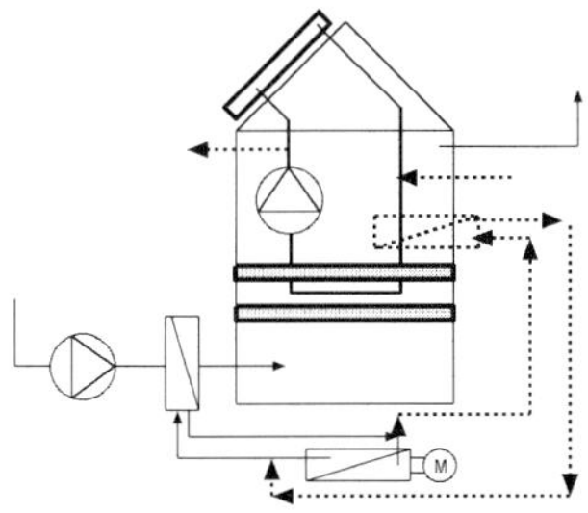 Figure III.2.5. Case B: night ventilation mechanically cooled if required and hypocaust mechanically cooled		Ventilation mechanically cooled if required plus hypocaust mechanically cooled Case B

Table III.2.5. Performance of system: ventilation with outside air only

Schematics of system	Performance

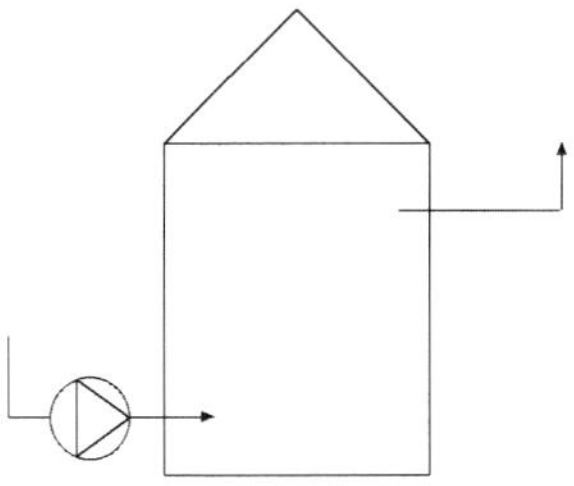

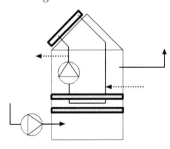

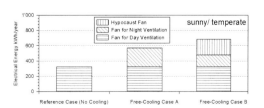

Figure III.2.1. Case A: Free cooling with outside air

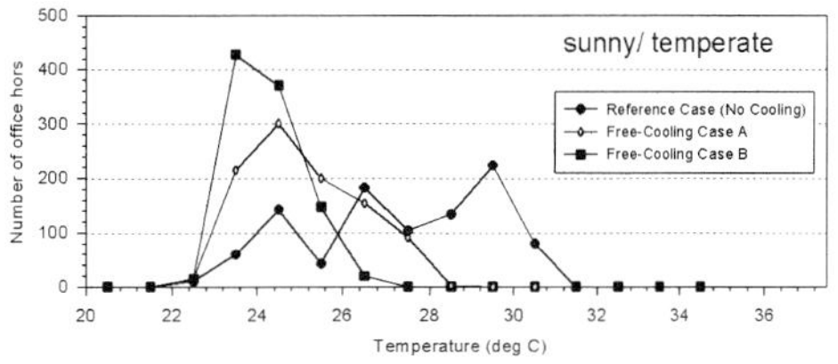

Figure III.2.6. Comfort performance: free cooling with outside air

Figure III.2.2. Case B: Free cooling and hypocaust cooling with outside air

Figure III.2.7. Energy performance: free cooling with outside air

Table III.2.6. Performance of system: ventilation with outside air and with air cooled in earth coupled channel

Schematics of system	Performance
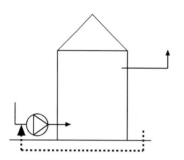 Figure III.2.1. Case A: free cooling with outside air	Figure III.2.8. Comfort performance: free cooling with air pre-cooled in earth coupled channel
Figure III.2.3. Case B: free cooling with outside air and with air cooled in earth coupled channel	Figure III.2.9. Energy performance: free cooling with air pre-cooled in earth coupled channel

Table III.2.7. Performance of system: combination with mechanical cooling

Schematics of system	Performance
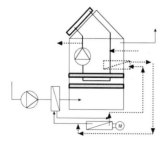 Figure III.2.4. Case A: night ventilation mechanically cooled if required	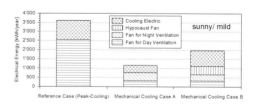 Figure III.2.10. Comfort performance: mechanical cooling
Figure III.2.5. Case B: night ventilation mechanically cooled if required and hypocaust mechanically cooled	Figure III.2.11. Energy performance: mechanical cooling

Thermal comfort
The performance of each system is presented as the number of office hours certain room temperature occurs. This indicates the overheating frequency.

Energy consumption
The energy consumption consists of the electrical power required for:

- fan (day ventilation, night ventilation, hypocaust cooling);
- mechanical cooling (assuming a chiller coefficient of performance of 2.5).

This system is shown for the sunny/temperate climate only. In sunny/mild climates it will lead to severe overheating and unacceptable discomfort. Without free cooling (reference case) severe overheating occurs. With free cooling with night ventilation air (Case A) the comfort is considerably improved, however, there are still approximately 250 office hours with temperatures above 26°C. Free cooling and hypocaust cooling (Case B) improves comfort to zero hours above 26°C with only marginal electrical consumption for the fan (to move the hypocaust air).

Using earth coupled channels to cool ventilation air results in a similar benefit as when using hypocaust cooling (see previous application). However, investment costs might be higher if the earth coupled channel was installed for cooling purposes only.

In sunny/mild climates the application of the earth coupled channel is not feasible. During the critical summer period the ground temperature is too high to provide enough cooling potential (see Figures III.2.13 and III.2.15)

Peak cooling with ventilation air was taken as the reference case. The ventilation air rate has to be increased to remove the cooling load. This leads to a high demand on electrical power for the ventilation fan. Ideal comfort cannot be achieved.

Mechanical cooling with night ventilation air (Case A) in addition to day ventilation consumes less energy but results in reduced comfort.

Best comfort results at relatively low energy costs result when utilizing the *hypocaust for cooling* during the night (Case B). An additional benefit (not appearing in the energy performance) is the lower required installed

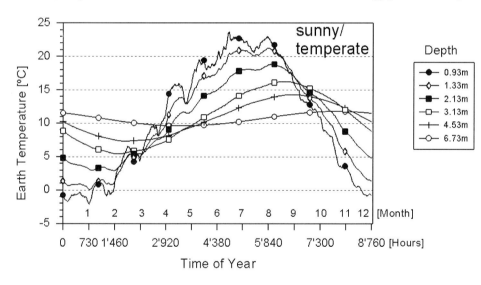

Figure III.2.12. Annual earth temperatures as function of depth: sunny/temperate

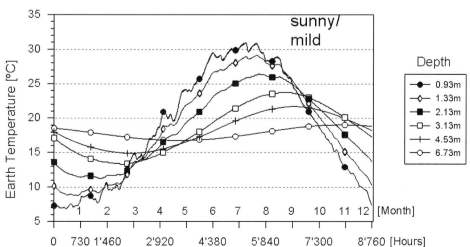

Figure III.2.13. Annual earth temperatures as function of depth: sunny/mild

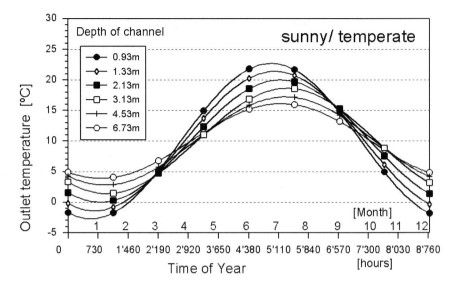

Figure III.2.14. Expected outlet temperature (smoothed) as function of depth for climate sunny/temperate (d = 30cm, L = 50m, maximum temperatures for cooling, minimum for heating)

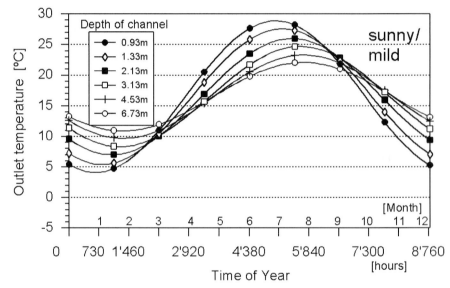

Figure III.2.15. Expected outlet temperature (smoothed) as function of depth for climate sunny/mild (d = 30cm, L = 50m, maximum temperatures for cooling, minimum for heating)

capacity of the chiller. This results in considerably lower investment costs.

SYSTEM DESIGN

Design procedures here focus on the earth coupled channel.

Sizing of hypocaust (see Chapter IV.6) Earth coupled channels

The following is a rough method for preliminary calculations of the system's depth, and the diameter and length of the earth channels, and predicting the system's performance.

Criteria
- duty: desired outlet temperature for:
  - pre-cooling of ventilation air in summer: T_{max} = 18°C
  - pre-heating of ventilation air in winter ?
  - hypocaust cooling?
- air flow rate: suitability for ventilation or/and hypocaust cooling
- earth conditions: type of soil, ground water level (still or moving?)

Earth temperatures in wet soil (no ground water movement)

The data shown here are evaluated only for no moving ground water. Figures III.2.12 and III.2.13 show typical patterns of the annual earth temperatures to be expected as a function of the channel depth for the two climate types.

With increasing depth the climate exerts less influence and the annual temperature amplitude decreases.

Figures III.2.14 and III.2.15 show the air outlet temperature of the earth coupled channels for a channel at 3m depth, 50m length and 30cm diameter. The curves show maximum temperatures to be expected during the cooling season and minimum temperatures to be expected during the heating season.

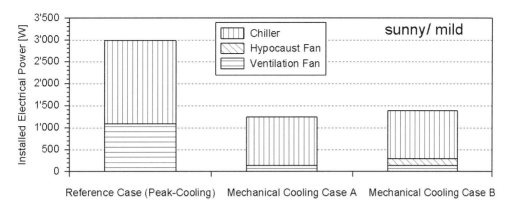

Figure III.2.16. Installed chiller power

Sizing of the chiller for mechanical cooling

Hypocaust cooling allows the chiller to be designed for lower cooling loads (see Figure III.2.16).

Controls

The right control strategy is important for cooling and difficult to determine because of:

- the storage capacity of the hypocaust and resulting thermal shift between the cooling load and cool air supply;
- limitations due to outside air temperature in the case of free cooling;
- constraints of avoiding condensation.

Do's and don'ts

- Do install channels with a slope to drain condensation water to the ground at the lowest point.
- For parallel channels use a minimum separation of one meter.
- Avoid air temperatures below 18°C in air channels or hypocausts. Condensation problems have to be avoided inside the building.
- Do not cool ventilation air below 19°C during office hours.
- Hygienic measures to be taken for design and operation of earth coupled channels:
  - elevate the air inlet above ground level, in an area free of dense vegetation which could hinder access for cleaning and at a distance from odour- or organism-emitting sources (i.e. composts, sewer access/vent points, loading docks)
  - install air screens at inlet and outlet of channels to prevent insects and small animals (children?) from entering the channels
  - do not use ribbed channels (they retain condensate and cause excessive pressure drops requiring more fan power)
  - use common air-filters used in ventilation systems, which should remove particle sizes of 2–5mm.

REFERENCES

Computer models:

TRNSYS Type160, by Karel Fort
　The hypocaust model can be used with some adaptations for the purpose of the earth coupled channels. This model does not take into account the latent heat of condensation/evaporation. This model was used for the simulations made in this chapter
　Type160 is available from TRANSSOLAR, Nobelstrasse 15, D-70569 Stuttgart.

TRNSYS Type 61: Hypocaust (Air-To-Soil Exchanger), by Pierre Hollmuller and Bernard Lachal, Centre universitaire d'étude des problèmes de l'énergie, Université de Genève, Route de Florissant 158, CH-1231 Conchres-Genève.

Reports:

Flückiger B, Lüthi P and Wanner H-U, 1997 *Mikrobielle Untersuchung von Luft-ansaug-Erdregistern*, Institut für Hygiene und Arbeitspsychologie, ETH-Zürich, Switzerland.

Chapter author: Karel Fort

III.3 Hybrid photovoltaic/heating system

INTRODUCTION

Description

Photovoltaic (PV) solar panels produce far more heat than electricity, so putting this heat to use improves the system's total efficiency and cost effectiveness. This chapter presents ways to capture and use this heat; it does not address the use of PV to power a fan or as a controlling strategy.

The main drawback to most PV systems has been the high initial cost and small electrical output compared to the solar input. Paying back the capital investment with the value of the electrical output could take decades, particularly for grid-connected systems, were it not for government subsidies or buy-back prices for electricity from utilities. This long payback can be reduced to an acceptable range if 'waste' solar heat is used, since the heat production per square metre can be as much as four times greater than the electrical energy produced.

Crystalline silicon PV panels are the most commonly used type in building integrated systems. Their efficiency ranges from 10% to 15%. Amorphous panels have lower efficiencies. Most of the non-utilized energy is thermal energy. The actual operating efficiency is invariably less than the rating provided by the standard tests, which are based on conditions which seldom occur in reality (operation by 1000 W solar radiation with the cell temperature 25°C).

For crystalline cells the performance drops as cell temperature rises. Removing the heat cools the cells and improves the operating efficiency. A PV cell can have a stagnation temperature of 50 K above the ambient if the heat is not removed. A solar air collector is more efficient, achieving 40% to 70% thermal efficiency and the panel temperature above ambient will be lower than the 50 K stagnation temperature since heat is being removed.

In a hybrid PV/thermal system part of the solar absorber is covered with encapsulated photovoltaic cells or modules. The PV cells are cooled by the continuous air stream behind the absorber. The end result is increased PV conversion efficiency and warm air available for building ventilation, space heating or even water heating via an air to water heat exchanger.

Variations

a   Conventional PV panels can be cooled by an air stream passing behind the panels as in a conventional back pass solar air collector.
b   PV cells can be integrated or mounted onto a thermal absorber designed to remove heat.
c   Two separate systems can be used, the first being a PV system with some form of heat collection with the warm air fed into a second higher temperature solar booster heating collector mounted above the PV panels.

Type (a)

Ambient air is drawn behind the PV array where it is heated, then enters the building's ventilation system as in System 1 designs (Chapter II.1). The design is very similar to a back pass solar air collector with the PV cells acting as the heat absorber. Figure III.3.1 shows a typical roof-mounted configuration. The thermal efficiency will be lower than with a typical solar air panel due to the less than ideal heat exchange between the back of the PV panels and the air stream. Overall thermal operating efficiencies should be in the 30% range while the PV efficiency is improved.

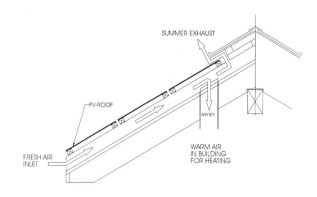

Figure III.3.1. PV panels mounted on roof with back pass heat removal

Type (b)
Here the PV cells are mounted on an unglazed collector absorber or configured in a way that allows the cooler ambient air to pass behind the PV cells in a uniform and controlled manner. Heat generated from the PV cells is transferred to the air, as shown in Figure III.3.2. Using the balancing features of the unglazed perforated panel (see Chapter IV.3) ensures good air flow around the PV panels. The whole system can be designed in accordance with Chapters II.1 and IV.3.

When used for preheating ventilation air, the thermal efficiency of the PV cell mounted onto perforated panels can achieve 50% to 60%. This design is suitable when there is a ventilation heating requirement and maximum PV output is desired.

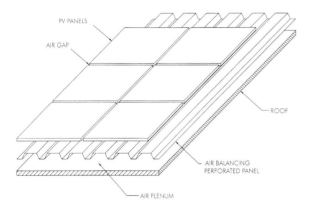

Figure III.3.2. PV modules mounted onto unglazed perforated solar collector

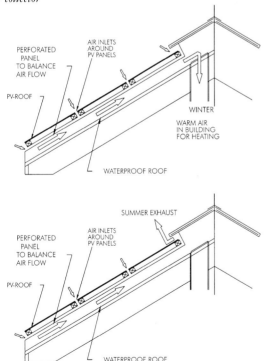

Figure III.3.3. Cross-section of PV panels on perforated collector

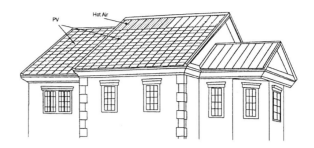

Figure III.3.4. PV panels and solar thermal panels in a two stage configuration

Type (c)
The third variation is a two-step system. Air warmed by a PV panel is fed into a conventional solar air collector located above the panel. Heat rising up the PV panel is drawn into the solar air collector. This combination offers more choices for selecting components and for using the heated air. Separating the two functions also allows design flexibility. Figure III.3.4 illustrates this concept.

Applications

By recovering heat from PV panels, the economy of the PV array will be greatly enhanced, improving the chances that the owner will approve the system.

Buildings in northern climates with a heating load are ideally suited for a combined PV and thermal concept. System 1 should be considered first; a second stage of solar heating should be considered when higher temperatures are needed, especially for commercial and multi-family residential buildings.

DESIGN

General selection criteria

PV panels integrated in walls and roofs must be detailed to dissipate the waste heat, otherwise overheating is probable. Overheating may damage the roof or wall construction. Designing a combined PV/thermal system requires two designs, one matching the electrical requirements and the other matching the heating needs. An electrical engineer or consultant normally designs the electrical PV array and that person should be advised of the requirement to integrate the PV panels into the building construction and to coordinate with the solar heating consultant.

Crystalline silicon PV panels, the most common PV panels, operate more efficiently when they are cooled, and drawing outside air across the panels performs this cooling function. Circulating room air behind the PV panels would not provide adequate cooling power. Crystalline panels should only be considered with System 1 (ambient air through solar panel). Summer heat gain can either be used for domestic hot water preheating as described in Chapter III.1 or excess heat can be dissipated to ambient.

Amorphous silicon panels actually perform better at higher temperatures so recirculating warm air behind them for space heating is possible. If air is recirculated from the building instead of using ambient air, the PV cell temperature will be higher, improving performance. Amorphous panels are less efficient than crystalline panels, require more surface area to achieve the same electrical output and would generally be recommended for milder climates where summer heat build-up may be a concern.

When designing the PV/thermal system either the electrical needs or the heating needs will dominate the design. If electrical generation is the primary motivation then the surface area for PV panels will be so determined and the thermal output will be determined by that area. The volume of air to cool the panels can vary widely, but is generally between 40 to 120 m^3/(h m$^2_{collector}$). Alternatively, heated air requirements may dictate the design, by the use of space heating, ventilation air heating, domestic hot water preheating or process air heating.

If the solar heating collector array is known, determine the PV coverage area possible or feasible for the PV modules. The PV surface area can then be given to the electrical consultant to calculate the electrical generation output.

For safety and so that the interconnections attain practical system voltages, the photovoltaic cells and/or modules must be electrically isolated from the metal panels. Therefore, cells (e.g. amorphous silicon) will not be deposited directly on the metal panels. Adhesives, if used, must be weatherproof and durable. Additionally, for the roof-mounted configurations, the panel or photovoltaic module design should withstand possible damage to the modules if people were to walk on the roof during installation or maintenance. Roof integration details of PV panels are available from PV suppliers. When designing the air flow, follow the advice given throughout this handbook.

Energy savings

The heat from combined PV thermal panels is normally used during the eight- or nine-month heating season,

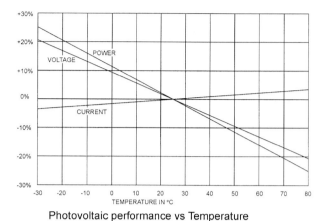

Figure III.3.5. Effect of cell temperature on electrical production

while the electrical output is useable over twelve months. Annual electrical production is generally in the 70–200 kWh/m^2 range depending on the climate.

Test results indicate that combining the PV panels with the thermal panels improves the efficiency of the PV panel and that the total combined efficiency can exceed the thermal efficiency alone. Figure III.3.5 shows the relationship between power generation and PV cell temperature. Lowering the cell temperature will increase the power output. The power output increases at a rate of between 0.4% to 0.5%/K of lower cell temperature. Lowering the cell temperature by 20 K by increasing the air flow rate around the PV panels can increase power output by 10%, e.g. heat removal can lower the temperature from 50°C to 30°C.

Ideally the PV panels will be roof-mounted to achieve the best sun angle in summer when the longest sunshine hours occur. High cell temperatures will result there because of the high roof temperatures, so heat removal is imperative, even if the heat is rejected in summer. But coupling this system with an air to water heat exchanger (see Chapter III.1) would allow even the summer heat to be put to use.

Chapter author: John Hollick

IV. Components

IV.1 Flat-plate air collectors

INTRODUCTION

Description

The solar air collector is designed to heat air when irradiated by the sun. It can be architecturally integrated into the building's skin or attached on top of it. The basic components are: cover, absorber, air passage and insulation, as shown in Figure IV.1.1. Solar radiation transmitted through the cover heats the absorber, which in turn heats the air in the air passage.

Efficiency

The relation between the useful gain of the collector and the solar radiation on the collector surface is defined as the efficiency:

$$\eta = \frac{\dot{Q}_u}{\dot{Q}_{sol}} = \frac{\dot{Q}_u}{A_c G} \qquad (IV.1.1)$$

$$\dot{Q}_u = \dot{m} c_p (T_o - T_i) \qquad (IV.1.2)$$

A_c: collector reference area, m²
c_p: specific heat of air, kJ/kgK
G: solar global irradiance, W/m²
$\dot{m}$: mass-flow rate, kg/s
T_i: collector inlet temperature, °C
T_o: collector outlet temperature, °C
$\dot{Q}_{sol}$: solar gain on the collector reference area, W
$\dot{Q}_u$: useful gain of the collector, W

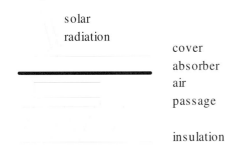

Figure IV.1.1. Typical collector elements

The collector reference area can be considered as: aperture area, absorber area or gross area.

Based on Duffie and Beckmann (1991) a general equation for solar collector performance is:

$$\eta_o = F_o [\tau\alpha - U_L (T_o - T_a) / \dot{Q}_{sol}] \qquad (IV.1.3)$$

where T_a is the ambient temperature and F_o is the collector heat removal factor in relation to T_o (the collector outlet temperature), and η_o is the efficiency when the outlet temperature is taken as reference.

F_o accounts for the fact that the absorber temperature is not the same as the outlet air collector temperature in either the horizontal or vertical direction.

The following physical factors influence the efficiency of the collector:

h: the heat transfer coefficient, which governs the heat transfer from the absorber to the passing air;
$\tau\alpha$: the glass transmission absorber absorption product;
U_L: the overall collector heat loss coefficient.

It appears that the efficiency η depends on the operating conditions of the collector. It decreases with increasing temperatures, because of increasing heat losses.

It is important to define the operating conditions of the collector given by the temperature difference between the 'overall collector temperature' T_k and ambient T_a.

An efficiency curve can be drawn as a function of a certain reference temperature, which corresponds to the collector temperature T_k. In a physically correct way one has to take a weighted mean temperature T_k of the whole collector box, but in practice this is not practicable. Accordingly, one of these three reference temperatures may be used: the inlet temperature (T_i), the outlet temperature (T_o) and a so-called 'mean' collector temperature (T_m) which is the arithmetic mean value between inlet and outlet temperature. Efficiency curves of solar collector corresponding to the three possible reference temperatures for a constant mass-flow rate are shown in Figure IV.1.2.

For solar *water* collectors it is customary to present the efficiency related to the mean collector temperature (T_m) representative for the heat losses of the collector. The temperature difference between the water inlet and outlet is very small (normally less than 10 K) and the heat

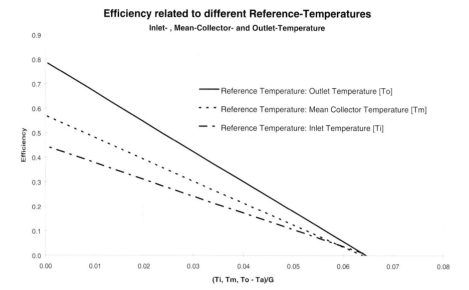

Figure IV.1.2. Efficiency curves related to different reference temperatures

transmission from the absorber to the fluid is high. Therefore, the arithmetical mean value (T_m) is in fact very close to the physical mean temperature (T_k) of the collector.

For *air collectors* the difference between inlet and outlet can be up to 30 K or 40 K, depending on the air mass-flow. Also important is the amount of heat transmitted from the absorber to the 'fluid', which in the case of air is not that much. Due to these effects there will no longer be a linear increase of the fluid temperature along the collector plate. Accordingly, the arithmetical mean value (T_m) is often not representative for the heat losses of the collector. Measurements indicate that the physical mean temperature (T_k) of the collector is often much closer to the outlet temperature (T_o) than to the arithmetical mean temperature (T_m). Therefore the presentation of the collector efficiency curve using the outlet temperature (T_o) often seems to be the best solution.

Presentation of efficiency curves

Solar air collectors usually operate in two different system types:

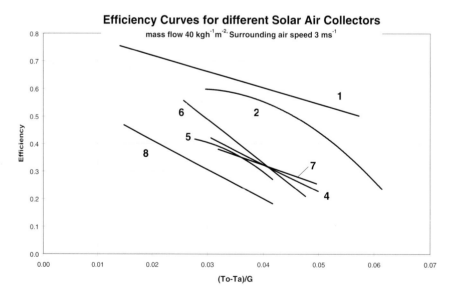

Figure IV.1.3. Collector efficiency related to outlet temperature for current collectors on the market

1 glazed collector (iron-poor), aluminum absorber with U-profiles, selective coating, underflow
2 glazed collector, black textile absorber
4 glazed plane absorber, black painted, facade element, underflow
5 glazed, rippled absorber, air flow on both sides
6 glazed plane absorber, black painted, facade element, air flow on both sides
7 glazed site-built collector, selective absorber, trapezoid profile, underflow
8 glazed plane absorber, black painted, facade element, underflow

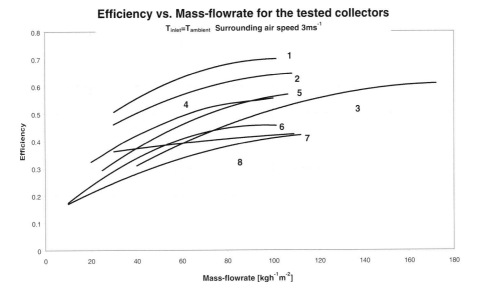

Figure IV.1.4. Efficiency versus mass-flow rate for selected collectors on the market
1 *glazed collector (iron-poor), aluminum absorber with U-profiles, selective coating, underflow*
2 *glazed collector, black textile absorber*
3 *unglazed perforated trapezoid absorber panel, aluminum, anthracite, strongly dependent on wind (curve for 3ms^{-1})*
4 *glazed plane absorber, black painted, facade element, underflow*
5 *glazed, rippled absorber, air flow on both sides*
6 *glazed plane absorber, black painted, facade element, air flow on both sides*
7 *glazed site-built collector, selective absorber, trapezoid profile, underflow*
8 *glazed plane absorber, black painted, facade element, underflow*

- Closed loop systems
- Open loop systems

CLOSED LOOP SYSTEMS

Here the inlet temperature can sometimes be much higher than the ambient temperature. For this mode of operation the presentation of the efficiency as a function of the difference between outlet and ambient temperature $(T_o - T_a)/G$, is adequate. Figure IV.1.3 presents measured efficiency curves for seven collectors on the market. Be aware of the fact that the efficiency values close to the y-axis only appear if the inlet temperature is below ambient, because if $T_o = T_a$, T_i is always below T_a.

OPEN LOOP SYSTEMS (SYSTEM TYPE 1)

The collector always draws ambient temperature ($T_i = T_a$). For these situations a presentation of the efficiency versus the mass-flow rate is better suited for engineering purposes, see Figure IV.1.4.

TEMPERATURE RISE OVER THE COLLECTOR

When designing solar air systems it is often useful to know which temperature level can be expected out of the collector. This is in direct relation to the efficiency and the mass-flow rate and can be calculated based on these. Figure IV.1.5 shows the measured temperature rise for the same eight collectors for which efficiency curves are shown in Figure IV.1.4.

Figure IV.1.6 shows the efficiency of the collectors modeled by Trnsair and used in the Trnsair simulations presented in Chapter II.1. The different areas are put together in series of 2 m² modules. Therefore the air speed will be higher in collectors consisting of more modules. The air flow within an air collector can be laminar or turbulent. The dynamic condition within the air channel depends on several facts, mainly the air velocity, the geometry of the channel and the air temperature. In the 2 m² module the transition between laminar and turbulent flow (which suddenly increases the heat transfer coefficient) happens at 40 m³/hm², whereas the flow is turbulent also at the lower flow rates for the larger collector areas.

Note that more efficient solar energy collection occurs at higher mass-flow rates for two reasons. The heat transfer is greater, and the higher the mass-flow rates, the lower the mean temperature of the collector and hence lesser the heat losses.

Variations

A number of different solar air collector designs have been built. They differ according to the construction of the three main collector features:

- cover (τ = transmission);
- absorber type and coating, α = absorptance, ε = emittance);
- air flow pattern.

The cover transmits sunlight and hinders heat loss by radiation and convection back to the ambient; it is the

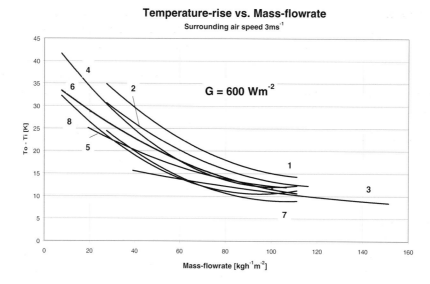

Figure IV.1.5. Temperature rise at G = 600 W/m² for selected collectors on the market (1998)

external weather-shield of the collector. The designer must therefore consider the characteristics of a candidate material for these functions.

The absorber converts the solar energy to heat. It must then give up as much of this heat as possible quickly to the passing air stream. Thus the absorber should be highly absorptive for the solar spectrum, and provide a high heat transfer to the air stream. The absorber's performance can be enhanced with a selective coating on the sun-facing surface. The laws of physics determine that the absorptivity (α) and emissivity (ε) are identical at the same wave-length. A selective coating is designed to have a high absorptivity at the wave-length of the solar spectrum and a low emissivity at the longer wave-length of the infrared (heat) radiation from the absorber. Therefore selective coatings are essential for high temperature applications.

Absorbers can also be a combination of a liquid and air absorber to serve two system applications (Jensen et al. 1987)

Four distinct air flow patterns exist:

- The air passes between the glazing and the absorber.
- Air flow passes below the absorber.
- Air flow passes on both sides of the absorber.
- Air flow passes through the (porous) absorber.

Characteristics, advantages and limitations

The choice of air flow pattern depends on the application. A very simple air collector can be constructed with the air flow between the absorber and the glazing. The efficiency will be quite poor due to the high convective heat transfer to the glazing, but it can be produced at low cost. The air flow behind the absorber is probably the most

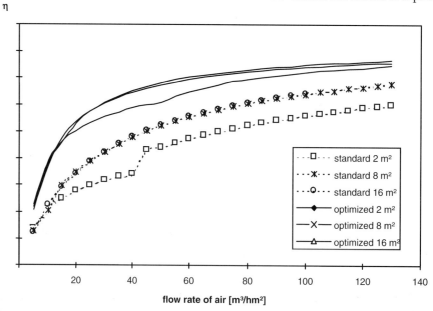

Figure IV.1.6. Efficiency curves for the collectors modeled with Trnsair, G = 800 W/m²

common solution, generally applicable and used with many different absorber types. Air flow on both sides of the absorber is often used in systems where the preheating effect of the outer air flow can be used, for example when heating outside air. Also the fourth air flow pattern offers the possibility of a cost-effective solution for medium temperature applications.

ADVANTAGES OF THE DIFFERENT AIR FLOW PATTERNS INSIDE A SOLAR AIR COLLECTOR

- The air passes between the glazing and the absorber: simple construction, inexpensive;
- The air passes below the absorber: the air gap between absorber and glazing provides insulation, and it is possible to geometrically increase the heat transfer surface area to the air stream;
- The air passes on both sides of the absorber: double effective heat transfer area;
- Air flow passes through the (porous) absorber: high heat transfer coefficient.

DISADVANTAGES OF THE DIFFERENT AIR FLOW PATTERNS INSIDE A SOLAR AIR COLLECTOR

- The air passes between the glazing and the absorber: only one surface transfers heat, high losses, especially at a high difference between absorber and ambient temperature, rapidly decreasing efficiency at high air velocities (double glazing reduces the heat losses but also decreases the solar input);
- The air passes below the absorber: only one surface is used as effective heat transfer area;
- The air passes on both sides of the absorber: poor efficiency at high differences between absorber and ambient temperature;
- Air flow passes through the (porous) absorber: high pressure drop, and depending on ambient air conditions (dust, pollution) the absorber may clog.

For these four air flow patterns the collector performance can be calculated using formulas derived by Duffie & Beckman.[1]

Another issue is the air flow direction. The collectors can be mounted and connected upwards, downwards or horizontally. Be aware of buoyancy forces, they are generally not a problem for fan-forced systems, but should not be ignored for low flow solutions.

A summary of different collector designs available on the market is shown in Figure IV.1.7.

DESIGN

The following paragraphs on design provide guidance for specifying either a prefabricated collector or a site-built collector. Important design criteria are:

- the transmission of the glazing;
- the absorber material, color and the coating and geometry;
- the overall collector geometry;
- the air flow pattern inside the collector;
- the way several collectors are connected;
- the required temperature levels;
- constructional constraints (tilt and orientation);
- initial costs and running cost.

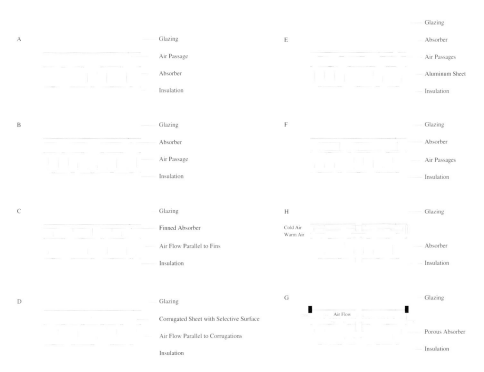

Figure IV.1.7. Examples of different air collector designs

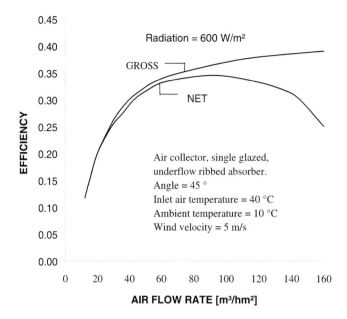

Figure IV.1.8. Collector efficiency curves. GROSS shows thermal efficiency, NET shows the efficiency when the fan electricity consumption has been subtracted

For example, for an industrial application of System 6 working at 60°C an efficient air collector with a high heat transfer rate designed for high air flow rates must be chosen, whereas for preheating of ventilation air, System 1, a simpler and less expensive air collector design is appropriate. For a system designed for direct room air heating or preheating of ventilation air in winter and domestic hot water heating in summer a medium-efficiency collector will be the best choice.

Collector design as part of a system

The collector is just one component of a system. The expected collector output/performance is needed for the design of the other components in the system, e.g. heat exchanger, fans, and to define the control strategy.

The performance of the air collector can be characterized by two interrelated parameters:

- temperature rise over the collector, $T_o - T_i =$
- collector air flow rate, $\dot{m}C_p$, and their product,
- the useful solar gain, $\dot{Q}_u = \Delta T \dot{m} C_p$

From these (and in accordance with equations IV.1.1 and IV.1.3) the following relations can be derived:

$$\dot{Q}_u = \eta G \qquad (IV.1.4.)$$

and

$$\Delta T = (T_o - T_i) = \frac{\dot{Q}_u}{\dot{m}C_p} = \frac{\eta G}{\dot{m}C_p} \qquad (IV.1.5.)$$

where A_c is chosen to be 1 for the normalized situation (per m²)

It is seen that ΔT is a product of two parameters η and $G/\dot{m}C_p$. The efficiency η is a function of G, $\dot{m}C_p$ and T_o for each collector design, and is generally shown in the form of efficiency curves (see above), given as product information or estimated, based on the actual collector geometry, etc. Based on the efficiency read off the efficiency curve, ΔT can be calculated by applying equation IV.1.5 for a specific combination of incoming solar radiation and collector air flow rate. Note that the flow rate for the efficiency read and the flow rate used in the equation need to be identical.

The air flow rate

The air flow rate is crucial to the overall system performance. Too high an air flow consumes excessive fan power and too low flow rates cause poor thermal performance of the system. In summary:

- The higher the mass-flow rates, the higher the efficiency of the collector.
- The electrical energy for the fan increases with the mass-flow rate.
- The effect of leakages increases with the air flow rate.
- For heating purposes a certain temperature level is often needed.

Altfeld *et al.* have defined an optimization procedure for the net energy including pressure losses in the optimization calculations.[2] This can also be expressed simply by saying that the electricity consumption for the fan has to be subtracted from the thermal output both converted to primary energy (tons of coal, litres of oil, m³ of gas) to give a true picture of the efficiency. When this is considered, the optimum efficiency occurs at lower capacity flow rates, as is seen on Figure IV.1.8.

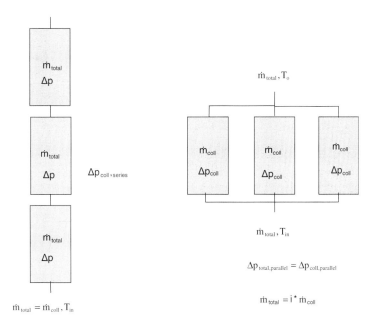

Figure IV.1.9. Air collectors connected in series and parallel

For a given application variable air flow rates may be chosen as the optimum solution and should be studied in detail for differing conditions during operation of the system in question. A drawback of this solution is the need for a more complex control system.

Constant flow or constant temperature outlet from the collector
The performance of the classic air system with a rockbed, (System 5) is greatly improved when the air flow in the collector circuit is varied to achieve a constant temperature of 40 or 50 °C out of the collector.[3] This assures a constant useful temperature level out of the rockbed for a longer time, when the air flow is reversed to extract the heat. Also fan electricity consumption is lowered. Corazza et al. have developed a set of performance equations for a collector operating under constant temperature conditions addressing the question: what percentage of the time can the collector cover the demand?[4]

Parallel or serial connection of collectors
Solar collectors may be connected either in series or in parallel as shown in Figure IV.1.9. When deciding on an air mass-flow rate, consider whether the collectors are to be connected in series or in parallel. If the same total air flow rate is assumed in a series or in a parallel connection, the thermal performance of the collector array is influenced as follows:

- The total heat transfer of the collectors in series is higher because the total air flow passes through all collectors
- The temperature rise in the collectors reduces the efficiency of each progressive collector in the series

where:
$$\Delta p_{total,series} = \Delta p_1 + \Delta p_2 + \Delta p_3 \ldots \quad (IV.1.6)$$

and
$$\Delta p_{coll,series} \gg \Delta p_{coll,parallel} \ldots \quad (IV.1.7)$$

$$\dot{m}_{total,parallel} = \dot{m}_{total,series} = \dot{m}_{coll,series} \ldots \quad (IV.1.8)$$

$$\dot{m}_{coll,series} = 3\dot{m}_{coll,parallel} \ldots \quad (IV.1.9)$$

Example
Three collectors are in series or in parallel heating ambient air. Each collector in the series connection has three times the mass flow of the collectors in the parallel connection. This means that the first collector of the series connection has a much higher efficiency than the collectors connected parallel (Figure IV.1.9). However, connecting in series leads to the fact that the second collector already has an inlet temperature above ambient, and the efficiency decreases according to Figure IV.1.10 below.

- Ambient temperature: 0° C
- Collector reference area: 1 m²
- Irradiation: 600 W/m²
- Cp: 1005 J/kgK

The total mass-flow of 33 kg/hm² for the parallel collector respectively:

- 100 kg/hm² for the series-collector gives
- for the parallel collectors: η = 52 % and

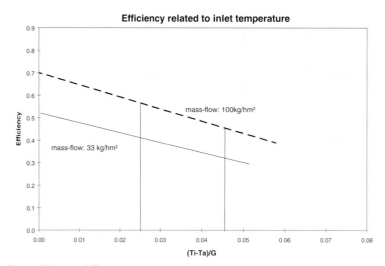

Figure IV.1.10. Efficiency related to inlet temperature

- for the first collector in series: $\eta = 70\,\%$.

From

$$\Delta T = \eta G A / m c_p \qquad (IV.1.10)$$

the temperature increase for the parallel collectors is 33.9 K and for the first collector in series 15 K. For the second collector in series, the efficiency drops down to 57% because of the increased temperature difference between collector and ambient (x-value: 0.025). For that reason, the further increase is 12.3 K, leading to a total of 27.3 K. For the third collector in series, the efficiency drops down to 46 % (x-value: 0.046). The temperature increase is 9.9 K, which leads to a total of 37.2 K.

Since the total mass-flow is the same in both connection types and the temperature increase in the series connection is 3.3 K higher than in the parallel connection, the energy gain in the series connection in this example is a little higher (9.7 %).

Since the electrical power needed for the fan directly depends on the product of volume flow times pressure drop, the pressure drop across the air collector is very important to the overall efficiency of the collector. The pressure drop increases drastically with the mass-flow rate, see Figure IV.1.11.

Connecting the collectors in parallel (33 kg/hm²) the pressure drop is about 5 Pa. Connecting in series (100 kg/hm²) the pressure drop is 18 Pa per collector, which gives a total of 54 Pa. Therefore the power consumption for circulating the air in the serial connection is much higher than in the parallel connection.

Comments

To avoid high pressure drops in serial connections, the area of the cross-section of the collector should be increased by widening the profiles, enlarging the height or choosing different absorber profiles.

A collector with a fixed air gap can only be used in the range of designed total flow rate. Therefore the question in this case is not how to connect collector modules

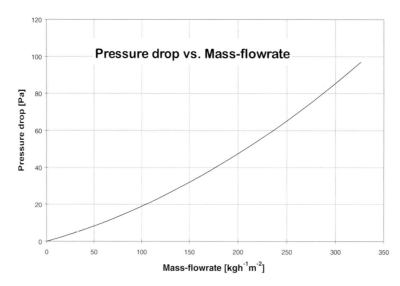

Figure IV.1.11. Pressure drop of an air collector

(parallel or serial) but how to connect the modules to reach the total flow rate and achieve best results.

On the other hand, if the air gap is an open parameter the array can be designed depending on architecture, costs of manifolds and operational costs. The air gap then has to be optimized due to heat transfer and pressure drop.

Connecting collectors in parallel needs more sophisticated manifold constructions to avoid uneven air flow. The final goal is to have an even flow from inlet to outlet from one edge to the other and to obtain the same air flow in each collector.

Design of a flat-plate air collector

Cover glazing

Glass and tempered glass are very durable covers even under the service conditions of a solar air collector. Special glass with a low iron content has significantly higher transparency than normal 'green' glass. Such glass is recommended for high temperature applications. To reduce the heat losses back through the glazing, a low-emissivity coating on the glass instead of the absorber might seem sensible. This is generally not recommended for several reasons: it is not as efficient as having the selective coating on the absorber, it reduces the transmission of the glazing and its low-emissivity characteristics are lost if there is condensation on the coating. Durability may also be an issue, as well as reusability of the glass.

Plastics and glass treated with a special coating may not hold up to the combined effects of intense UV exposure and extreme temperature variations. Such inexpensive glazing may however be suitable for low temperature applications, such as System 3 (double envelope).

Special attention should be drawn to the air- and rain-tightness of the cover and its framing. Several issues should be considered:

- It is possible to choose polycarbonate sheets in one, two or more layers. More layers reduce the heat loss but also the solar transmission coefficient.
- The solar transparency of the covers should not deteriorate appreciably during their service life, the covers should be resistant to UV radiation, air pollution, high humidity and condensation.
- Acrylic covers suffer from embrittlement when they age. This means that they will easily crack and will have to be replaced. A reduction in the tensile strength or impact strength of a cover material may lead to a failure of the collector cover.
- Plastic expands more as it is heated than metal. A good rubber seal should be used to separate a plastic cover from a metal casing to take up this difference.

Gasketing

Flange frames and high temperature resistant sealant should be used. If an efficient collector with the air flow behind the absorber has been designed it is important to prevent air from circulating around the absorber and between the absorber and cover, as this will increase the losses considerably.

Absorbers

These are typically made of steel or aluminum. Different sections are used and the surface is sometimes roughened to optimize heat transfer. Metal absorbers may be painted in dark colours or have selective coating (the efficiency of different dark-coloured absorbers is often not much less than black). Also porous cloth materials are used for absorbers.

The characteristics of the absorber surface governs the absorption of short-wave radiation and emittance of long-wave radiation. Typical values for the absorption of short-wave radiation lie between 0.9 and 0.95 and emissions range from 0.1 (for selective coatings) to 0.95 for standard black painted absorbers.

Ingenious solutions have been developed to increase high heat transfer without creating excessive pressure drops. The heat transfer is directly proportional to the heat transfer area. This can be increased by using a corrugated V absorber, by adding fins or even complete ducts to the rear of a flat absorber. The absorber might also be roughened which would result in the transition from laminar to turbulent flow at a lower air velocity. Other absorber types used and described in the literature are porous cloth, capillary structures, jet plate, matrix type, packed bed. The reader is referred to the literature for further reading.[5-7]

The absorber should be made from materials which can survive the mechanical, thermal and chemical conditions of the application. The stagnation temperature can exceed 200° C in the case of high performance collectors. Absorber coatings should retain their optical properties under high temperature, high humidity and condensation and sulphur dioxide at high humidity.

Insulation

At stagnation temperatures the insulation should not melt, shrink or outgas (a frequent cause of fogging of the cover glazing). Water or humidity absorption by the insulation material may reduce shortly or permanently the insulation performance.

Manifolds

The manifold should be designed to ensure an equal flow distribution over the absorber. To ensure a uniform flow distribution all the pressure losses for parallel areas have to be the same from inlet to outlet. Connections must be air-tight. For a large collector array it is worthwhile to spend some time and effort on designing a manifold which is easy to mount and seal.

Checklist

- Covers, thermal insulation, seals and other collector components should withstand stagnation temperatures as well as thermal shocks (a shower of cold rain pouring down on a collector in stagnation).

- Differential thermal expansion due to different thermal expansion coefficients of the materials exposed to extreme temperature variations should be accommodated.
- Components exposed to the ambient should withstand snow loads combined with wind and hail impact. If a collector box is used, it should be tight to prevent water penetration.
- Transparent insulation or Teflon films to create additional insulating air layers should not distend or become clouded during the service life due to UV radiation, high temperature and humidity.
- Solar collectors can be exposed to airborne fire or radiant heat. The use of non-flammable materials is preferred.
- If the absorber is directly anchored to the frame without a thermal break, when the absorber becomes hot the heat losses through the frame become enormous. Such constructions are only suitable for very low temperature applications. In such cases, the frame might be painted black to achieve as high a temperature as possible.
- Solar collectors should be environmentally friendly: the materials should be recyclable, the production process not polluting and the amount of energy used for manufacturing should be minimized.

Hydraulic aspects

Pressure drop
The pressure drop in a collector depends on the air flow rate to the power of 2.21.[8] It is therefore absolutely necessary to consider how much air will be passing through the collector in order to keep fan power requirements minimal. The first issue is whether the collectors are to be connected in series, parallel or a combination of series and parallel.

Rule of thumb: keep the pressure drop over the collector below 8–10 Pa for a 2 m² collector and the air flow below 70 m³/hm². This corresponds to a yearly fan energy consumption of less than 4 kWh/m². The manifold must have a considerably lower pressure drop than the absorber. This is important not only for the pressure drop but also to ensure an equal flow distribution over the entire absorber.

Morhenne *et al.* followed the energy approach and have showed how optimum collector geometries of the standard rectangular fin types are influenced by the overall geometry and pattern of the collector array.[9]

The pressure drop in the duct system can be significantly higher than the pressure drop in the collectors (especially for collectors connected in parallel) and should not be neglected.

Leakage
Although air leakages are less important than in water systems, the air-tightness of the collector is none the less important. Cold air drawn into the system or solar heated air lost out of the collector reduces its efficiency. Especially in the case of using only one fan (at the inlet or at the outlet) the effect of leakage can be significant.

A Swiss testing institute has determined the air-tightness of a number of air collectors and estimated the consequent performance reductions.[8] Depending on the air flow rate, air leakage rates were found to be as high as 10%, resulting in a significant decrease in collector efficiency. Accordingly, although it is difficult to construct absolutely air-tight air collectors, this should be given importance. Not only thermal performance is affected, but also durability. Dust and particles entering the collector may be deposited on the absorber and accumulate in a heat exchanger. Preventing leakage on site-built collectors is more difficult than for prefabricated. To minimize the leakage the length of joints should be as short as possible and temperature-resistant sealants should be used. The amount of air leaking is mainly dependent on the static pressure in the system, and thus the location of the fan(s) within the system.

PRESSURIZED OR UNDER-PRESSURIZED COLLECTOR CIRCUIT
Where to mount the fan can be decisive for the following aspects:

- If the fan is mounted before the inlet to the collector the heat dissipated to the air stream from the fan is used less efficiently, than if it is mounted after the outlet of the collector.
- If the fan is mounted after the outlet of the collector, high temperature tolerant fans must be used.

CONSTRUCTION AND OPERATION

Building integration

The opaque air collector is well suited for building integration. There is no risk of freeze damage to other building components due to air stream leakage. Another advantage is its relatively lightweight construction, which simplifies installation in the roof or facade (i.e. in a modern curtain wall).

Site-built collectors

Site-built collectors potentially cost less than factory-built units. This advantage can be maximized by using standard components produced for other purposes (e.g. Al trapezoidal external wall panels). Construction of site-built collectors requires knowledge about heat transfer and hydraulics. Prototypes should be built and tested in advance of the project construction.

Standardized sub-components
Air collectors may be constructed on site using standardized sub-components, see Figure IV.1.13.

Figure IV.1.12. Photos of integrated collectors

Practical questions

During construction it is important to protect the collectors against building dust as this may degrade the performance of the collectors as well as the fan and the heat exchanger in the collection circuit. Especially, selective coated absorbers must be mounted with care since dust, rain, high humidity and even finger prints can degrade the selectivity of the surface. Therefore it is important to close the collector immediately after constructing and to mount temporary filters at both sides of the collector.

Operation modes – controls

Most air collectors have a low heat capacity. This means they heat up and cool down rapidly in cadence with the ambient conditions. To prevent the system from cycling on and off the controls must allow for these temperature swings. Also the control considers the time it takes for the air collector to heat up all the air in the ducts (see Chapter II.6) and to compensate for the heat losses of the ducts.

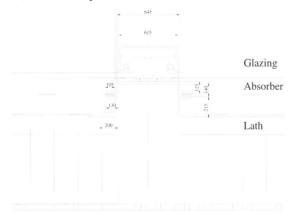

Figure IV.1.13. Section showing air collector based on standardized sub-components

The location of the sensors in the air stream behind the absorber plate and not on the plate itself is essential.

In many applications there will be a summer and a winter mode with different corresponding air mass-flow rates. A slow air mass-flow rate means a higher temperature rise over the absorber but the whole system including ducts will heat up more slowly. This needs to be taken into consideration in the control system by including a time-delay before the system is switched off, allowing for the system to heat up.

Maintenance

Optical degradation

Condensation on the cover is sometimes not easy to clean but decreases the solar transmittance of the cover only slightly. Moreover it promptly vanishes with increasing temperatures.

Snow

Snow often slides off, if the slope is at a minimum of about 25° and no barrier-like framing or roof-element hinders the sliding process. As with conventional roof construction, consider the hazard of roof avalanches.

Filter

An open system requires a filter to protect the absorber. The filter has to be replaced at certain intervals. A pressure indicator or alarm may be necessary in larger installations to avoid excessive fan energy consumption if the filter has not been changed as needed.

Maintenance scheme

A maintenance scheme should be worked out during the design phase. This should include yearly inspections of the cover, the absorber and the connections from the manifold to the duct system. Every fifth year the system should be tested for leakage, and sealings repaired if necessary.

Do's and don'ts

- Avoid thermal bridges connecting the absorber with the ambient.
- Block thermal circulation of air around the absorber.
- Specify high temperature resistant sealing between collector and duct system.
- Specify a control sensor mounted in the air stream path of the absorber before delivery.
- Choose a sensor to withstand high (stagnation) temperatures.
- Protect the wires from the sensors against animals (martens, weasels).

Acoustics – noise

Problems with noise are not likely to stem from the collectors. However, collector designs should be carefully checked for sharp bends or any other obstacle to the air flow, which might cause an unpleasant noise and also imply large pressure losses. To isolate fan noise these should ideally be located in the attic or basement.

Checklist

- Collector has a reasonable efficiency at the operating temperatures desired.
- Collector efficiency is given or tested for different flow rates.
- Temperature sensor type and location are specified.
- Insulation level is optimized.
- Air-tightness of the collector is assured.
- Collector is constructed from durable and environmentally friendly materials.
- Pressure drop for operation mode is known.
- Tilt and orientation are optimized.
- Checks for shading during the whole year have been done.
- Materials can withstand the stagnation temperatures.

COMPUTER MODELS SPECIFIC TO THE OPAQUE AIR COLLECTOR

Soleff: A program specifically developed to model different air collector designs. Institute for Building and Energy, Building 118, Technical University of Denmark, 2800 Lyngby.

Trnsys: A transient systems simulation program with a modular structure. Solar Energy Laboratory, University of Wisconsin-Madison, 1500 Engineering Drive, Madison, WI 53706, USA.

Trnsair: Trnsys-based program developed specifically to model the six solar air system types described in this handbook. Transsolar Energietechnik GmbH, Nobelstr. 15, 70569 Stuttgart, Germany.

EMGP3: A modular simulation program for thermal systems with an interactive pre-processor. Katholieke Universiteit Leuven, Institut voor Mechanica, B-3030 Heverlee, Belgium

AVAILABLE PRODUCTS

- Grammer, Werner-von-Braun-Str. 6, D-92224 Amberg, Germany. Tel.: +9621 601 151; Fax: +9621 601 260; Email: solar.info@grammer.de; Website: http://www.bs.grammer.de/sit
- Secco Solar Air-Chimney, Secco Sistemi SPA, via Terraglio 195, I-31022 Preganziol, Italy. Tel.: +39 422 490316; Fax: +39 422 490713
- Solahart World Corporate Headquarters, 112 Pilbara Street, Welshpool, Western Australia 6106. Tel: +618 9458 6211; Fax: +618 9351 8034; Email: solahart@solahart.com.au; Website: http://www.solahart.com.au
- Aidt Miljø, Kongrovensbrovej, Aidt DK-8881 Thorsø, Denmark. Tel.: +45 86 96 67 00; Fax: +45 86 96 69 55; Email: aidt@email.dk; Website: http//www/aidt.dk
- Solarwall, Conserval Engineering Inc., 200 Wildcat Road, Ontario, Canada M3J 2N5 Tel.: +416 661 7057; Fax: +416 661 7146; Email: conserval@globalserve.net; Website: http://www.solarwall.com
- ABB, The Friendly Wall, ABB Miljø, PO Box 6260, N-0603 Oslo, Norway; SUNLAB, Mr H N Røstvik, Alexander Kiellandsgt. 2, N-4009 Stavanger, Norway. Tel.: +47 51 53 34 42; Fax: +47 51 52 40 62

(see also *Solar Air Systems Product Catalogue*, James & James Science Publishers Ltd, London, 1998)

REFERENCES

1. Duffie J and Beckman W (1991) *Solar Engineering of Thermal Processes*, J Wiley, New York.
2. Altfeld K, Leiner W and Fiebig M (1998a) Second Law optimisation of flat-plate solar air heaters, Part 1: The concept of net exergy flow and the modelling of solar air heaters. *Solar Energy* **41**, 127–32.
3. Abbud I A, G O G Löf and Hittle D C (1995) Simulation of solar air heating at constant temperature. *Solar Energy*, **54** (2), 75–83.
4. Corazza A, et. al (1998). Design, development and performance studies of a large sized solar air heater in nonconventional mode of operation. *Int. Conf. Alternative Energy Sources Today and for the 21st century*. Brioni, Oct. 5–8, 1988. QUERY 5
5. Dai, Hui and Li (1991) Fully developed laminar flow and heat transfer in the passages of V-corrugated solar air heater, *ISES'91, Denver*, proceedings. QUERY 3
6. Gupta D, Solanki S C and Saini J S (1997) Thermohydraulic performance of solar air heaters with roughened absorber plates. *Solar Energy*, **61** (1), 33–42.
7. Matrawy K K (1998) Theoretical analysis for an air heater with a box-type absorber. *Solar Energy*, **63** (3), 191–98.

8. Keller J, Kyburz V and Kölliker A (1988), *Untersuchungen an Luftkollektoren zu Heis- und Trockungszwecken, 1988*. Schlussbereicht des KWF-Projektes Nr. 1296. Paul Scherrer Institut, Wrenlingen und Villingen.
9. Morhenne J and M Fiebig (1990) *Entwicklung und Erprobung einer Baureihe von optimierten, modularen Solarlufterhitzern für Heizung und Trocknung*, Ruhr-Universität Bochum.

FURTHER READING

Altfeld K, Leiner W and Fiebig M (1998b) Second Law optimisation of flat-plate solar air heaters, Part 2: Results of optimisation of and analysis of sensibility to variations of operating conditions. *Solar Energy*, **41**, 309–317.

Biondi P, Cicala L and Farina G. (1988) Performance analysis of solar air heaters of conventional design, *Solar Energy*, **41** (1), 101–107.

CE-Standard of solar collectors, Thermal solar systems and components – Collector – General requirements, CEN TC 312 N164, a draft paper by CE TC 312-PT1. European Committee for Standardization, Central Secretariat, rue de Stassart 36, B-1050 Brussels, Belgium

IEA Status Seminar on Solar Air System, March 1-3, 1988, Utrecht, The Netherlands, Part 1: Report and Part 2: Technical papers.

Jensen S Ø (1987) *Roof Space Collector, Validations and simulations with EMGP2*. Institute for Energy and Building, Technical University of Denmark, Report No. 87–15.

Jensen S Ø, Olesen O and Kristiansen F (1987) *Luft/væskesolfangere*. Solar Energy Centre Denmark, DTI Energy.

Lo, S N G, Deal C R and Norton B (1994) A school building reclad with thermosyphoning air panels, *Solar Energy*, **52** (1), 49–58.

Mørck O and Kofod P (1993) *Udvikling af luftsolfanger*. Cenergia Energy Consultants, Denmark.

Fechner H (1999) Investigations on several manufactured solar air collectors, Working Paper within IEA Solar Task 19, Arsenal Research and Testing.

Chapter authors: Ove Mørck and Hubert Fechner

IV.2 Window air collectors

INTRODUCTION

Description

A window air collector is made of a double window with a moveable blind that acts as the solar absorber between the inner and outer glass panes. Air is circulated through this gap, is warmed by the blind and then extracted from the window (Figure IV.2.1). The hot air can be transported to a room or storage where the heat can be used. Thereby, the window serves as a collector and provides overheat and glare protection, as well as still being a window offering a view to the outside and admitting daylight.

Modes of operation

A window air collector is generally operated in four different modes according to the season, the level of solar radiation and the time of the day. These are:

- The Active Collection mode: When solar radiation exceeds approximately 400 W/m² the absorber blind is lowered and the fan activated to transport the hot air to either thermal storage or a north-facing room. Only a small amount of radiation, together with some heat from the warm inner glazed surface, enters the room directly.
- Direct Gain mode: When the radiation does not reach a certain level, i.e. when direct sunlight to the heated space is acceptable, the blind is raised and the fan stops.
- Night-mode: The blind is lowered to reduce heat loss through the window and electrically operated dampers are closed to prevent reverse flow from the storage through the collector.
- Summer-mode: The blind is lowered with its reflection side facing outside. The outdoor air cools the blind, preferably by free convection so as not to consume electricity.

Variations

The window air collector and the exhaust air window belong to the family of air flow windows illustrated in Figure IV.2.2.

The exhaust air window corresponds to type B in Figure IV.2.2 and always runs in a single operational

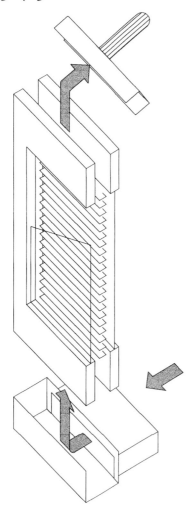

Figure IV.2.1. The principle of the window air collector

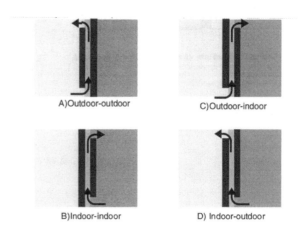

Figure IV.2.2. Types of air flow windows

mode: Air is distributed by a ventilation system throughout the building and leaves each room through a narrow slot at the bottom edge of the inner window pane. The room air is drawn up between the panes of the exhaust air window and is collected into the air recirculation ducts. The air, which is warmed both by the sun and by the heat losses of the room, is then transferred to a heat-recovery unit or extracted from the building in summer. Exhaust air windows are used in external walls of all orientations, especially in northern countries. Winter solar energy is scarce in these countries, which demonstrates that these windows provide comfort even at low radiation levels and outdoor temperatures.

Applications

Air flow windows may be used in dwellings as well as in office buildings. Many examples of exhaust air windows can be found in North European countries,[1] whereas window air collectors are more common in Central Europe,[2] where they should preferably face south (west-facing units are prone to summer overheating).

Characteristics, advantages and limitations

Air flow windows not only save energy; they also improve comfort in both winter and summer by keeping the inner glass pane at comfort temperatures and by reducing glare.

DESIGN

Thermal design

There are two ways to consider the losses from a window air collector, namely as a collector with losses to the front and back and as a window with an overall U-value.

Heat loss coefficient of the collector to the front and back
The heat loss is best determined by means of the efficiency curve (see Figure IV.2.4). The heat loss coefficient of the collector is defined by the negative slope of the efficiency curve. Typical heat loss coefficients for a window air collector for an air flow of 65 m^3/hm^2 range from 5.5 to 7.5 W/m^{2}K.

Heat loss from an air flow window
The window air collector is also characterized by the losses of the entire construction from the room to the ambient. For the case of single glazing on the room side and double glazing on the outside the equivalent U-value is determined as shown in Table IV.2.1.

Figure IV.2.3 shows how the equivalent U-value depends on the air flow rate in the air gap of an exhaust air window. At zero air flow the curves tend towards the static U-value of 1.5 W/m^{2}K for a triple-glazing win-

Table IV.2.1. Energy flow

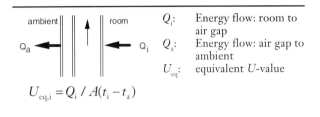

$U_{eq,i} = Q_i / A(t_i - t_a)$

$U_{eq,a} = Q_a / A(t_i - t_a)$

Q_i:	Energy flow: room to air gap
Q_a:	Energy flow: air gap to ambient
U_{eq}:	equivalent U-value
	The total U-value is the inverse of the sum of the inverse values of $U_{eq,i}$ and $U_{eq,a}$

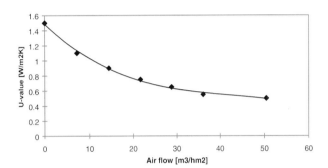

Figure IV.2.3. Equivalent U-values for different air flows in an exhaust air window

dow. From Figure IV.2.3 it is obvious that, to produce a significant effect, air flows should exceed 15m^3/hm^2 per window. U-values less than 1 W/m^2 K can be reached with special glass with low-emissivity coating. The effect of the blinds additionally reduces the U-value.

The overall heat loss from the window air collector is comparable to that of an exhaust air window: Daytime U-values range from 1.4 to 0.9 W/m^2 K, whereas during nighttime values range from 1.1 to 0.7 W/m^2 K as a result of the lowering of the absorber blind.

The performance of a window air collector
The efficiency curve of the window air collector (Figure IV.2.4) compares to that of low performance flat-plate air collectors. See Chapter IV.1 for a comparison of efficiencies and discussion of the different representations of the collector efficiency.

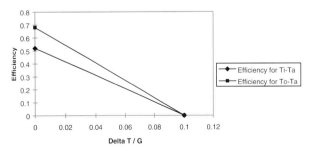

Figure IV.2.4. The efficiency of a window air collector at 65 m^3/hm^2

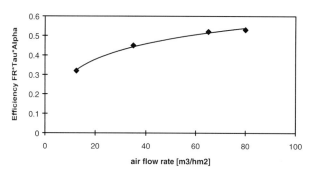

Figure IV.2.5. Maximum collecting efficiency as a function of air flow rate

Figure IV.2.7. View of a two-storey window air collector in the Solar Trap House in Switzerland (Architect Th. Kurer, Zürich)

Figure IV.2.5 shows the maximum collecting efficiency (zero heat loss) as a function of the air flow rate. For the window air collector an air flow rate of 65 to 80 m^3/hm^2 is appropriate. It can be clearly seen that there is no net improvement in the collecting efficiency if high flow rates are used. In addition, the overall system efficiency is strongly dependent on the pressure drop in the collector loop.

Temperature rise over the collector
The typical temperature rise over the window air collector ranges from 5 to 25 K, depending on the solar radiation (Figure IV.2.6), the air flow rate and the ambient temperature. The colour of the absorber is assumed to be dark but not necessarily black. The **stagnation temperature** for zero air flow rate may reach 80°C at high solar radiation levels (800 W/m²).

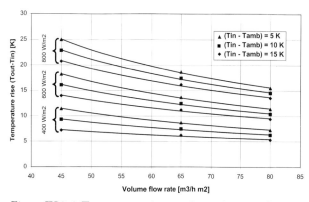

Figure IV.2.6. Temperature rise over the window air collector as a function of solar radiation

Shading
With the blind lowered and the louvres tilted 30°, the heat gain by the window collector is reduced to approximately 20% of that of a normal clear double window (shading factor of 0.20). With the blind louvres horizontal (tilt = 0°) the shading factor is approximately 0.50, while for the fully closed position (90°) it reduces to approximately 0.08.[3]

Hydraulic aspects

The pressure drop in the window air collector ranges from 0.5 to 3 Pa/m². Because of the wide air gap and low flow velocities through the collector compared to a commercial flat plate, the pressure drop is also significantly lower than in flat-plate collectors.

Construction

Building integration
Window air collectors are always site-built or site-assembled. The window air collector will preferably be a one- or two-storey element. Window air collectors may easily be combined with ordinary windows, thus giving a coherent look to the facade (Figure IV.2.7).

Assembling the components into a system is a critical phase of the construction (Figure IV.2.8). The varying cross sections of collector, ductwork and storage are very often the cause of pressure losses that are too high.

Construction details
Window air collectors are far from being standardized. Figure IV.2.9 offers a detailed section of a typical collector.

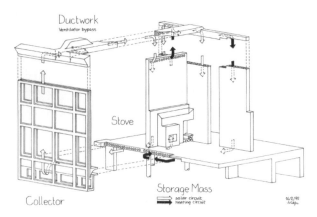

Figure IV.2.8. Connecting a window air collector to the duct system and the storage (Architect U. Schaefer)

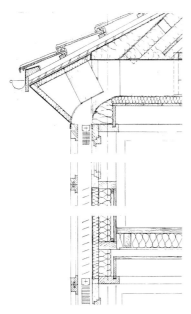

Figure IV.2.9. Vertical section of a window air collector (Architect U. Schaefer)

Materials
Glass pane: Normal tempered glass and double glazing on the inner and outer surface are used. Ideally the glazing towards the room should have the lowest U-value (i.e. 1.3 W/m^2 K) to keep the room-facing glass surface from becoming too hot during collection. The glass towards the ambient can have a worse U-value with a correspondingly better solar transmittance.

Absorber: The absorber may be a moveable venetian blind or a dark-coloured transparent roller blind. A venetian blind has the advantage that the blades can be tilted to adjust for full daylight without direct sun penetration. It is important that the air flow passes both the front and the back of the absorber to guarantee optimum heat exchange.

For absorber coatings, a non-selective dark surface with an absorption coefficient of approximately 0.9 can be used. Dark colours can be used instead of black.

Frame: Wooden frames are most common for window collectors.

Solar protection: For optimal overheat protection in summer, additional shading outside the window collector can be added.

Operation and maintenance
Hand operation of the blind/absorber in the window collector requires less maintenance. Remember that the enclosed space in which the blind hangs is subject to temperature extremes. These are hostile conditions, not only for an electric motor, but also for the strings supporting the venetian blind.

Ideally, one sash of the window collector should be openable to allow access for cleaning and maintenance. Obviously, this window should be gasketed so that it seals tightly.

REFERENCES

1. Seppanen O (1981) 'Windows as cost effective solar collectors'; *Proceedings of the 6th ASES conference, Portland, Oregon*, Volume 6, pp. 621–625. ASES, Boulder, CO 80301.
2. Kurer T, Filleux Ch; Lang R and Gasser H (1982) Research project SOLAR TRAP; Final report (in German), Zürich. For information, e-mail: chfilleux@bhz.ch.
3. Rheault S and Bilgen E (1990) 'Experimental study of full-size automated venetian blind windows'. *Solar Energy* **44**, 157–160.

FURTHER READING

Choudhury C, Andersen S L and Rekstad J (1988) 'A solar air heater for low temperature applications'. *Solar Energy* **40**, 335–343.
Hastings S R (Principal editor) (1994) *IEA Task 11: Passive Solar Commercial and Institutional Buildings (A Sourcebook of Examples and Design Insights)*. John Wiley & Sons, Chichester.

Chapter author: Charles Filleux

IV.3 Perforated unglazed collectors

INTRODUCTION

Description

The principle of this collector is simple: outside air is drawn through perforations in a dark-coloured, sun-warmed metal facade. As the air passes over and through the facade, it is warmed. A ventilation fan creates negative pressure in the wall cavity to draw air through the holes.

The perforated unglazed collector (Figure IV.3.1), also referred to as a transpired collector or by the trade name *SOLARWALL*, is most commonly used in System 1. Because the concept is so simple, it is highly reliable and very economical. Low costs are achieved by reducing the material in the collector to a single metal sheet mounted away from a wall or roof to form an air cavity. The heat delivered can pay back the installation costs in only a year in some cases. The unglazed perforated collector looks like a conventional metal wall, but collection efficiency can exceed 70% at high air flow rates.

Applications

Building types
This technology is now North America's leading solar collector for heating ventilation air. Ideal building types for the unglazed solar collector are industrial buildings with large wall surfaces (Figure IV.3.2), sports halls, schools, hospitals, offices, maintenance buildings and parking garages. In apartment buildings the panels are used to heat common areas and as window ventilators. Another proven application for the perforated collector is providing low-temperature process heat for agricultural or industrial purposes (i.e. crops and drying cardboard, textiles or paint). Drying applications are economical since the solar heat is also usable in summer.

Space heating
The panels can also be used as a space heater in milder climates or in the warmer spring and fall months. Whenever the fresh air is heated above 20°C, the solar heat will provide a space heating benefit.

Summer cooling
In summer the dark-coloured solar panels may actually feel cooler to the touch than a dark piece of unperforated metal in the sun. This is explained by the cooling effect of ambient air entering the bottom half of the wall, rising by convection, then exiting through the top half, removing the solar heat. The main wall is spared the direct heat from the sun, reducing the cooling demand of the building. If the walls and roofs are uninsulated, the reduction in solar load can be very significant. In fact, in some climates, the summer cooling benefit can exceed the winter heating benefit. Figure IV.3.3 illustrates the summer cooling effect.

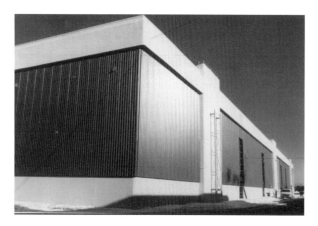

Figure IV.3.1. An unglazed perforated solar collector

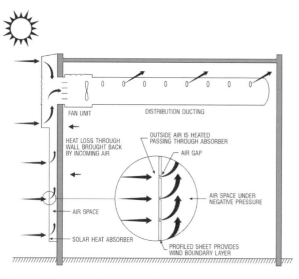

Figure IV.3.2. A perforated panel integrated into a wall and connected to an interior fan

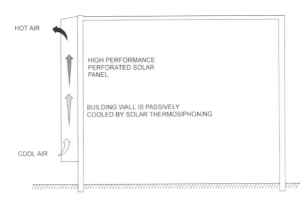

Figure IV.3.3. Summer cooling with perforated unglazed panels

Variations

Wall mounting – canopy
The main variation in wall mounting is the method of collecting the solar-heated air at the top of the solar wall array. For small air volumes all of the air can normally be accommodated in the air gap between the perforated absorber and the wall. For larger air volumes, however, either multiple air connections to the fan(s) or a larger air space is needed. The larger air plenum is called a canopy and can be built either on the face of the wall as an architectural feature, behind the wall or on the roof. The canopy can also be glazed to provide an extra heat charge before the ventilation air enters the building.

In some cases, the canopy will be built above a wall that is not perforated to collect the solar-heated air as it rises up the exterior of the wall (Figure IV.3.4). This would be suitable for walls with numerous windows or doors. The canopy would be made from the perforated panel and air entering the perforations would have been preheated by the lower wall surface.

Two-stage solar heating
The unglazed solar heater has a limited ability to raise the air temperature, the limit being 25 K above ambient. If a higher temperature is desired, a two-stage system can be designed. The first stage is the unglazed perforated panel and the second stage is a glazed solar panel, which would be supplied with preheated air from the first stage. This tandem configuration can elevate temperatures an additional 10 to 20 K or more. The glazed section can be the surface of the canopy so that air entering the canopy would be heated again as a back-pass collector when it travels horizontally to the nearest fan intake.

Roof mounting
Roof-mounted systems are equally efficient but less common. Walls are preferred in northern latitudes owing to the low angle of the winter sun and the added reflection of sun off the snow in front of the wall. Roofs are also less desirable because of possible snow build-up impeding the air flow. Roof panels are better for countries closer to the equator or where drying or process applications are operated in the summer.

Advantages and limitations

Advantages are:

- Low cost: the collector consists of a single sheet of metal and uses the building wall for support.
- High efficiency at high flow rates: with no glazing, panels receive 100% of sunlight and operating temperatures are low, minimizing heat losses.
- No maintenance: the wall has no moving parts and the coating is designed to last for decades.
- Attractive appearance: similar to other building wall facades and architecturally attractive; choice of colours increases consumer acceptance.
- Improves indoor air quality: more air can be heated for free so no need to skimp on fresh air.
- Savings can be substantial as ventilation air can represent 50% of a building's heating needs.
- The absorber recovers heat that would otherwise be transmitted through the wall to the ambient. Heat losses are picked up by the air stream and returned to the building when the fans are running.

Disadvantages are:

- Application limited to heating ventilation air, not space heating.
- Is not suitable for applications where a high temperature rise is required; ideal design temperature rise is 10 to 20 K with a maximum design of 25 K over ambient.

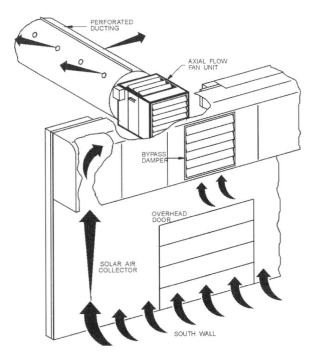

Figure IV.3.4. A canopy constructed over the wall captures the rising heat

- Requirement for ventilation air is a prerequisite for this system.
- Low efficiency at low flow rates

Energy performance

The collector is surprisingly efficient. This is explained by several facts:

- If there is no glazing in front of the absorber, it receives 100% of the sun's energy. Heat loss with no protective glazing is less than intuition might indicate. The boundary air layer in front of the metal is pulled into the collector through the perforations before the heat can be lost to the ambient. Accordingly, the collection efficiency is highest at high air flow rates. Even on cloudy days, the unglazed panels can still generate a few degrees of heat and act as a preheater for the air before it reaches the auxiliary heater (see Figure IV.3.5).
- The efficiency of any solar collector is highest when the temperature of the air entering the solar panel equals the ambient temperature. This occurs with the perforated-plate collector. Most solar efficiency curves show the panel efficiency based on a formula that includes the ambient air temperature and the temperature of the air entering the solar panel. The efficiency drops as the difference between the two temperatures increases. With the perforated panel, the two temperatures are always the same and the solar panel operates at maximum efficiency (see Figure IV.3.6). Typical temperature rises are shown in Figure IV.3.7.
- In space-heating designs, the building air enters a solar panel to be heated above room temperature. On cold, overcast days, there may be insufficient solar energy to achieve this, whereas the perforated panel generates heat above ambient, whether it be a temperature rise of 2 or 20 K, and this gain is useful energy.

DESIGN

Design parameters

Wall or roof mounting
The perforated cladding absorbs most sunlight when facing within 20° of south. If the south wall is not

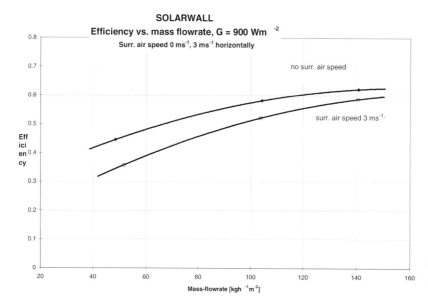

Figure IV.3.5. Efficiency of a perforated absorber as a function of air flow

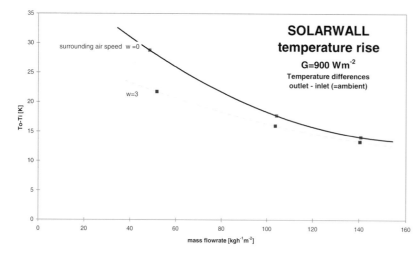

Figure IV.3.6. Temperature rise of a perforated absorber as a function of mass flow rate

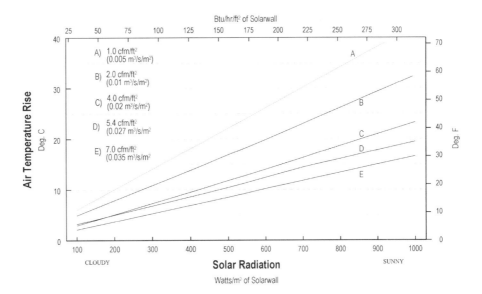

Figure IV.3.7. Solar temperature rise for various air flow rates and solar radiation levels

suitable, consider either or both of east and west walls. If a large volume of air is to be heated, all three walls can be utilized. Only the solar contribution is affected by the collector orientation; the heat recovery benefit remains the same.

The wall area to be considered does not have to be blank. Wall surfaces around doors and windows may be suitable if they can be connected together or to fans that deliver air inside the building. This is an important consideration, since many buildings will not have large wall surfaces without windows and doors. Consider using the perforated panels as the building material for covering the wall around doors and windows and do not be overly concerned about optimizing the design. Rather, cover as much of the wall as possible and use as much of the free heat as possible. If parts of the wall are shaded or not readily accessible for uniform air flow distribution, again, do not sacrifice appearance for highest efficiency, instead design the air flow for a range of heating curves, as long as some air is moving through the perforations. In new construction, the capital cost of a perforated metal cladding system is similar to that of conventional walls. In retrofit situations, the perforated solar cladding can be applied over most existing walls of block, metal, glazing or precast concrete (see Figure IV.3.8).

If no wall is suitable or available, consider using a south-facing roof. The slope of the roof should be at least 30° and preferably more. If snowfall often occurs at the proposed site, the minimum slope should be 45° to allow the snow to slide off.

Required ventilation air volume
The volume of air to be heated and the desired temperature are dictated by the use of the building. In Figure IV.3.7 curves A and B represent flow rates for applications where higher temperatures are needed and the panels can provide some space heating needs of the building as well as heating the ventilation air. Curves C and D are typical of most ventilation heating designs.

Curve E and higher air flows are used in industrial applications where large volumes of air must be heated and only a small temperature rise is needed.

If the quantity of outside air to be heated is low in proportion to the wall area, then a perforated canopy design may be the most cost-effective choice. The capital cost would be less than covering an entire wall. A wall with numerous shipping doors, windows or other obstructions may not be suitable for other styles, so a canopy may be the only option. In such a case, if the doors are a dark colour, they will collect heat when closed and the heat would rise to the canopy as illustrated in Figure IV.3.4. The face of the canopy would be constructed from the perforated cladding to increase collection efficiency.

Air flow direction
Air behind a perforated panel system normally travels in two directions, vertically up the wall to the top

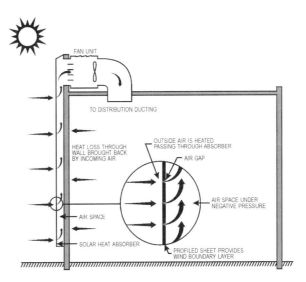

Figure IV.3.8. Solar cladding mounted on an existing wall and connected to a roof fan

plenum or canopy and then horizontally in the canopy to the nearest fan intake. The two directions simplify balancing to ensure that air is drawn through the entire panel surface; otherwise some of the solar heat may be lost.

Air mixing

Industrial buildings usually require a large amount of fresh air and the usual practice is to supply separate fans with mixing dampers and ducting, which distribute the solar air at ceiling level as far into a building as practical. The mixing dampers are temperature controlled and will mix hotter ceiling air with the solar-heated air. The mixing is necessary for night use or cloudy days. For non-industrial buildings, most solar heating projects use the perforated panels as a preheater to the conventional ventilation fan, with a provision for bypassing the solar panels in the summer months when heating is not necessary.

Design guide

The method of sizing a perforated solar air heater is relatively simple and not as precise as one might think, since more or less air can be heated by the same panel area. The velocity is calculated by dividing the air volume to be heated by the solar panel area.

If a high temperature rise is needed or low volumes of ventilation air are to be heated, design for lower velocity across the solar collectors (lowest solar collection efficiency). Use:

18 to 54 m/h
0.005 to 0.015 m/s
1 to 3 feet per minute.

For ventilation air heating in schools, offices, factories, design for (typical selection criteria):

54 to 108 m/h
0.015 to 0.03 m/s
3 to 6 feet per minute.

For higher air volumes, preheating of air and low temperature rise (highest solar efficiency), design for:

108 to 200 m/h
0.03 to 0.055 m/s
6 to 11 feet per minute.

Perforated panels can be roof-mounted, provided the main roof is waterproof. The solar sheets directly over the air-intake opening must not be perforated, so as to prevent water from entering air intake.

Pressure drop through the perforated metal panels is approximately 20 Pa and the total pressure drop through the solar wall panels and canopy is approximately 50 Pa. The pressure drop from the inside of the solar wall to the ventilation system should be calculated in the normal manner, with duct and other losses taken into account.

The air velocity behind the perforated metal panels should not exceed 3 m/s when the air enters the canopy. The maximum velocity in the canopy is 5 m/s. If multiple fans or duct connections are made to the canopy, then the canopy can be smaller or even eliminated.

If the solar collector area is known, calculate the volume of air that can be heated by multiplying the panel area by the desired temperature-curve flow rate from Figure IV.3.7 or by using the velocity guide shown above. The curves show the temperature rise of air at different flow rates.

Example

The perforated panel area is 500 m^2 and a large temperature rise is desired. Pick Curve B at 36 $m^3/h\ m^2$ (m/h) to give an air flow rate of 18,000 m^3/h.

If both panel area and volume of outside air to be heated are known:

panel area = 500 m^2
air flow = 50,000 m^3/h,

the air flow through panel is:

(50,000 m^3/h)/(500 m^2) = 100 $m^3/h\ m^2$.

From Figure IV.3.6, 100 $m^3/h\ m^2$ (m/h) corresponds to Curve D for the temperature rise of the air.

Step-by-step design procedure

1. Decide on solar panel size and location. Is the south wall suitable? If not, consider east or west walls. Note that a south wall may actually be south-west and the east wall would then be south-east. In this case use both walls for maximum benefit.
2. Determine the volume of outside air required in the building. Heat as much fresh air as possible. This will improve the indoor air quality without increased fuel costs.
3. Calculate the volume of air per area of solar heater and then refer to the temperature chart to determine the expected temperature rise.
4. Select the colour.
5. Determine the spacing of the solar cladding from the main wall and whether to use a separate or internal canopy plenum.
6. Locate the fans as close to the solar panels as possible. Position the solar-fan connections at a maximum spacing of 30 m apart. Closer spacing will mean a smaller canopy plenum.
7. Industrial buildings can save additional energy by destratification. Determine the amount of ventilation or make-up air required and then position the

ducting to distribute the air throughout the building. The distribution ducting should be located in the areas where the ceiling temperature is the hottest so as to disperse the heat and save energy.

CONSTRUCTION

Building Integration

Panels
The metal panels are perforated with very small holes or slits and resemble a conventional metal facade. Panels are available in many colours, including black and dark shades of brown, gray, red, blue and green. They are usually overlapping panels of varying lengths, one meter wide and installed to give a continuous appearance along the entire wall. To add structural strength and rigidity, the material is processed through rollers to form corrugations. The corrugations are 20–50 mm tall and spaced 150–300 mm apart. A large number of cross sections are available. Figure IV.3.9 shows a close-up photograph of a typical unglazed perforated collector. The standard solar panel profile is shown in Figure. IV.3.10.

The perforated panels are made of either aluminium or galvanized steel. Initially the panels were made from aluminium because there was concern about possible corrosion around the holes if steel was used. Corrosion experts have examined galvanized panels that have been in use since 1989 and found no rust. The galvanizing protects the steel from rusting and the air movement through the holes dries any water that may be present. Water runs off the wall and the holes are so small that the surface tension prevents most water from entering the holes. The two materials perform the same. In North America and most developing countries galvanized steel is the preferred material because of the lower cost. Aluminium is more expensive, but is still popular in Europe and Japan.

Colour
The perforated panels can be any dark colour. The darker the better, as dark colours absorb more sunlight. Black is best and then dark brown. Other colours that have been used are dark shades of grey, green, blue and red.

Coating
The coating need not be selective. As the solar panels are part of the building, it is important that a durable and proven paint system be used, which will last for

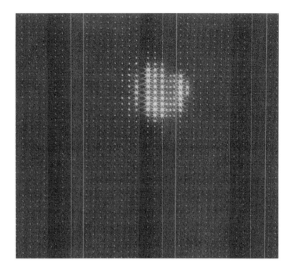

Figure IV.3.9. Photograph of an unglazed perforated collector

decades without maintenance or repainting. The solar absorber coating on the panels is the same premium coating used extensively in the building industry. The dark coatings are designed to last for decades and may come with an extended non-fade warranty similar to the metal building-industry warranty.

Air cavity
A certain air gap is necessary to allow the heated air to travel up the wall and across to the nearest fan intake. The air gap can be reduced if air is drawn off at several locations. The depth of the air cavity between the wall and solar facade must provide an upward air flow rate not exceeding 3 m/s. The canopy at the top of the wall is built as a plenum and is sized for a flow rate not exceeding 5 m/s. Connecting the fan or fans at multiple points reduces the required size of the canopy since the volume of air going through the canopy to each connection is lower.

The cladding and canopy can be mounted onto a wall in the following different ways (see also Figure IV.3.11):

1. The entire cladding mounted the same distance from the wall, but not more than 300 mm.
2. The cladding tapered with a larger spacing at top and a smaller spacing at bottom.
3. The cladding mounted away from the wall with a canopy at the top.
4. The cladding mounted directly on the wall with a canopy at the top.
5. The cladding mounted against the wall with a collection duct behind solar panels on the roof.

Figure IV.3.10. Standard profile used for an unglazed collector

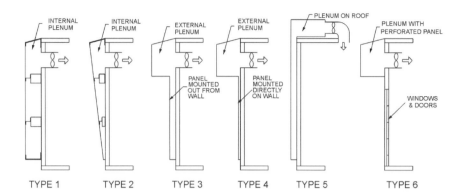

Figure IV.3.11. Wall designs to accommodate different air volumes

6. The cladding on a canopy over dark section of wall, doors or windows

The wall and canopy designs are based on the volume of air, cost and appearance. The volume criteria were discussed earlier. The cost and appearance issues are related, because a canopy is more expensive than a flat wall but it can also enhance the appearance of a building. If a separate canopy can be omitted and an internal canopy used, construction costs will be lower.

The air space within the cladding profile may be sufficient for low-flow designs, but not sufficient for higher air volumes. If more air space is needed, the solar panel must be mounted further from the main wall. The method of securing the panel away from the wall will vary depending on the option selected. The lowest-cost options are the internal canopy types 1 and 2, followed by type 4. Type 4 saves costs on the main wall because the panels are mounted directly on the wall, but a canopy is necessary to collect the air. The most expensive designs are types 3 and 5, since both require separate support structures for the solar cladding and the canopy. Option 5 is suitable for designs where the fan is on the roof or the existing air intake to a fan is above the roof line of the wall. Type 6 is used when a facia or canopy is built over windows or doors and can serve as a decorative feature.

If the architect is planning to include an architectural feature along the top of the wall, such as a canopy or mansard roof, then it should be designed to act also as the plenum to collect the solar-heated air. A glazed canopy is also possible for higher temperatures and may blend in with architecture requiring glazing.

Construction details

The unglazed panels are installed in a similar manner to other metal wall facades, although there is one difference. The panels are mounted away from the main wall to create the cavity for collecting the solar-heated air. The wall must be structurally sound and comply with local building codes.

If the main wall is masonry, attaching the perforated panels with an air gap is relatively simple, since the clip and support system can usually be fastened anywhere on the wall. If the main wall is a metal wall with support bars or grits spaced two or three meters apart, the supports for the solar wall panels must be connected to the structural supports and not to the metal sheets. Figures IV.3.12 and IV.3.13 show typical construction details for a masonry wall and a metal wall. The masonry wall example also includes a canopy.

Air will be flowing against the main wall, which must have a waterproof and non-combustible surface such as bricks, blocks, metal, glass, stucco or foil-covered insulation. If masonry is used, it may act as a limited heat storage and provide some heating benefit for up to a few hours after the sun sets.

AVAILABLE PRODUCTS

SOLARWALL perforated panels are available from:

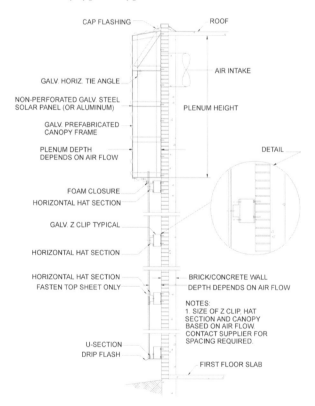

Figure IV.3.12. Typical construction details for a panel on a masonry wall

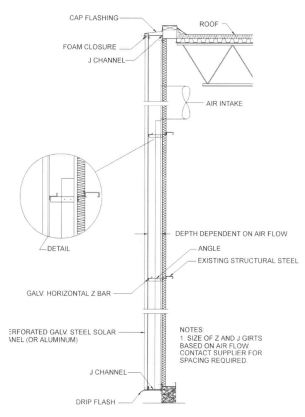

Figure IV.3.13. Typical construction details for a panel on a metal wall

- Conserval Engineering Inc., 200 Wildcat Road, Toronto, Ontario, M3J 2N5 Canada; Email: Conserval@solarwall.com
- Conserval Systems Inc., 4254 Ridge Lea Road, Unit 1, Buffalo, NY 14226, USA; Email: Conserval@aol.com.
- Solarwall International Ltd, Gottingen, Germany; Email: Solarwall@compuserve.com; Web site: http://www.solarwall.com.

BIBLIOGRAPHY

Publications

Canmet, Natural Resources Canada (1994) *Performance of the Perforated Plate Canopy Solarwall at GM Canada, Oshawa. Monitoring Report.* NRCAN, 580 Booth St, Ontario, K1A 0E4 Canada

Conserval Engineering Inc. *Descriptive information and test results from solar panel supplier.* http://www.solarwall.com.

Kutscher C F (1996) 'Transpired Solar Collector Systems: A Major Advance in Solar Heating'. *Proceedings 19th World Energy Engineering Congress*, Atlanta, GA. *Association of Energy Engineers, Fairmont Press, 700 Indian Trail, Lilburn GA 30247, pp. 481–489*

US Department of Energy (1998) *Transpired Collectors (Solar Preheaters for Outdoor Ventilation Air).* Federal Technology Alert: April. US Department of Energy Federal Energy Management Program, 1000 Independence Ave. SW, EE-92, Washington, DC. 20585

Computer models

SWIFT 99, Solarwall International Feasibility Tool. Simulates solar energy savings with SOLARWALL, uses hourly or monthly weather data. Produced by Canmet, Natural Resources Canada, available from Enermodal Engineering Inc., Kitchener, Ontario, Canada; Web: http://www.enermodal.com.

ESAP2, Energy Savings Analysis Program (1992). Simulates SOLARWALL energy savings; uses monthly weather data. Available from Conserval Engineering Inc., 200 Wildcat Road, Toronto, Ontario, M3J 2N5 Canada.

RETScreen (1998). Prefeasibility analysis of ventilation solar air heating using Excel software. Produced by Canmet, Natural Resources Canada, available from http://retscreen/.

Chapter author: John Hollick

IV.4 Double facades and double-shell facades

INTRODUCTION

By adding a glass facade in front of the main facade a sun-tempered space is created. The gap between the facades, typically between 60 and 80 cm, reduces thermal losses from the building and the air may be supplied to the ventilation intake. This construction is often referred to as a 'double facade'. Non-energy benefits are often decisive in this design being chosen. For example, sun shading can be located in the gap, protected from wind and the elements. The outer glass skin may serve as a barrier to the street noise at urban sites. Not surprisingly, the double facade has become popular among architects and clients in recent years.

This chapter describes various applications of the double facade and quantifies the influences of key design parameters.

In general, a double facade separates the different functions of the classic facade into the different layers, providing more flexibility for the specific solutions. The weather protection given by the external glazing, for example, reduces the water- and air-tightness requirements for the inner facade. Some of the functions a double facade can serve are (Figure IV.4.1):

- creation of a buffer zone;
- solar preheating;
- saving energy;
- weather protection for inner facade and any shading device;
- natural ventilation;
- wind protection for manual ventilation by windows;
- sound reduction;
- air pollution reduction.

Winter

In winter, the buffer effect leads to an intermediate climate in the double facade and therefore to higher surface temperatures of the inner glazing and better comfort in the building zone. Solar gains and transmission losses from the building into the buffer can be used to reduce the losses or to cover some of the ventilation losses via a preheating of fresh air. Especially in commercial buildings, direct solar gains in winter are in general not useful because of the problems of glare and the illumination conditions for on-screen working. With a double facade, these blocked solar gains can be used via the intake air, which has to be preheated anyway to prevent draft problems.

Natural ventilation

In combination with a natural ventilation concept, the double facade can provide wind and weather protection, allowing openings in the inner facade. On noisy sites and for high-rise buildings, the noise and wind protection provided by the external glazing are prerequisites for ventilation. Depending on the openings in the external facade, the sound reduction varies between 0 and 30 dBA through the additional layer. On the other hand, the air space can be used for transportation of fresh or exhaust air and as an additional installation space. This allows manual ventilation on a building facade that looks out onto heavy traffic or a highly polluted road, bringing the fresh air from another side of the building via the double facade.

Summer

To prevent a glazed building from overheating, external shading is essential. To avoid damage, shading devices are installed with a wind guard, which closes the device into the park position. On windy sites or in high-rise buildings, this can lead to a loss of shading during

Figure IV.4.1. Benefits of a double facade

Figure IV.4.2. Different ventilation concepts for double facades

many sunny hours, causing overheating. Sun-protection glazing is no solution, since it blocks solar gains throughout the year and reduces daylight to unacceptable levels, while internal shading devices in combination with good insulation glazing are unable to reflect enough radiation back to the outside to avoid overheating. Therefore, wind and weather protection of the external shading device by the double facade is a valuable asset. The shading in the cavity of the double facade leads to higher temperatures than outside. This differential can be used as the driving force for a ventilation system that uses the double facade as a solar chimney, sucking exhaust air out of the building. The weather protection also allows devices for daylight control. Different ventilation concepts for double facades are shown in Figure IV.4.2.

Advantages and disadvantages

The advantages of double facades are:

- a thermal buffer, using solar gains and transmission losses;
- weather and wind protection for the shading device and the inner facade;
- sound reduction;
- reduced requirements for the inner facade;
- reduced heating and cooling demands;
- additional usable space.

In contrast, the disadvantages are:

- a temperature increase in the double facade due to solar gains on the shading device and depending on the ventilation rate and concept;
- reduced wind pressure and temperature difference as a driving force for natural ventilation;
- reduced daylight due to the additional glass layer;
- an acoustic connection between neighbouring zones through the facade corridor;
- additional cost of extra construction;
- reduced visual contact with the ambient.

VARIATIONS

Closed buffer (closed to the inside)

For the closed-buffer concept (Figure IV.4.3), the building space has no connection with the air space of the double facade. Therefore, the function of the double facade is reduced to a thermal buffer in winter and ventilated weather protection for the shading device in summer, together with a good sound insulation by two

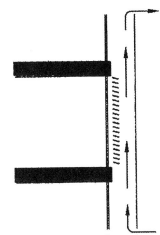

Figure IV.4.3. Closed buffer

glazing layers. Depending on the necessary sound protection and ambient air quality, the summer ventilation is driven in a natural or mechanical way.

The summer ventilation of the buffer zone can be made permanent (to keep costs low) or can use mechanical flaps that are controlled as a function of the temperatures in the double facade.

Open buffer (permanent)

This concept (Figure IV.4.4) is often constructed by leaving open gaps between the glass panes. The cracks are small enough to keep the rain out but allow ventilation in summer. Another concept uses overlapping panes with an air gap. As a result of the effect of reduced insulation caused by the infiltration, open buffers are mostly single glazed. Depending on the gaps, the sound reduction drops from 30 dBA for the closed construction down to 12 dBA.

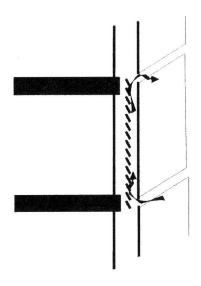

Figure IV.4.4. Open-buffer facade

Open buffer (controlled)

A better solution is ventilation controlled by moveable flaps, closed in winter for the best buffer effect and open in summer for good exhaust ventilation. Opening in summer may lead to reduced sound insulation.

Mechanically ventilated buffer

To keep the sound reduction at a certain level (which may be necessary for specific sites), summer ventilation can be done mechanically with exhaust fans and intake air ducts.

Ventilation buffer

In this concept, the inner facade can be opened for natural ventilation into and/or out of the buffer space. To solve acoustic and fire-protection problems, most of the projects built have horizontal separators at each floor (Figure IV.4.5). For multi-storey open spaces, the temperature increase in the buffer space has to be considered. The height may be beneficial for the intake air, but air quality in the upper floors may suffer. Higher ventilation rates can partially solve this problem, but will cause higher losses in winter.

Open ventilation buffer (permanent)

For the permanent open ventilation facade 1:1 model measurements have been made by the facade company Gartner. Figure IV.4.6 shows the ventilation of the

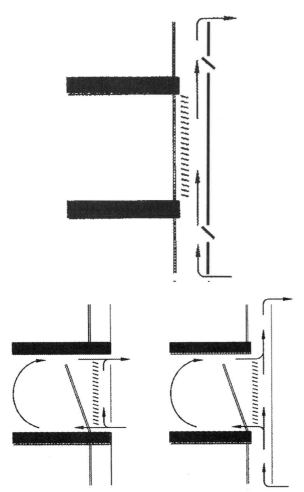

Figure IV.4.5. Ventilation concepts including the twin-shell facade with and without floor separators

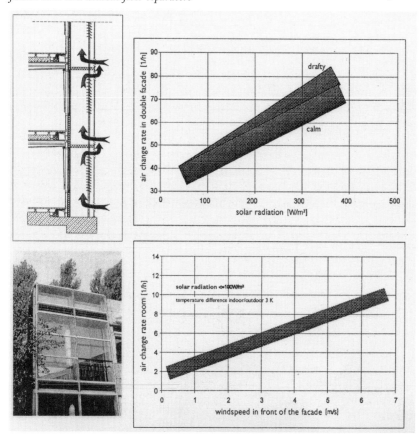

Figure IV.4.6. Ventilation rates depending on radiation and wind speed for a 1:1 mock-up

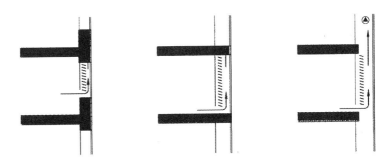

Figure IV.4.7. Exhaust facades from window, through each storey to the building height

buffer space to the ambient and of the occupied space to the buffer, the basic values for the fresh air requirements for the rooms and the exhaust ventilation for the buffer in summer. The values are valid for permanent openings with a height of 20 cm at the bottom and top of the external glazing. The realized office examples do not all have 100% natural ventilation and have mechanical ventilation and cooling in addition. The measurements in the mock-up have shown a good climate for 70% of the year. For the rest of the time, mechanical ventilation is used. Better activation of the thermal mass of the building and night flushing could possibly eliminate totally the need for mechanical ventilation.

Ventilation buffer (controlled)

If the opening in the external skin can be controlled, the ventilation rate, and therefore the temperature of the buffer, can be regulated in response to the external temperature, solar gains and wind effects on the facade. This will allow the extended use of natural ventilation for the adjacent spaces, as has been achieved in the just-finished office building Stadttor Düsseldorf, with its corridor buffer that has separators on each floor.

Exhaust air facade

This is a reversal of the venting direction described above. In this case, room air is exhausted into the double facade and hence reduces the transmission losses (Figure IV.4.7). The outer shell should be double glazed to avoid condensation on this shell, while the inner pane can be single glazed without causing concern about comfort. This concept has been implemented on a small scale in Scandinavia and Switzerland in the form of exhaust-air windows, while in Germany this system is marketed in conjunction with decentralized ventilation systems in the spandrel panels. The disadvantage that the solar gains in the facade cavity cannot be utilized becomes an advantage in the summer. The risk of overheating is overcome through an active venting of the twin-shell facade with internal air. Compared to the reference facade with internal sunshades, the cooling requirement can be significantly reduced through night-time ventilation and thermally active building mass.

Compared to mechanical supply and exhaust systems with heat recovery and a single facade, a pure supply or exhaust facade without heat recovery will increase the heating requirement by about 15%. However, a proper comparison should not ignore the higher electricity consumption for the supply- and exhaust-air plant, weighted to account for the primary energy source.

Intake air facade

A supply-air facade (Figure IV.4.8) enables solar preheating and partial recovery of the transmission losses in the twin-shell facade. The savings in the heating can reach 30% compared to heat-protected glass and external ventilation. For interior comfort, the glazing of the twin-shell facade should employ single glazing for the outer shell and double glazing for the inner one. In summer, exhaust-air vents in the outer shell must be opened while maintaining the direction of venting in order to guarantee temperatures within the facade in the region of the ambient temperature. Here, the sun shading placed between the panes should be as reflective as possible. If the venting is ignored, then simulation studies, checked by measurements on a test facade, show that, compared to a version with internal shading, the cooling requirement almost doubles and can offset the savings in the heating requirement.

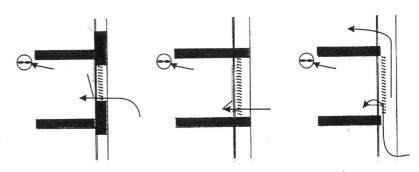

Figure IV.4.8. Intake facades from window, through the storeys up to building height

Figure IV.4.9. Twin-face ventilation principle

Combined exhaust and intake concepts

A mixture between intake facade and exhaust buffer is the patented twin-face concept (Figure IV.4.9). Separated floor by floor, supply-air windows allow the users to obtain fresh air from this buffer zone. The used air is vented the same way via the window in this space, but exhausted by a vertical exhaust chimney that collects the waste air, using its heat and solar gains as the driving forces. The problems with this concept are the permanent openings in the outer skin, sized to be not too large to destroy the buffer effect, but reducing the wind pressure and temperature differences between room and ambient as driving forces for natural ventilation.

THERMAL PERFORMANCE AND DESIGN RECOMMENDATIONS

Two modes of ventilation of double-facade systems were chosen for parametric studies:

- Air-intake facade: Using a double facade to preheat air, which is mechanically extracted for ventilation. This variation yields the maximum heating savings.
- Ventilation buffer: Natural ventilation through the double facade either during the entire year or in spring and fall. The benefit of this variation is that no electrical power is required to move the air.

Parametric studies illustrate the potential of double-facade technology. Results are given for typical and extreme conditions.

Method

The thermal performance of the double-facade building is compared with the performance of the same building without the double facade. Performance factors were:

- heating demand (winter performance), calculated in kWh/m² floor area;
- 'surplus temperature degree hours' (summer performance), which are analogous to heating degree days; at room temperatures above the tolerable temperature maximum, the difference between the room temperature and this tolerable temperature is calculated and summed over the hours that such conditions exist; the tolerable temperature is set at 27°C;
- running hours of the ventilation system; the criterion for switching on the ventilation system is the air temperature at the intake in the buffer space; two tolerable temperatures are given for winter (5°C and 12°C) and one for summer (22°C).

The computations were done with TRNSYS.

Reference building

The shoebox (see Appendix H: Nomogram assumptions) was modified as follows:

- The two floors were put side by side so that only one zone had to be simulated. Thus, the net floor area remained the same, although the proportion of the enclosing surfaces was different. Furthermore, the floor was considered to be adiabatic. This reduces the number of thermal zones to one for each orientation.
- The northern zone of the building was omitted. Along the inner wall, an adiabatic boundary condition was assumed, reducing the number of thermal zones to one. Because of the orientation parameter, the north facade was still included in the computation.
- Forced ventilation was assumed to run every night during summer in order to achieve maximum cooling. Throughout the year, forced ventilation was started at temperatures exceeding 26°C.
- The double low-E glazing with argon gas filling was substituted with krypton gas-filled glazing (same g-value, U-value = 1.1 W/m²K), since this was considered to be the current standard.

All other factors (such as air changes, internal gains, control strategy, time schedules, etc.) were taken from the shoebox model.

Standard configuration

Four standard types of double-facade building were selected for analysis:

- the standard office (new construction and retrofit);
- the standard dwelling (new construction and retrofit).

The standard parameters of these configurations are shown in Table IV.4.1.

In the ventilation buffer case only the office (new construction) in a cloudy, temperate climate was calculated. Ventilation was assumed always to be on unless there was a trigger: when the temperature in the buffer space is between a certain range (case 1: 5–22°C, case 2: 12–22°C), the inhabitants ventilate their spaces naturally and the mechanical ventilation system is switched off.

Modelling the double facade

The outer facade was a distance of 60 cm from the building, with a standard configuration as given in Table IV.4.2. The shading elements were set to reduce the direct solar gains to 50% of the incident amount during winter, and 27% during summer. In the case of the air-intake facade, the flaps in the outer facade were completely open during summer conditions and closed in winter. The pressure drop created by the fan pulled outside air through the double-facade spaces and into the building. The air change in the double-facade buffer with opened flaps driven by buoyancy was calculated with a simple approximation,[2] which defines the speed of the outlet flow as a function of the flap area and the temperature gradient.

The modelling of the inner glazing area in TRNSYS needed to be specially developed, since this component was not included in the standard program. Though the specified model represents the real energy flow patterns fairly well, the following considerations should still be noted:

- Direct solar radiation gains within the building that are reflected back into the double facade are ignored. This would have amounted to less than 1% of the solar gains in the buffer space.
- The shading elements are defined as blinds that reflect direct and diffuse radiation equally. This simplification was justified because use of the shading is highly user-dependent.
- The convective heat transfer coefficient was set to 4.6 W/m^{2}K along the external side of the inner wall between building and buffer. This corresponds to a maximum wind speed during winter of 0.15 m/s.
- Thermal bridging through construction joints and anchors was neglected, since these can be largely avoided with appropriate construction details.
- The intensive night flushing in summer, assumed continuous, was not actually possible because of the need for wind and weather protection, especially in the case of conventional buildings with natural ventilation.

Parameter variations

The system behaviour of the air-intake facade was analysed with respect to the following variables:

Table IV.4.1. Double-facade construction

Parameter	Standard configuration
Insulation level	Retrofit: U = 1.39 W/m^{2}K New construction: standard U = 0.31 W/m^{2}K Conductance other areas: 20.1 W/K
Mass	Heavy
Outer glazing	Single glazing U = 5.8 W/m^{2}K
Inner glazing	Retrofit: double glazing U = 2.8 W/m^{2}K New construction: low-E glazing U = 1.1 W/m^{2}K
Inner glazing area	9.6 m^2
Absorption coefficient: opaque elements in buffer	0.75
Room depth	5 m
Infiltration/leakage	Three per hour buffer space Retrofit/reference: 0.4 per hour (dwelling); 0.4 per hour (office) Retrofit/double facade: 0.1 per hour (dwelling); 0.1 per hour (office) New construction/reference: 0.2 per hour (dwelling); 0.1 per hour (office) New construction/double facade: 0.1 per hour (dwelling); 0.05 per hour (office)
Location, climate	Zurich, cloudy/temperate
Double-facade orientation	South

Table IV.4.2. Envelope construction

Standard configuration: air intake facade	Area (m^2)	Material	U-value (W/m^{2}K)
Outer window	36.7		
Glazing	90%	Single light	5.8
	10%	Thermally separated profiles	2.3
Roof	6.25	Light construction	0.5
Floor	6.25	Light construction	0.5
Flaps (× 2)	2		
Ventilation buffer openings (× 2)	0.5		

- For retrofit buildings, the inner facade was left in its former condition and only a second glazing layer was attached. All other components (such as the roof and the other exterior walls) are conventionally constructed. The reference building in this case was the original – i.e. not remodelled – building, as well as the conventionally retrofitted one.
- Utilization: All cases were calculated for both an office and a dwelling. Since occupant time schedules and internal gains are use-dependent, infiltration is the variable parameter.
- Location: The four solar similarity indices (SSI) were computed (kWh/m^{2}Kd).
- Orientation: The main facade was turned from south (standard) to east, west and north. The parameter used was orientation (°).
- Room depth (relation between facade area and floor area): Whereas solar gains and leakage losses re-

main the same with increasing depth, effective thermal mass is increased and internal gains due to increases in the amount of equipment and the number of people rise; this also raises the hygienic ventilation demand. The thermal conductance of the envelope was kept constant for room depths of 5 to 10 m.

- Thermal conductance/surface to exterior: The external areas, with the exception of the double-facade area, are varied from the maximum (roof, two walls) to minimum (only double-facade wall to exterior). The parameter used was thermal conductance (thermal coupling coefficient), 20–55 W/K.
- Thermal conductance: The thermal quality of the inner facade of the double facade was varied with respect to the reference model (shoebox). The average U-value of the standard configuration was modified to reflect both the highly and poorly insulated cases. The parameter used was thermal conductance (thermal coupling coefficient) (W/K).
- Glazing area of the inner facade: The standard size was changed in two cases to about 9/10 and 1/10 of the facade area, respectively. The parameter used was glazing area, 3–27 m^2.
- Quality of glazing of inner and outer facades: The quality of the inner-facade glazing for the standard configuration (double-isolation glazing) was changed to single glazing and to double low-E glazing. The outer single glazing was substituted with a low-E glazing filled with krypton gas. The parameters used were the U-value of the glazing, 5.8 and 1.1 W/m^{2}K, and the g-value, 0.85 and 0.60.
- Absorptance of opaque components in the buffer: Absorption coefficients were chosen to reflect cases of more extreme colours than in the standard configuration (red, blue or grey), whereby dark grey was exchanged for black and the standard colours for brighter shades. The parameter used was the absorption coefficient, 0.6–0.9.
- Thermal mass: As with the shoebox model, two cases relating to the building, one with heavy and the other with light construction, were calculated. The parameter used was the total mass, 380–790 kg/m^2.
- Shading strategy: The shading device efficiency was varied between 0 and 0.5. The parameter used was the shading factor, 0–0.5.
- Night flushing: The night flushing was varied between a high air change rate and no possibility for air change during the night. The parameter used was air change rate, 0–10 litres/h.
- Infiltration: The infiltration rate (a.c./h) may be either improved or worsened. The parameter used was the air change rate, 0.6 relative to reference infiltration for old buildings and 0.2 to reference infiltration for new or retrofitted buildings [litres/h].

For the ventilation buffer system, the set temperatures for natural ventilation were varied. Two temperature intervals, between 5 and 22°C and between 12 and 22°C, were used for natural ventilation:

Only the case of new office construction in a cloudy, temperate climate was considered.

Results and design recommendations

The results of this study answer such questions as how high the actual savings potential is with respect to heating energy or running hours of the mechanical ventilation systems and to what extent uncomfortable interior conditions must be anticipated in summer.

In general, it can be said that parameter variations affect the heating energy demand of the air-intake facade more than the unregulated ventilation buffer cases. The minimum and maximum heating demand, as well as surplus temperature degree hours, were plotted for the evaluated parameters of every design situation under scrutiny. The heating energy demand is on the whole moderately sensitive to variations in location, to the glazing area of the inner facade and, to a lesser extent, orientation and room depth. For surplus temperature degree hours, a non-linear relationship is generally the case. Thermal conductance, absorptance and infiltration showed roughly a linear relationship with overheating in the investigated range.

New construction: office
Since new office buildings generally have small heating loads, the absolute effect of energy-saving measures is minimal under winter conditions. The range of possible savings for all cases remains within 10 kWh/m^2 floor area per year for large buildings and 15 kWh/m^2 for smaller buildings.

Figure IV.4.10 shows the results of the thermal performance of a double-facade new office building in a cloudy, temperate climate. The first column in each case shows the building with double facade, while the second column is the reference building. The main issue here is summer overheating. This becomes critical if the inner facade is largely glazed and night air flushing and/or the thermal storage mass are insufficient.

In sunny, temperate climates a lack of night air flushing results in over 2000 Kh per annum surplus temperature degree hours for every orientation. In comparison, large glazing areas, i.e. more than 50% of the facade area, are not quite as critical (about 500 Kh per annum for a south-oriented facade and >1000 Kh per annum for a west-oriented facade). The south-oriented cases that show acceptable levels of surplus temperature degree hours (less than 20 Kh per annum) are characterized by the following:

- a glazing area of inner facade < 30%;
- the total ventilation flap area > 0.03 m^2/m^2;
- thermal storage mass >~1200 kg/(m^2 floor area),
- air change rate of night flushing > 1.5/h;

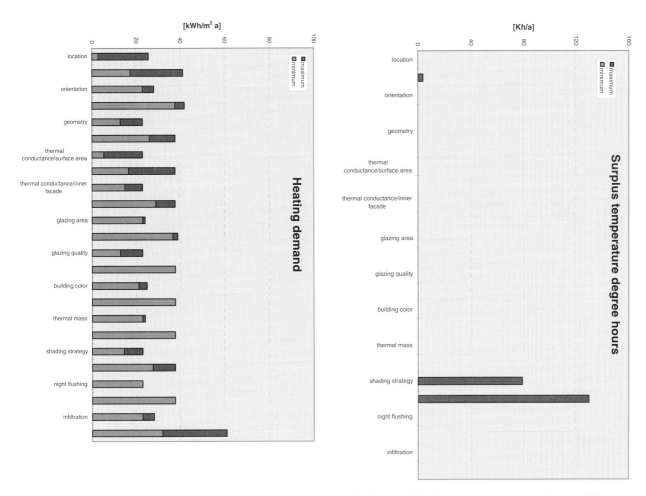

Figure IV.4.10. New office building in a temperate climate (column 1: building with double facade; column 2: reference building)

- absorptance of opaque exposed surfaces < 0.6 (e.g. middle grey, yellow, pink);
- effective shading coefficient for the facade (combination of outer glazing and shading devices).

In cloudy, temperate climates, the likelihood of overheating is greater for all cases. Acceptable levels of thermal comfort (below 20 Kh per annum) can be achieved without mechanical cooling only if:

- the glazing area of inner facade < 15%;
- the total ventilation flap area > 0.05 m²/m²;
- thermal storage mass > 1700 kg/(m² floor area);
- air change rate of night flushing > 3/h.

The ventilation buffer primarily serves to enable natural ventilation during the transition periods and, therefore, reduce the total running hours (energy demand) of the mechanical ventilation system.

The annual frequency distribution of exterior and buffer temperatures for an office in a cloudy, temperate climate is presented in Figure IV.4.11. The ventilation system is assumed to achieve 60% heat recovery, the buffer space is permanently open (open area 0.05m²/m²). The running hours of the ventilation system, the heating demand and the surplus temperature degree hours are given in the Table IV.4.3.

In this particular case, the heating energy demand is relatively insensitive to increased natural ventilation during the winter, since heating occurs largely while the mechanical ventilation systems are assumed to be running. Furthermore, the effective reductions in recovered heat are compensated by air preheating and reduced infiltration losses (under the ideal assumption that the naturally ventilated air change is limited to the hygienic rate).

In sunny, cool climates, a reduction of up to 1900 hours could be reached, which corresponds to an annual energy saving of approximately 5 kWh/(m² floor area). This is a real energy saving, if energy value of electricity is considered.

New construction: dwelling
For the case of new residences, the heating demand is higher than for the office building as a result of lower internal gains and longer occupancy. Energy savings ranged from 15 kWh/m² per annum for large buildings to 40 kWh/m² per annum for smaller buildings, as can be seen in Figure IV.4.12.

For dwellings the problem of overheating can be avoided in almost all cases. Slight overheating occurs in the situation with an inner shading device, as well as for the light constructed building with outer low-E glazing. In sunny, temperate climates the orientation

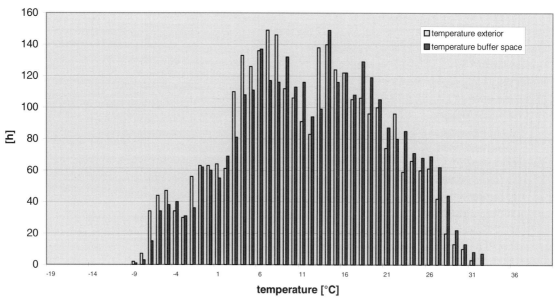

Figure IV.4.11. Annual frequency distribution of ambient and buffer temperatures (cloudy, temperate climate)

strongly influences heating demand (south-oriented: 16k Wh/m², north-oriented: 60k Wh/m²). The following measures optimize the thermal behaviour in winter:

- south-oriented double facade (at least east/west oriented);
- glazing area of inner facade > 25%;
- controlled buffer ventilation;
- heavy construction, i.e. > 1500 kg/(m² floor area);
- absorptance of opaque exposed surfaces > 0.75 (e.g. grey, red),
- inner shading device for the winter.

In an overcast, temperate climate the thermal performance is not as pronounced as in a sunny, temperate climate, but energy savings are still possible.

Table IV.4.3. Relationship between the operation of the ventilation system, heating demand and comfort

	Running hours of ventilation system (h per annum)	Heating energy demand (kWh/m²a)	Surplus temperature degree hours (Kh per annum)
Reference system	3132	25.3	16
Double facade: natural ventilation when buffer temperature between 5 and 22°C	1192	26.1	52
Double facade: natural ventilation when buffer temperature between 12 and 22°C	2019	24.0	53

Retrofit: Office

For the case of retrofitted office buildings, the savings from a double facade are comparable to those of conventional energy-saving measures such as adding insulation. Figure IV.4.13 shows the thermal performance of double-facade retrofit to the office building in a sunny, temperate climate, while Figure IV.4.14 shows the performance in an overcast, temperate climate. The first column is the building with double facade; second column is the reference building.

The simulations show a reduction in annual heating demand, ranging from 30 to 75k Wh/m², depending on the application as well as on the thermal quality of the reference case. Generally, the best performance is achieved with high-quality glazing of the outer facade (low-E), coupled with strongly reduced shading coefficients. The benefit of this is most pronounced for sunny, cold climates and inner facades with large glass areas.

Additional improvement can be achieved by means of darker colouring; however, these savings would be offset by summer overheating. Indeed, under summer conditions, the critical cases are similar to those outlined for new construction (see above). General optimization for both winter and summer is only feasible for sunny, cold climates, in which the overheating tendency with low-E glazing can still be controlled without mechanical cooling.

In sunny, temperate climates a lack of night air flushing results in over 2000 Kh per annum surplus temperature degree hours for every orientation. In comparison, large glazing areas, i.e. more than 50% of the facade area, are not quite as critical (about 500 Kh per annum, south-oriented and >1000 Kh per annum,

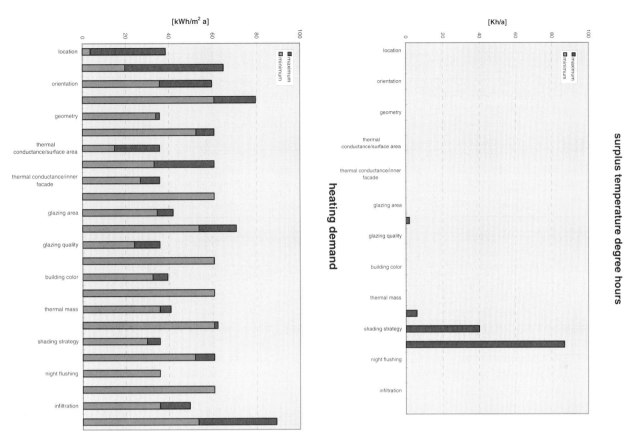

Figure IV.4.12. Performance of a double facade on a new dwelling in a cloudy, temperate climate. (column 1: building with double facade; column 2: reference building)

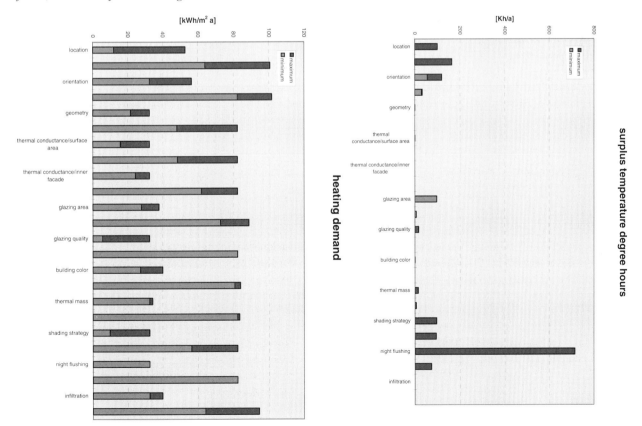

Figure IV.4.13. Thermal performance of double-facade office building, retrofit, sunny, temperate. (column 1: building with double facade; column 2: reference building)

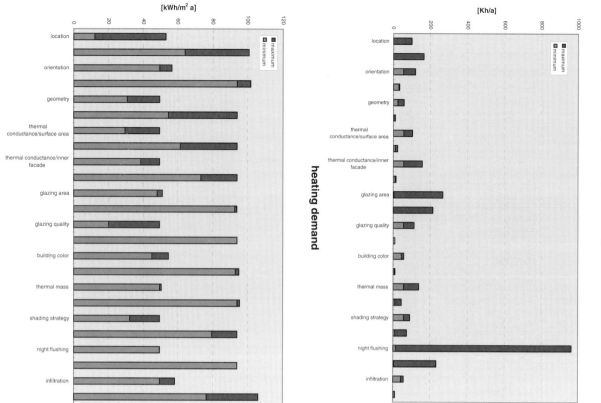

Figure IV.4.14. Thermal performance of double-facade office building, retrofit, cloudy, temperate. (column 1: building with double facade; column 2: reference building)

west-oriented). The south-oriented cases that show acceptable levels of surplus temperature degree hours (less than 20 Kh per annum) are characterized by the following concurrent parameters:

- the glazing area of inner facade < 30%;
- the total ventilation flap area > 0.3 m²/m²;
- thermal storage mass > 1200 kg/(m² floor area);
- air change rate of night flushing > 1.5/h;
- absorptance of opaque exposed surfaces < 0.6 (e.g. middle grey, yellow, pink);
- an effective shading coefficient for facade system (combination of outer glazing and shading devices).

In cloudy, temperate climates the likelihood of overheating is greater for all cases. Acceptable levels of thermal comfort (below 20 Kh per annum) can be achieved without mechanical cooling only if:

- the glazing area of inner facade < 15%;
- the total ventilation flap area > 2 m²/m²;
- thermal storage mass > 1700 kg/(m² floor area);
- air change rate of night flushing > 3/h.

Retrofit: dwelling
For the retrofitted dwelling, the energy saved by a double facade is comparable to that of conventional energy-saving measures (improved U-value, low-E glazing, or savings of about 40–50%).

The simulations show a reduction in annual heating ranging from 50 to 100 kWh/m² per annum, depending on the application as well as on the thermal quality of the reference case (Figures V.4.15 and V.4.16). Optimal configurations for winter conditions are generally characterized by high-quality glazing of the outer facade (low-E), coupled with strongly reduced infiltration losses of the inner and outer facades. The results are thus most pronounced for sunny, cold climates (e.g. Denver) and for largely glazed inner facades.

Some additional improvement can be achieved by means of darker colouring without worsening of the summer situation. Under summer conditions, even the critical thermal behaviour configuration shows no tendency to overheat. For these reasons, the optimization of thermal performance focuses on the winter case.

In sunny, temperate climates the impact of orientation on heating demand is significant. The following concurrent parameters enable a good winter performance:

- south (or at least east/west) orientation;
- low-E glazing of outer facade;
- air tightness < 0.2;
- thermal storage mass > 1500 kg/(m² floor area),
- absorptance of opaque exposed surfaces > 0.6 (e.g., grey, red).

In cloudy, temperate climates (e.g. Zurich), the summer situation is similar to that in sunny, temperate

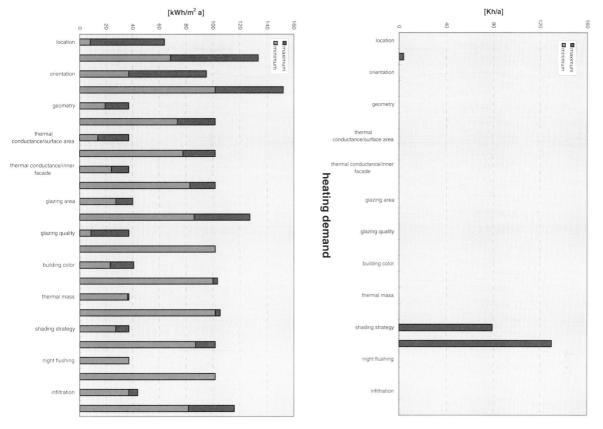

Figure IV.4.15. Performance of a retrofit double facade on a dwelling, in a sunny, temperate climate. (column 1: building with double facade; column 2: reference building)

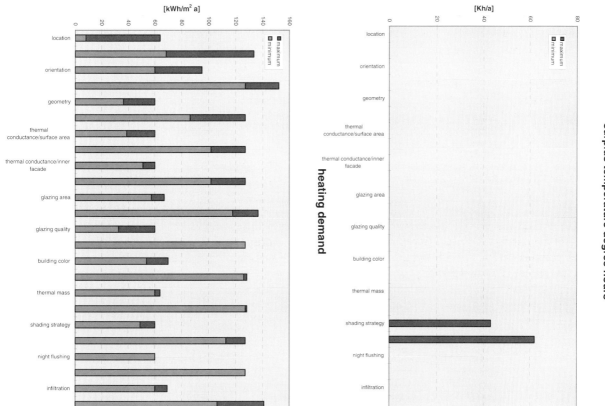

Figure IV.4.16. Performance of retrofit double-facade dwelling, in a cloudy, temperate climate. (column 1: building with double facade; column 2: reference building)

climates. However, the parametric influence on heating demand is somewhat weaker, especially regarding the solar influences.

Considerable savings can be achieved with:

- south (or at least east/west) orientation;
- low-E glazing of outer facade;
- infiltration control < 0.2;
- thermal storage mass > 1500 kg/(m² floor area);
- absorptance of opaque exposed surfaces > 0.6 (e.g., grey, red).

Cost/benefit

The costs reported below are typical for Austria and Germany, whereby the assumptions of Table IV.4.4 are made.

Based on the estimates in the table, it is apparent that the basic cost of constructing a double facade lies within the present-day range of common facade solutions. Low-cost double facades are not more expensive than typical single facades. Compared to the investment in erecting a typical office building, the double facade itself constitutes only a fraction of the overall construction costs. Thus, for certain applications, saving energy by means of a double facade may be economically quite feasible. Improved comfort (e.g. better sound protection) is an added benefit.

Table IV.4.4. Cost data

Facade area/floor area ratio	0.5	
Investments costs (excluding VAT)		
Facade	€350–700/(m² facade area)	Curtain wall and conventional construction
Additional costs with double facade (new construction)	€150–350/(m² facade area)	Single glass with fixed openings
Double facade (retrofit)	€300–700/(m² facade area)	Single glass with controlled flaps
Double facade (retrofit)	€180–550/(m² facade area)	Single glass with fixed openings
Total construction costs	€1000–2000/(m² floor area)	
Mechanical ventilation system with heat recovery	€35–70/(m² floor area)	Good planning
Exhaust-air ventilation system	€5–20/(m² floor area)	Good planning
Power consumption of ventilation system		
Mechanical ventilation system with heat recovery	€0.5–1.5 W per unit of ACR (m³/h)	
Exhaust ventilation	€0.1–0.5 W per unit of ACR (m³/h)	
Energy costs (system efficiency included)		
Heat production costs	4.1 cent(€)/kWh	
Costs of electrical energy	12.0 cent(€)/kWh	

Table 15.5. Cost analyses for double facades on different building types

	Office new	Office retrofit	Office retrofit conventional	Dwelling new	Dwelling retrofit	Dwelling retrofit conventional
Annual savings (€/m²)						
Air-intake facade	0.65	1.65	1.65	1.07	2.15	1.93
Ventilation buffer	0.36					
Additional costs (€/m² per annum)						
Both Systems	90–350	100–370	70–115	90–350	100–370	70–115
Simple payback period (years)						
Air-intake facade	See discussion	60–230	45–70	See discussion	45–170	40–60
Ventilation buffer	See discussion					

Five cases are presented to illustrate costs and savings:

- Office and dwelling, Zurich, new construction, air-intake facade;
- Office and dwelling, Zurich, retrofit, air-intake facade;
- Office, Zurich, ventilation buffer.

The evaluations were carried out with a simple cost/benefit calculation. Maintenance and repair costs, as well as other potential benefits of a double facade (e.g. better sound or weather protection), were not considered. The results are shown in the Table IV.4.5 for a building with double facades in two modes: air-intake facade with exhaust-air ventilation and permanently opened ventilation buffer with mechanical ventilation (heat recovery 60%).

The payback period varies between 45 and >100 years and depends largely on the costs of the double facade. If energy savings alone are considered, new buildings with a double facade are less economical than conventional office buildings. The economic performance of a double facade as a retrofit measure is better, particularly for dwellings, which proved to be the best of all the analyzed applications. With a service life of 50 years, only simple double facades promise a positive payback over the system lifetime, even in the case of dwelling retrofit. Nonetheless, the double facade also can serve other important design objectives, such as formal aspects or noise protection. Taken together with its energy-saving potential, a seemingly 'expensive' double facade may actually prove more economical overall than conventional systems with lower apparent costs.

FURTHER READING

Faist A *et al.* (1998) *La façade double-peau*. EPFL,-LESOPB, CH 1015-Lausanne.

Hauser G (1997) 'ESVO 2000 - ein Konzeptvorschlag'. Internationaler Bauphysik-Kongress, Bauphysik der Aussenwände, pp.11-25. Fraunhofer IRB Verlag, Berlin.

Meletta A *et al.* (1995) Fallstudie Sanierung des ZTL-Gebäudes, Luzern. Hochschule für Technik und Arch., Technikumstrasse, CH 6048-Lucerne.

Recknagel *et al.* (1997) *Heizungs- und Klimatechnik*. 68th edn. Oldenbourg Verlag, Munich.

Schittich Staib, Balkow, Schuler, Sobek (1999) *Glasbau-Atlas*. Birkhäuser Verlag, Basel, CH.

Thiel D. (1995) 'Doppelfassaden - Bestandteil einergetisch optimierter Bürogebäude'. *Facility management* 1/95. Technopress Fachzeitschriftenverlags-GesmbH; Iglaseegasse 21-23, Pf. 176; A-1191 Wien.

Chapter authors: Thomas Zelger, Matthias Schuler and Alexander Knirsch

IV.5 Spatial collectors

INTRODUCTION

Description and principles

Spatial collectors are glazed habitable spaces from which sun-warmed air can be extracted. While glazed atria, sunspaces[1] or glazed balconies[2] are not as efficient at heating air as are flat-plate collectors, they do offer the benefit of being habitable part of the time. Atria can provide a pleasant access and circulation route and a social gathering place. Similarly, sunspaces are brightly lit spaces extending the living space of a residence for a considerable part of the year. This chapter describes concepts to make active use of the warm air that accumulates at the top of such spaces. The primary use is as a source of fresh air for ventilation.[3]

While the methods of construction and the shapes and sizes of glazed spaces are practically limitless,[4,5] certain features must be common to all variations if they are to save energy and be comfortable for as much of the year as possible, namely:

- clear, low U-value glazing in the roof and walls (e.g. double low-emissivity glass);
- vertical glazing oriented within ±20 degrees of south, with good solar access;
- exterior ventilation openings of about 10% of the floor area with maximum height difference between the lower and upper openings to provide natural ventilation;
- moveable sun shading overhead, for west- and south-facing walls;
- thermal mass in the floor or back wall if evening occupancy is desired;
- no auxiliary heating except, perhaps, as freeze protection for plants;
- larger windows to the building than would be the case were there no sunspace or atrium (to compensate for light loss due to the framing and glazing of the glazed space).

Tall glazed spaces are particularly suitable as a 'solar air collector', since the air at the top of the space may reach temperatures sufficiently high to be used for ventilation purposes (the energy benefit must exceed the electricity input for the fans). Examples can be found in schools,[6] offices and houses.[7] Invariably, the configuration is that of System 1 (see Chapter II.1), that is, ambient air enters the glazed space, is sun warmed, rises and is extracted at the top of the space.

Advantages
- Energy savings by using solar gains and recovering heat otherwise lost by the building to the ambient.
- Only minor modifications are needed to a well designed atrium or sunspace to create a spatial collector and these do not alter the appearance of the glazed space or the parent building.
- Minimal cost for ducting the sun-warmed air into the building, especially if a mechanical ventilation system already exists.

Limitations
The habitability of the atrium/sunspace must not be unduly compromised, e.g. temperatures at floor level during the hours of occupation should be between 15 and 25°C. In smaller spatial collectors (sunspaces and glazed balconies) this means that the temperature of the air from the collector is limited to 25°C. However, occupancy densities in dwellings are low, so required fresh air volumes are low. Therefore, the habitability of an appropriately sized sunspace is only marginally affected by its use as a spatial collector (see below). Tall glazed spaces allow warm air to accumulate at the top for extraction, while at ground level it is cooler. Even in such cases, temperatures are still likely to be lower than those achievable by a flat-plate collector. Accordingly, the output of a spatial collector is best used directly for ventilation purposes and not fed into thermal storage. It is unlikely that the cost of storage would be paid back.

Unlike opaque collectors, it is difficult to quantify the instantaneous heat output of spatial collectors as a function of solar irradiance and ambient temperatures. This is because the output temperature is influenced by the thermal mass of the sunspace, transmission gains and losses from the building and occupant use of sunshading and ventilating devices. While some thermal simulation programs[5] can estimate the energy and comfort performance of a spatial collector, such computations are time-intensive. If a sunspace or glazed atrium must be built for non-energy reasons, energy saving is a fringe benefit and may not necessitate precise calculations.

Energy saving capabilities

Pfrommer conducted a systematic study of the habitability of various sunspaces using dynamic thermal simulations with London and Stuttgart climate data.[5] He assumed that sunspaces were designed as above, so that summertime extremes could be curtailed by ventilation and shading. The sunspaces were deemed habitable when the temperature exceeded 15°C between the hours of 07:00 and 23:00.

When the sunspaces were used as spatial collectors, the annual hours of habitability fell by only about 4%.

If it is assumed that a sunspace has an area of about 1/10 of the floor area of the house (and the house is insulated to current UK (1990s) standards), then its use as a solar air collector reduces the energy consumption of the building by at most 20%. In practice, however, the reduction may be less as a result of unforeseen air leakage, thermal bridges and occupant behaviour.

Computer modelling of a four-storey atrium in Neuchatel, Switzerland substantiated the conclusions of Table IV.5.1,[8] namely that maximum energy saving benefits and the largest periods of habitability occur from 08:00 to 17:00. Thermal comfort (between predicted mean votes (PMV) of –1 and +2) was achieved in the unheated atrium, which has a double-glazed south-facing wall, high thermal mass and moveable internal blinds vented at the bottom and top by windows that open to the exterior.

DESIGN

Most well-designed sunspaces and atria can also serve as solar air collectors. However, atria and sunspaces that cover a large area of the building's facade perform better since they recover more of the heat otherwise lost to the outside. This extends the period of habitability and improves the potential for extracting warm air.

In taller spaces stratification of air occurs, producing much higher temperatures at roof level than in the occupied zone close to the floor. For example, on a summer day, when the ambient temperature was 25°C, a four-storey atrium with shading had a floor-level temperature of about 25°C, whilst at roof level the temperature was as high as 45°C.[8] On the day in question the solar radiation intensity peaked at about 700 W/m^2.

A checklist for designing a glazed habitable space for warm-air extraction

- Warm air from the spatial collector should be extracted as high as possible to ensure that the hottest air is taken into the building. In the case of internal sunshading, air can be extracted at the top of the gap between the shading and the glazing, thereby gaining the hottest air and improving the effectiveness of the shading.
The Grafenau Building in Zug, Switzerland (Figure IV.5.1) has three, five-storey atria, which operate as solar air collectors. Ducts are positioned below the atrium ridge to extract air at the highest possible point with automated internal translucent fabric blinds below to prevent overheating and glare.[9]
- Locate outside air inlets to the glazed space away from seating areas to avoid discomfort. These should be closeable, so that when no warm air is extracted, the glazed space cools down more slowly. To improve comfort, ground-level openings should not be used when air is being extracted. Openings higher up are preferable so that some mixing of the warmer air higher up occurs, reducing the likelihood of cold drafts.
Each of the atria in the Grafenau Building has mechanically operated windows in the glazed south-walls to admit fresh ambient air;[9] the other three sides are bounded by offices.
- The air extraction should be stopped when the temperature at floor level is below a pre-set value (e.g. 15°C during occupied hours). This requires a temperature sensor and control logic that can also shut air inlets and possibly the duct inlets

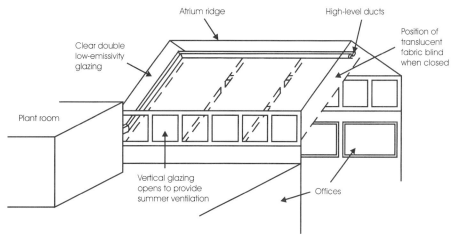

Figure IV.5.1. Schematic diagram of the air collector system of the Grafenau Building atrium

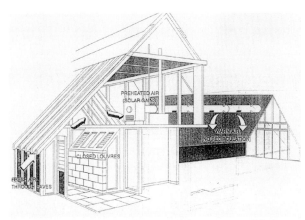

Figure IV.5.2. Schematic diagram of the sunspace at Netley Abbey Infant School

- Especially in non-domestic buildings, a bypass to admit fresh air into the ventilation system should be provided to prevent the glazed space from becoming too cold in winter and to avoid admitting hot air into the ventilation systems in summer.
- The stale exhaust air from the building should not be circulated back into the glazed space.

  Netley Abbey Infant School is naturally ventilated in summer, but incorporates a simple combination of sunspace collector and warm-air heater to deliver solar-warmed fresh air in the heating season (Figure IV.5.2). The option of air intake from ambient was not provided, nor was there a dedicated exhaust air outlet path.[6] Not surprisingly, occupants complained that the building was stuffy.
- Sunspaces and atria are invariably designed as architectural elements first, then, if appropriate, used for solar air heating. A residential sunspace with a floor area of about one tenth of that of the house could deliver almost all the pre-heated ventilation required. In the case of commercial buildings with higher ventilation requirements, this coverage is smaller. Each five-storey atrium in the Grafenau Building (see above) has a floor area of about one fifth of that of the offices that it serves. In the heating season, about 35% of the required fresh air comes from the atrium and the remainder from the outside.
- High-level open walkways must be avoided in tall atria that are performing as effective spatial collectors because they may be uncomfortably warm. By raising the roof above the parent building, overheating of adjacent spaces can be avoided.
- It can be advantageous to position air-handling units close to the spatial collector to minimize duct runs. In the Netley Abbey Infant School the heater was actually in the collector and in the Grafenau Building the HVAC plant is on the roof adjacent to the atria.

REFERENCES

1. Ove Arup & Partners (1988) *Passive Solar Design Studies for Non-domestic Buildings, Case Studies*, for ETSU on behalf of the Department of Energy, Harwell, Didcot, Oxfordshire OX11 0RA, England. Report ETSU-S-1157-P2.
2. Hastings R (ed) (1999) *Solar Air Systems: Built Examples*, Chapter III.4. James & James (Science Publishers), London, for the International Energy Agency.
3. Baker N V (1985) *The Use of Passive Solar Gains for the Pre-heating of Ventilation Air*. For ETSU on behalf of the Department of Energy, Harwell, Didcot, Oxfordshire OX11 0RA, England. Energy Technology Support Unit (ETSU) Report No. ETSU-S-1142, 242pp.
4. Saxon R (1986) *Atrium Buildings, Development and Design*, 2nd edn., The Architectural Press, London.
5. Pfrommer P (1995) *Thermal Modeling of Highly Glazed Spaces*. PhD Thesis, Institute of Energy and Sustainable Development. De Montfort University, Leicester, England.
6. Hobday R A, Trollope M, Palmer J, Shaw P (1992) *Netley Abbey Infant School*, EPA Domestic Technical Report, Databuild Ltd. For ETSU on behalf of the UK Department of Trade and Industry, Harwell, Didcot, Oxfordshire OX11 0RA, England. Report No. ETSU S 1160/12.
7. Energy World (1986) *An International Exhibition of 50 Energy-Efficient Houses, Official Guide*. Milton Keynes Developement Corporation, Saxon Court, 502 Avebury Boulevard, Central Milton Keynes, England.
8. Seager A (1991) *Passive and Hybrid Solar Commercial Buildings, Advanced Case Studies Seminar, IEA Solar Heating and Cooling Task XI*. Energy Technology Support Unit (ETSU), Harwell, Didcot, Oxfordshire OX11 0RA, England.
9. Lomas K J (1995) The Grafenau Building, Zug, Switzerland. Report to ETSU under contract S/N6/00235/00/00 Atrim Studios. Available from the author at the institute of Energy and Sustainable Development, De Montfort University, Seraptoft Campus, Leicester LE7 9SU, England.

Chapter author: Kevin Lomas

IV.6 Hypocaust and murocaust storage

INTRODUCTION

Description

A **hypocaust** as defined here is a massive floor with channels through which solar warmed air is circulated. The mass of the floor then radiates the heat to the room with a time delay. A **murocaust** is a massive wall with channels serving a similar function.

Room heating by these components is comfortable since it is a low-temperature radiant heat from a large surface area. The storage is short term, with a delay of few hours to maximum one to three days. This may be desirable in applications where there is a large direct gain through a window and the aim is to distribute solar gains in the evening.

Applications

Suitable applications depend on:

- *The type of air circulation*: Solar-heated air can be circulated through the hypocausts and murocausts by natural or forced convection. Forced circulation is more common and results generally in higher efficiency of the system. Natural convection systems must be carefully designed hydraulically to avoid excessive pressure drops:
  - In applications with natural convection the hypocaust/murocaust provides a heat sink that creates the natural driving force of the system. The difference in height between collector and the storage will guarantee air flow due to gravity.
  - The temperature drop along the channels must be calculated to ensure that both the hypocaust and the murocaust are being charged/discharged.
- *The type of discharge*: Discharge may be passive by radiation or active as a result of a fan forcing air through the storage. Passive discharge is more common and simpler. Active discharge is suitable for applications where the building has heat demand during the night. It is therefore not suitable for buildings with night setback.
- *The type of system loop*: Closed loop and radiant discharge or open loops are possible. Hypocausts and murocausts can be connected in parallel or in series.

System 1. Solar heating of ventilation air
This system (Figure IV.6.1) is suitable for buildings with a high ratio of window area to floor area. At night the system complements the large daytime direct gains.

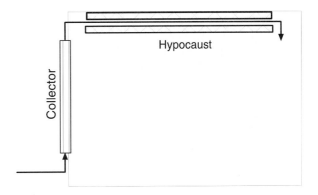

Figure IV.6.1. Hypocaust in System 1

System 2. Open collect loop with radiant discharge
For natural convection systems very low air velocities (0.1–0.5 m/s) are imperative and the hypocaust can provide the necessary heat sink and thus driving force required. With this system insulation should not be used on ceilings (hypocaust). Figure IV.6.2 shows the system and the properties and material dimensions are given in Tables IV.6.1 and IV.6.2.

System 4. Closed loop with radiant discharge
Here (Figures IV.6.3 and IV.6.4), it is necessary to design a hypocaust so that the return air to the collector is not too hot; otherwise collector efficiency will be

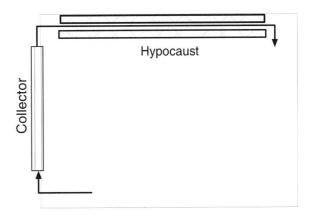

Figure IV.6.2. Hypocaust in System 2

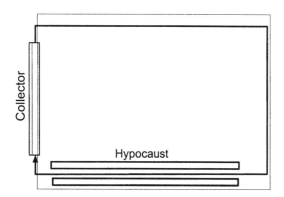

Figure IV.6.3. Hypocaust in System 4

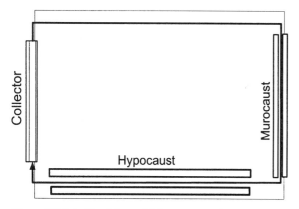

Figure IV.6.4. Closed loop with hypocaust and murocaust

reduced. The floor temperature of the hypocaust should be within the comfort range (see Do's and Don'ts at the end of the chapter). The pressure drop should be limited by keeping air velocities in the range of 1–3 m/s for forced circulation and below 1 m/s for natural convection.

System 5. Closed loop charge, open loop discharge
This system (Figure IV.6.5) is used in applications where a controlled discharge is required to avoid overheating. However, to avoid uncontrolled cooling down of the storage, dampers are needed.

Variations – an overview of component types

Selected component types are shown in Figures IV.6.6 to IV.6.9. There are of course far more design possibilities for different hypocausts and murocausts than can be presented here.

Figure IV.6.5. Closed loop charge, open loop discharge

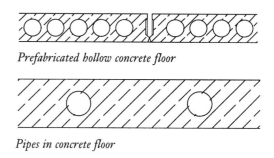

Figure IV.6.6. Pipes in concrete floors and walls – type A

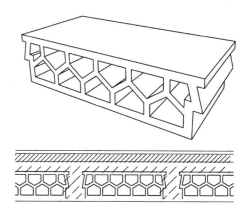

Figure IV.6.7. 'Spannton' (above) and 'Hourdis' (below) floors – type B

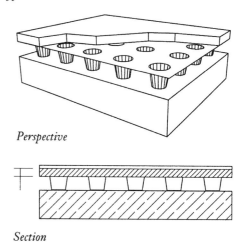

Figure IV.6.8. 'Harbo' hollow floor (air gaps between bearing floors and screed concrete) – type C

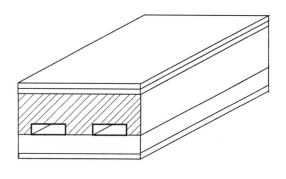

Figure IV.6.9. Barra-Constantini floor – type D

DESIGN

Design options and design parameters

Table IV.6.1 gives some characteristic parameters for different types of hypocaust components. The variables in the table are as follows:

- HSC is the specific heat storage capacity (kWh/m²K).
- The time constant τ is defined as the time in which 63% of the energy absorbed from the air into the hypocaust is transferred into the room (see Figure IV.6.10). This is important for estimating how fast the heat stored in the hypocaust penetrates the room.
- The phase delay Φ is the time difference between the maximum air inlet temperature to the hypocaust and the maximum heat flux from the surface of the hypocaust (see Figure IV.6.10). This gives an idea how long the heat transferred to the room peaks after the peak of collector output. Phase delay and time constants depend on the surface insulation and on the specific storage capacity. A greater phase delay and a longer time constant can be found with hypocausts with insulation on the surfaces.
- The air velocity in a channel is important for pressure drop and heat transfer (recommended air velocities are given in Table IV.6.1).
- The outlet temperature (not shown in the table) is important for the next hypocaust in series or for the collector efficiency. If two hypocausts are in series, the outlet temperature of the first hypocaust has to be sufficiently high for the second hypocaust to be heated. The exit temperature of the second hypocaust should be a maximum of 2 K above the room temperature.

It is recommended that, for comfort reasons, the maximum surface temperature does not exceed 26°C in dwelling areas that are occupied for long periods. In areas like bathrooms, where people stay for only short periods, higher temperatures (up to 33°C) are acceptable. Considerations regarding thermal comfort are described by Fanger.[1,2]

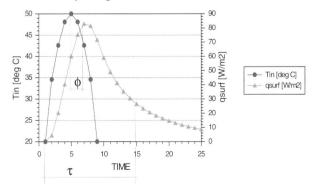

Figure IV.6.10. Definition of phase delay and time constant

Table IV.6.1. Characteristic values for various hypocaust types (the variables are defined in the text)

Type of hypocaust*	Short name	HSC (kWh/m²K)	τ (h)	Phase delay (h)	Recommended range of air velocity (m/s)
A1111111	A1	0.138	10	4	1–3
A1111122	A2	0.139	10	3	1–3
A2322333	A3	0.249	176	14	1–3
A1111223	A4	0.142	17	3	1–3
A2322231	A5	0.229	83	10	1–3
A1111221	A6	0.133	16	3	1–3
A1111131	A7	0.115	7	2	1–3
C1121	C1	0.181	13	3	0.1–2
C1122	C2	0.181	25	3	0.1–2
C1211	C3	0.144	9	3	0.1–2
C1213	C4	0.145	27	3	0.1–2
C1311	C5	0.144	9	3	0.1–2
C1333	C6	0.218	42	3	0.1–2
D11111	D1	0.103	5	3	0.1–0.5
D11223	D2	0.151	26	2	0.1–0.5
D21121	D3	0.165	12	4	0.1–0.5
D13111	D4	0.104	16	3	0.1–0.5

*Refers to different dimensions; see also Figures IV.6.14, IV.6.16, IV.6.18 and Tables IV.6.2 to IV.6.4.

Selection

Single- and double-sided radiant heat distribution
Either one or both sides of a hypocaust may be used for radiant discharge (Figure IV.6.11). Typical examples of double-sided discharge are a floor between two heated zones and a wall between two rooms. The advantage of this is that the building element distributes heat over twice the surface area, compared to a single-sided hypocaust. Good design practice is to oversize hypocausts relative to what the system calculations suggest.

Floors
Floors can be constructed as shown in Figure IV.6.11. The absence of thermal insulation guarantees the distribution of heat both downwards and upwards. Acoustic insulation can be provided by placing a small piece of rubber below each support.

Walls
Masonry double walls may be substituted for ducting and at the same time serve as murocausts. In such cases the thermal conductivity of the wall is not critical. Brick cavity walls can also be used. If the system only uses murocausts without hypocausts, materials with higher conductivity such as concrete may be preferable.

Thermal distribution and layout – walls or floors?
Distributing heat evenly to several rooms of different sizes and locations in a building is difficult. Even distribution is more easily achieved by using floor elements, because the area of the hypocaust is automatically proportional to the size of the room.

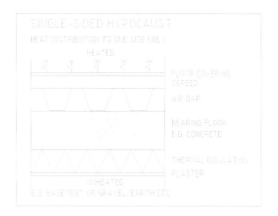
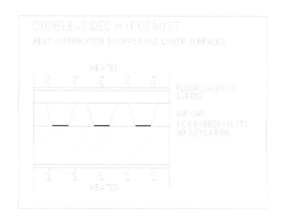

Figure IV.6.11. Single-sided and double-sided hypocausts

Heavy or light-weight components – implications
This chapter is mainly focusing on heavy hypocausts, which both store and distribute heat. There are also light weight murocausts, usually made of plaster-board, the main function of which is the distribution of radiant heat. A comparison of the properties of light and heavy hypocausts as follows:

- *Heavy components/hypocausts with high thermal mass*: Hypocausts made of heavy materials undergo only small and slow temperature changes. There is little need for controls since the systems tend to be self-regulating. Masonry, concrete or screeded flooring can be constructed to be reasonably air-tight.
- *Light components/murocausts with little thermal mass*: The temperature of murocausts made of plaster(gypsum)-board changes rapidly. This makes it difficult to control the room temperature, so that better control systems are required for the charging temperatures. Such light constructions can also be leaky.

Hypocausts typically have a greater surface area than murocausts.

Design procedure

The design procedure is iterative with reference to the design of the whole air system. The heat output and the air exit temperature of the hypocaust are determined when the inlet temperature and air flow rate are known.

Thermal design

By the time this stage is reached, the components may have been specified but they may change, so only the main variables are considered.

Definitions
Figure IV.6.12 shows how the variables are defined. The fully unsteady (real) situation is too complicated to be treated here. A simplified quasi-steady method using time averages has therefore been developed.

Transient variables, which have been *averaged over the total time*, are shown in the following definitions with an asterisk *, for example q_a^*. The inlet and outlet temperatures of the air have to be *averaged over the active hours* (hours when hot air is moving through the system) and are referred to as $T_{av,in}$ and $T_{av,out}$.

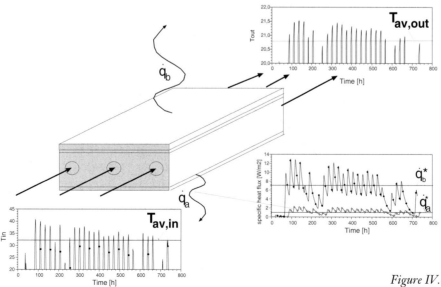

Figure IV.6.12. Definition of variables

Variables are defined as in the following formulae:

$$T_{av,out} = \frac{\left(\sum T_{out}\right)_{active\ hours}}{\sum_{active\ hours}}$$

$$T_{av,in} = \frac{\left(\sum T_{in}\right)_{active\ hours}}{\sum_{active\ hours}}$$

$$q_a^* = \frac{\left(\sum q_a\right)_{total\ hours}}{\sum_{total\ hours}}$$

$$q_b^* = \frac{\left(\sum q_b\right)_{total\ hours}}{\sum_{total\ hours}},$$

where:

T_{out} (°C) is given by $T_{out} = T_{out}$(time);
$T_{av,out}$ (°C) is the averaged outlet temperature of air during the *active* hours;
T_{in} (°C) is given by $T_{in} = T_{in}$(time);
$T_{av,in}$ (°C) is the average inlet temperature of air over the *active* hours;
q_a (W/m²K) is the specific heat flux on surface *a*; $q_a = q_a$(time);
$q_{av,a}^*$ (W/m²K) is the averaged specific heat flux over *total* hours;
q_b (W/m²K) is the specific heat flux on surface *b*; $q_b = q_b$(time);
$q_{av,b}^*$ (W/m²K) is the averaged specific heat flux over *total* hours.

$$X = \frac{U_{tot}^*}{Flow/m \times c_p \times width}$$

$$NTU = \frac{U_{tot}^* \times A}{Flow/m \times c_p \times width} = X \times length$$

$$LMTD^* = \frac{\left(T_{av,in} - T_{av,out}\right)\frac{\sum_{active\ hours}}{\sum_{total\ hours}}}{\ln\left(\frac{T_{av,in} - T_{av,room}}{T_{av,out} - T_{av,room}}\right)} = LMDT \times \frac{\sum_{active\ hours}}{\sum_{total\ hours}}$$

$$q_a^* = LMTD^* \times U_a^*,$$

where:

NTU is the Number of Transfer Units, a dimensionless number; NTU is the ratio of heat transfer capacity of the hypocaust (product of heat transfer area and heat transfer coefficient) to the heat capacity (product of mass flow and specific heat of the air); with increasing NTU the hypocaust will transfer more heat from the air to the room;
U_{tot}^* (W/m²K) is the total heat transfer coefficient averaged over *total* hours;
A (m²) is the heat transfer area of the hypocaust surface in the room; A = length × width;
Flow/m (m³/h m) is the volume flow rate of air per metre of the hypocaust face width;
c_p (J/kg K) is the specific heat of air;
LMTD* (K) is the Logarithmic Mean Temperature Difference averaged over *total* hours;
U_a^* (W/m²K) is the heat transfer coefficient for surface *a* averaged over *total* hours;
$T_{av,room}$ (°C) is the room temperature averaged over *active* hours;
X (1/m) is the intermediate result used in General Nomogram G200;
Width (m) is the total hypocaust face width (perpendicular to the flow direction).

Note that under this definition the U^* values are defined from the heating medium (air) to both surfaces. The total U^* value consists of two parallel values rather than a series of U^* values, as is the case for a one-dimensional wall.

Design of the hypocaust
The following values are already assumed to be known from system-considerations:

- type of air system selected (1, 2 or 4) (see the system chapters II.1–II.6);
- total air flow rate;
- average collector output temperature, which is the inlet temperature to (the first) hypocaust, $T_{av,in}$; preferably use the monthly average for March, i.e. assume that the daily average of a sunny March day will be optimistic concerning the heat flux achieved.

1. Select the hypocaust type A, C or D that is suitable for the System, using the hints in this chapter.
2. Select geometry (cross section) depending on the desired storage capacity and the time constant; check the recommended air velocity in the channel by referring to Table IV.6.1.

Thermal design
3. Determine Flow/m by dividing the total flow rate by the hypocaust face width. This will be the main input value for the nomograms.

Use the type-specific nomograms in Figures IV.6.13 to IV.6.16 (depending on type) to determine values of X, U_a^* and U_b^*. These values will be needed in the next step to determine the air outlet temperature and the heat output in nomograms G100 and G200 (Figure IV.6.16).

4. Determine the air outlet temperature $T_{av,out}$ from nomograms G100 and G200:
Enter nomogram G100 using the value of X (the result of step 3) and, using the length of the hypocaust as the parameter, determine the value of NTU (Number of Transfer Units).
Enter nomogram G200 using the value of air overtemperature ($T_{av,in} - T_{av,room}$) and, using NTU as the parameter, obtain the overtemperature of the exit air ($T_{av,out} - T_{av,room}$).
Is this $T_{av,out}$ reasonable as the inlet temperature for the next hypocaust (it must be high enough above the room temperature) or for return to collector (it must be close enough to room temperature)? If not, return to Step 2 and change the design (geometrical) parameters (channel dimension, insulation).

5. Determine the average specific heat flux q over the surface using the nomograms G300 and G400 in Figure IV.6.16, which are common for all hypocausts. This value will be used to determine the auxiliary heat required.

Hydraulic design

6. Determine the pressure loss across the hypocaust and the complete hydraulic design of the system.

Tables IV.6.2 to IV.6.7 give the material properties and the dimensions of types A, C and D hypocausts, while the variables used are shown in Figures IV.6.17 to IV.6.19.

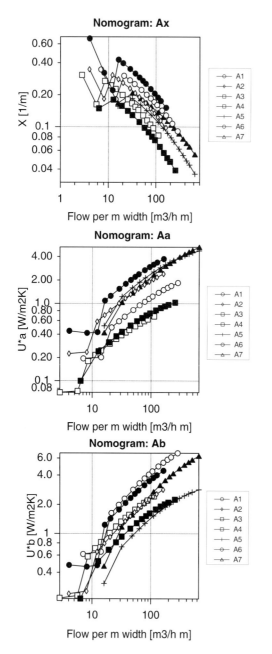

Figure IV.6.13. Nomograms for type A

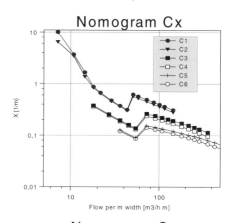

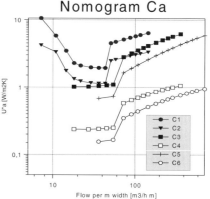

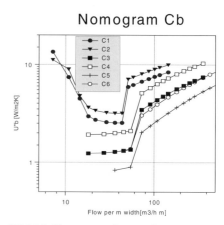

Figure IV.6.14. Nomograms for type C

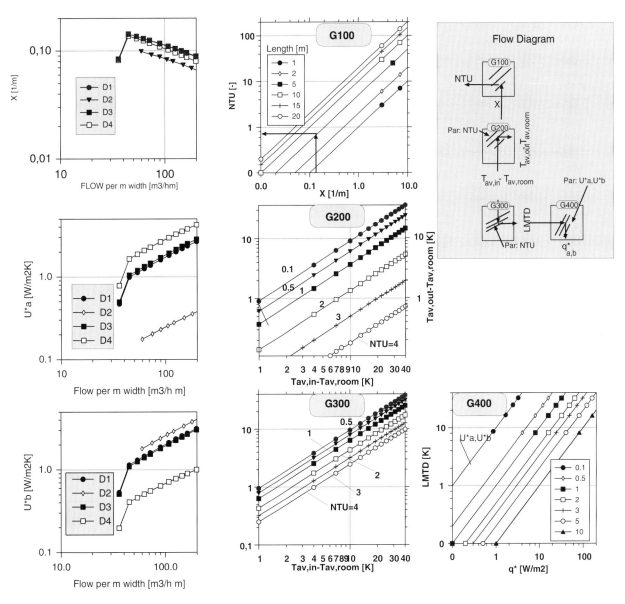

Figure IV.6.15. Nomograms for type D

Figure IV.6.16. Nomograms for heat output and exit air temperature; common to all types of hypocausts

Table IV.6.2. Properties of materials used in the type A hypocaust (Figure IV.6.17)

Material	Density ρ (kg/m³)	Conductivity λ (W/m²K)	Specific heat c_p (J/kgK)	$\rho \times c_p$
Concrete (E)	2200	1.5	1100	2,420,000
Insulation (Rb, Ra)	60	0.04	600	36,000
Concrete (A, B)	2400	1.8	1100	2,640,000
Air channel (D)	1.1	0.2	1000	1,196

Table IV.6.3. Dimensions for the type A hypocaust (Figure IV.6.17)

Name	Short name	E (cm)	Rb (cm)	A (cm)	B (cm)	Ra (cm)	D (cm)	lr (cm)
A1111111	A1	0	0	10	10	0	7.6	40
A1111122	A2	0	0	10	10	0	10	70
A2322333	A3	6	4	15	15	15	15	100
A1111223	A4	0	0	10	10	0	10	100
A2322231	A5	6	4	15	15	15	15	40

Table IV.6.4. Properties of materials used in the type C hypocaust (Figure IV.6.18)

Material	Density ρ (kg/m³)	Conductivity λ (W/m²K)	Specific heat c_p (J/kgK)	$\rho \times c_p$
Floor covering (Rb*)	1400	0.22	1470	2,058,000
Floor covering (Rb**)	800	0.12	1470	1,176,000
Hypocaust (C)	2200	1.2	1100	2,420,000
Insulation (Ra)	60	0.04	600	36,000
Concrete (A)	2400	1.8	1100	2,640,000
Air channel (B)	1.19	0.2	1000	1,196

Table IV.6.5. Dimensions for the type C hypocaust (Figure IV.6.18)

Name	Short name	Rb (cm)	C (constant)	B (cm)	A (cm)	Ra (cm)
C1121	C1	0.009	5	4	20	0
C1122	C2	0.009	5	4	20	2
C1211	C3	0.043	5	5	15	0
C1213	C4	0.043	5	5	15	10
C1311	C5	0.043	5	10	15	0
C1333	C6	0.043	5	10	25	10

Table IV.6.6. Properties of materials used in the type D hypocaust (Figure IV.6.19)

Material	Density ρ (kg/m³)	Conductivity l (W/m²K)	Specific heat c_p (J/kgK)	$\rho \times c_p$
Slab E (concrete)	2200	1.5	1100	2,420,000
Concrete (B)	2200	1.2	1100	2,420,000
Insulation (Ra, Rb)	60	0.04	600	36,000
Concrete (A)	2400	1.8	1100	2,640,000
Air channel	1.19	0.2	1000	1,196

Table IV.6.7. Dimensions for the type D hypocaust (Figure IV.6.19)

Name	Short name	E (cm)	Rb (cm)	C (cm)	D (cm)	F (cm)	B (cm)	A (cm)	Ra (cm)
D11111	D1	0	0	5	40	20	4	15	0
D11223	D2	0	0	8	40	20	4	25	10
D21121	D3	6	0	8	40	20	4	15	0
D13111	D4	0	4	5	40	20	4	15	0

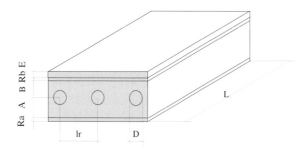

Figure IV.6.17. Dimensions for the type A hypocaust

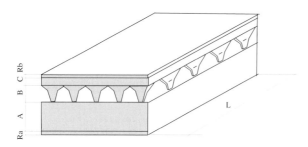

Figure IV.6.18. Dimensions for the type C hypocaust

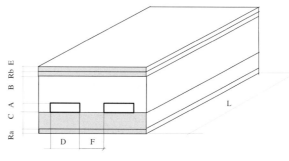

Figure IV.6.19. Dimensions for the type D hypocaust

Hydraulic design

Pressure drops are shown in the nomograms (see Figure IV.6.20 for type A) for hypocausts/murocausts only (excluding manifolds, change of direction etc.). For calculation of pressure drops in other system parts see Chapter V.1.

Example/Worksheet

An example of a worksheet is shown as Table IV.6.8

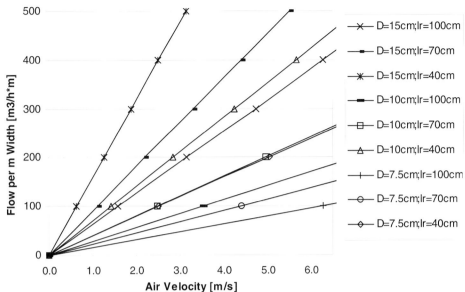

Figure IV.6.20. Type A nomogram; pressure loss per m length in a type A hypocaust

Table IV.6.8. Example of a worksheet

	Units	Result	
Air flow rate	m³/h	150	Total air flow rate for this hypocaust
Average air inlet temperature $T_{av,in}$	°C	40	If possible, use an average of the active hours over a few days; see Chapters IV.1 and IV.2 on air collectors
Desired outlet temperature	°C	< 25	The hypocaust outlet temperature is the air collector inlet temperature
Time delay desired	h	< 5	For higher time delays install more mass and insulation on the boundaries
Room temperature $T_{av,room}$	°C	20	Use average temperature in room
Hypocaust overall dimensions			
Width	m	4.8	
Length	m	5.0	
Selected hypocaust type		A1	
Diameter of channel	m	0.075	
Channel distance *lr*	m	0.4	
Estimated velocity	m/s	1.7	See the section on *Selection*
Check recommended velocity	m/s	1–3	See Table IV.6.1 => OK
Thermal design			
Flow/(m Width)	m³/(h m)	31.2	(150 m³/h)/(4.8 m)
Determine X	1/m	0.35	Use Nomogram Ax (Figure IV.6.13)
Determine U_a^*	W/m²K	1.7	Use Nomogram Aa (Figure IV.6.13)
Determine U_b^*	W/m²K	1.7	Use Nomogram Ab (Figure IV.6.13) – in this case the hypocaust is symmetrical: $U_a^* = U_b^*$
Determine NTU	–	1.75	Use Nomogram G100 (Figure IV.6.16) using X and parameter length 5 m
Determine $(T_{av,in} - T_{av,out})$	K	4	Use Nomogram G200 (Figure IV.6.16) with $(T_{av,in} - T_{av,room}) = 40 - 20 = 20$ K
Determine q_a and q_b	W/m²	~15	

CONSTRUCTION

Building integration

Using load-bearing or non-load-bearing components
Both load-bearing and non-load-bearing components may serve as hypocausts. The hypocaust can be added onto the load-bearing floor or put in the air gap between the load-bearing floor and the screed concrete. Load-bearing hypocaust floors can be prefabricated elements or poured concrete slabs containing channels at regular intervals. If it is preferred to leave the load-bearing structure unchanged, an alternative is to create an air cavity between the load-bearing floor and screed concrete on top.

A murocaust can also be added onto a load-bearing wall.

Layout and air flow
- *Directional air flow.* A unidirectional hypocaust consists of parallel air cavities and needs inlet and outlet manifold ducts at the beginning and the end of the hypocaust. This is usually the case when prefabricated load-bearing floors are used. To obtain an even air flow through all air channels it is recommended that the inlet and outlet ducts be placed at diagonally opposite corners of a rectangular hypocaust. If for some reason this is impossible, an even flow distribution can also be achieved by varying the openings of air cavities. The duct at the inlet

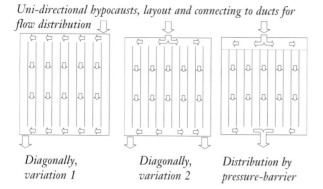

Uni-directional hypocausts, layout and connecting to ducts for flow distribution

Diagonally, variation 1   *Diagonally, variation 2*   *Distribution by pressure-barrier*

Figure IV.6.21. Various possibilities of flow distribution in unidirectional hypocausts

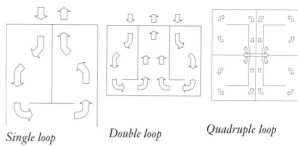

Air-(gap)-floors, layout and connecting to ducts for flow distribution

Single loop   *Double loop*   *Quadruple loop*

Figure IV.6.22. Possibilities of flow distribution in bidirectional hypocausts

end can have an enlarged sectional area. This will even out the air pressure at the ends of all the air cavities. Be careful not to increase the pressure drop across the system. Various options are shown in Figure IV.6.21.

- *Internal loops*. Another layout is to use a hypocaust within which the direction of the air flow may vary. Then loops may be formed so that only vertical ducts are used at the inlets and outlets. Another option is to substitute vertical ducts by a murocaust. This avoids 'bottle-necks' between horizontal and vertical components, which would cause large pressure drops. Various options are shown in Figure IV.6.22.

Do's and Don'ts

- Size the hypocaust to cover the whole floor rather than make it the minimal size indicated by the calculations.
- Keep pressure drops low; duct connections are the most critical points.
- Floor hypocausts may be better for distribution than wall murocausts.
- Use murocausts in combination with hypocausts to help reduce ducting.
- Ample thermal mass yields better distribution and may eliminate the need for sophisticated controls.

- Ensure air-tightness (more easily achieved in concrete and masonry than in light-weight 'dry' constructions).
- Place the fan on the hot side of the hypocaust. This causes less trouble with leaky hypocausts. A fan at the cold end of a leaky hypocaust will extract air from the room. Furthermore, only a part of the collector air may reach the hypocaust!

REFERENCES

1. Fanger PO (1970) *Thermal Comfort*. Danish Technical Press, Copenhagen.
2. International Standard ISO 7730 (1992) *Moderate thermal environments – Determination of the PMV and PPD indices and specification of the conditions for thermal comfort*. International Organization for Standardization, Geneva.

FURTHER READING

Fort K, Gygli W. *TRNSYS – model TYPE 160 for Hypocaust Thermal Storage and Floor Heating*. Available from the authors at Volketswil, Switzerland and from TRANSSOLAR, Nobelstrasse 15, D-70569 Stuttgart, Germany.

Morhenne J, Langensiepen B (1995) *Planungsunterlagen für solarbeheizte Hypokausten*. Projekt der AG Solar NESA, available from Ing.Büro Morhenne GBR, Schülkestrasse 10, D 42277, Wuppertal, Germany.

Chapter author: Karel Fort

IV.7 Rockbeds

INTRODUCTION

General principles

Since the early days of heating air by the sun, rock storage has been a common method of storing excess solar heat. Rockbeds (Figure IV.7.1) can either be thermally coupled to heated spaces or isolated from them. Discharge from isolated systems can be in both open and closed loops, driven by a fan or by thermosyphon.

A rockbed has two main features. The large surface area of the stones serves as an excellent heat exchanger, while a properly designed rockbed lets air pass through easily. The desired pressure drop can be determined from simple calculations to determine both the size of the stones and the size and geometry of the rockbed.

This chapter focuses on the fundamentals of rockbed use with active solar air heating systems. The same method of calculating the rockbed characteristics is used for all system variations. The rockbed size, pressure drop and thermal insulation vary with the system type.

Variations

A rockbed may be varied according to its containment, its insulation and the direction of the airflow. The containment should have a simple shape with a uniform cross section perpendicular to the airflow. Irregularly shaped containments with variable cross sections are not suitable. The most practical cross section is rectangular. Depending on the system type, the rockbed is either insulated or thermally coupled to the heating space. The direction of the airflow through the rocks can be vertical, usually downward, or horizontal. Figure IV.7.2 shows the types of rockbed.

Applications

Rockbeds may be used in various ways:

- solely as a storage unit;
- as a storage unit with radiant heat release;
- as a unit to delay the distribution of solar-preheated fresh air;
- to even out the temperature in rooms with direct gain.

An overview of possible applications in a system environment is given in Table IV.7.1.

DESIGN

Determining the application type and the necessity for storage

The necessity for storage depends on the strategy of the applied system for solar gain (see Applications section above). After the system has been roughly configured and the airflow temperature and rate from the collector have been determined, the rockbed can be dimensioned.

Figure IV.7.1. A rockbed under construction

Figure IV.7.2. Rockbed types: Insulated with vertical airflow, thermally coupled with vertical airflow, thermally coupled with horizontal airflow.

Table IV.7.1. An overview of rockbeds and systems

System 1. Ventilation air heating
A. Delay of yield
The rockbed is used to delay the heat yield of a system for solar preheating of fresh air. This may be useful for coordinating direct solar gains through glazing and the solar gain from the fresh air system, thus avoiding overheating during the hours of sunshine.

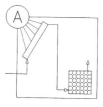

System 4. Closed loop collection/radiant distribution
B. Single loop for collection and distribution
The solar heat is stored in a rockbed, thermally coupled with the heated space, and supplied to the room by radiant heat from the surface of the containment. Either horizontal or vertical rockbeds may be used.

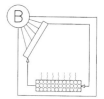

C. Two-loop with integrated auxiliary heating source
A second loop is used, with forced circulation or thermosyphon, for discharging the heat to the rooms via hypocausts. An auxiliary heat source also uses the second loop. All heat distribution is by radiant heat. The rockbed application is conditional on the hypocausts being shared by solar and auxiliary heat. This saves the cost of a separate auxiliary heating system.

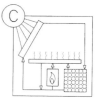

System 5. Closed-loop collection/open distribution
D. Thermally coupled rockbed/air from rockbed to room
The rockbed is thermally coupled to the room and provides radiant heat. A fan provides additional heat distribution by blowing air from the rockbed into the room.

E. Thermally insulated rockbed/auxiliary heating integrated into open-loop air distribution
A rock bin is used to provide a conventional air heating system, such as is commonly used in North America, with solar heat. The rock storage container is insulated and the heated air enters the rooms through a duct system.

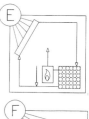

Saving passive solar gain for later use
F. In rooms with excess passive gain, warm room air is blown through the rockbed and enters the room again at a lower temperature. After the sun has stopped shining, the fan continues to circulate air, retarding the cooling of the room. The rockbed temperature is typically lower than in other applications.

Parameters

The main design parameters for rockbeds are storage capacity, the pressure drop, the size of the stones and the length of the airflow through the rockbed needed to transfer 95% of the heat to the stones.

Storage capacity
The storage capacity of a rockbed must be sized according to the heat output from the solar air collectors (i.e. from 0.1 to 0.50 m^3/m^2 of air collector area) and the desired longevity of storage, usually from one to three days. An undersized rockbed achieves higher temperatures, reducing the efficiency of the system. An oversized rockbed will provide better system efficiency, but for most of the time only a small part of the rockbed will have useful temperatures. The specific volumetric heat capacity of rocks is typically 0.37 kWh/m^3K and varies little with most mineral types. Some porous volcanic rocks have a lower density and are inappropriate for use in a rockbed.

Rockbed temperatures
Both vertical and horizontal rockbeds stratify thermally along the direction of airflow. The inlet end nearer the collector is obviously hotter and, as already noted, the cool end of the rockbed should not greatly exceed the room temperature during the main heating

season. The return of air that is too warm to the collector decreases its efficiency. The difference in temperature between the highest average rockbed temperature and room temperature ΔT (K) is used when calculating the highest storage capacity of the rockbed. The return air to the collector has the same temperature as the cool end of the rockbed.

Air flow
The airflow rate of an air collector may be between 40 and 100 m^3/m_{coll}^2. High airflow rates increase the collector efficiency, but deliver lower temperatures and demand considerably higher fan power. The overall system efficiency will typically be best at approximately 50 m^3/m_{coll}^2.

Rockbed geometry
The geometry of the rockbed is important. The cross section (usually rectangular) must be uniform. Two opposite sides (or the top and the base for vertical flow) are open to a plenum so as to allow the air to enter and to exit the rocks. These 'face areas' and the depth of the rockbed, through which the air has to pass, influence the pressure drop.

Face velocity
The face velocity is the volumetric airflow rate divided by the cross-sectional area of the rockbed. It represents the velocity of the air entering the rocks. Increased face velocity increases the pressure drop across the rockbed.

Stone size
Small stones produce a high pressure drop, but they also provide a larger surface area for heat transfer from air to stone and, consequently, a shorter length of air flow for heat transfer. The average size of the stones is the main parameter for determining pressure drop and heat transfer within a given containment. The smallest stones should not be smaller than half the size of the largest. Otherwise the pressure drop will be too high and the calculations will be invalid. The stones should be preferably round, as found in riverbeds, and not crushed.

Rockbed depth
This is the distance between the two face areas.

The critical depth
This is the length required to achieve 95% heat exchange from the air to the rocks. This value must always be less than the total depth. As a rule of thumb, it is advisable not to let the critical depth exceed one third of the total depth, although in most cases it is considerably less.

Pressure drop
A low pressure drop saves fan energy, whereas a high pressure drop helps to maintain an even temperature distribution in the rockbed. The ideal value will vary

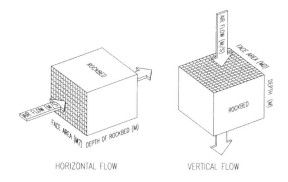

Figure IV.7.3. Standard horizontal and vertical rockbeds with the main parameters shown

according to the system variation. Open-loop systems may have higher pressure drops than closed-loop systems. Figure IV.7.3 shows the main parameters.

Calculation steps

System-related values
Refer to the system appropriate chapter to determine:

1. **Storage capacity C (kWh).** Establish the desired storage capacity.
2. **Storage temperature.** Find the temperature range of the rockbed depending on what can be delivered (by the air collector). The difference in temperature between the average rockbed temperature and room temperature is used. As noted, the cold end of the rockbed should not greatly exceed the room temperature. The return air to the collector has the same temperature as the colder end of the rockbed.
3. **Airflow.** Determine the airflow rate (m^3/h) of the system. The recommended airflow rate for an air collector might be 50m^3/m^2h. See the example below.
4. **Volume.** The volume of the rockbed, $V = C\,[kWh]/(\Delta T\,[K] \times 0.37\,[kWh/m^3K])$. Depending on the system type and the available space, it might be feasible to exceed the calculated volume.
5. **Pressure drop.** The total pressure drop across all the components except the rockbed, e.g. ducts and air collector, should be determined in advance.

Rockbed values
6. **Geometry.** Choose the geometry of the rockbed, thus determining the length and the face area.
7. **Face velocity.** Calculate the face velocity (m/s).
8. **Rock size.** Select the average rock size.
9. **Pressure drop across the rockbed.** Use the nomogram in Figure IV.7.4 to determine:
   Y axis: Pressure drop per metre. Multiply by the depth of the rockbed.
   X axis: Critical depth for 95% of heat exchange.
10. **Total pressure drop across the rockbed.** This is the pressure drop per metre multiplied by the depth of the rockbed.

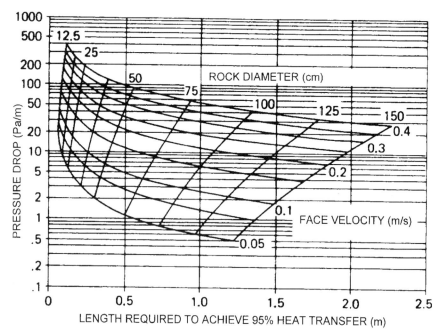

Figure IV.7.4. Nomogram to determine the pressure drop per metre of airflow and the critical depth depending on the size of the stones and the face velocity[1]

11. **Total pressure drop of the system.** Determine the pressure drop across the whole loop by adding the pressure drop across the ducting, air collector etc.
12. **Fan.** Select a fan that matches the airflow and the pressure drop of the system.

Optimization
Change geometry and rock size if necessary. To reduce the pressure drop:

- Use larger rocks.
- Enlarge the face area.
- Shorten the depth of the rockbed.

To increase the pressure drop do the opposite.

The appropriate pressure drop depends on the application. Closed-loop systems (Applications B and C; see Table IV.7.1) may profit from a very low pressure drop to reduce fan power. Systems needing an even temperature distribution within the rockbed require higher pressure drops. Application E, with an open-loop heat distribution as in a conventional North American air heating system, depends on even storage temperatures. Other open-loop applications, such as A, D and F, are less sensitive to even temperature distribution within the rockbed and priority may be given to a low pressure drop and low fan-power consumption.

Check the critical depth of the rockbed
The length through the rockbed required to provide 95% heat exchange should at no time exceed a third to a half the total depth of the rockbed. A short critical depth helps to make use of as much of the volume of rocks as possible. Remember that the air should be cold after leaving the rockbed. If the critical depth has to be reduced, choose smaller rocks.

Finally, recalculate rockbed values, steps 6 to 12, until satisfactory results are achieved.

Calculation example

A solar air heating system is to be installed in a new single-family house. The system has 40 m² of air collector. The recommended airflow rate for the collector is 50 m³/m_{coll}^2. A vertical rockbed will be installed in the basement and will have a depth of 2 m.

It has been determined that the peak capacity of the rockbed should be at least 70 kWh. The rockbed temperatures at a peak load are expected to reach 55°C at the top and 25°C at the bottom, giving an average temperature of 40°C and ΔT above room temperature of 20 K.

1. Total airflow = 40 m² × 50 m³/m_{coll}^2 = 2000 m³/h.
2. Volume of the rockbed, V = 70 kWh/(20 K × 0.37 kWh/m³K) = 9.5 m³.
3. Geometry of the rockbed. This gives a ground area of 2.18 m × 2.18 m = face area of 4.75 m². However, a width of 1.9 m fits better into the ground plan and a length of 3.0 m fits perfectly with the structure of the house, resulting in the following dimensions for the rockbed:
4. New volume and face area, V = 1.9 m × 3.0 m × 2.0 m = 11.4 m³.
   Face area = 5.7 m².
5. Face velocity = airflow rate/face area:
   v = 2000 [m³/h]/(3600 [s/h] × 5.7 [m²]) = 0.1 m/s.
6. Select the average rock size. Diameter = 30 mm.
7. Use the nomogram, the rockbed performance map, in Figure IV.7.4. From the face velocity and the rock size the following values are obtained:

relative pressure drop = 12 Pa/m;
total pressure drop = 12 Pa/m × 2 m = 24 Pa;
critical depth to require 95% heat transfer = 0.2 m.

This value is only one tenth of the depth of the rockbed and therefore quite safe. The critical depth usually only becomes significant with larger stones, a short depth and a higher face velocity.

8. Find the total pressure drop by adding the pressure drop across the ducts and the collector.
9. Select a fan according to the fan characteristics, as given in a fan catalogue. Check how much power the fan consumes. Power consumption for fans may vary greatly and it is good practice to optimize carefully. The two parameters influencing fan power are the airflow rate and the pressure drop across the system.
10. Changing the pressure drop of the rockbed: if the pressure drop is too high or too low (see criteria according to system type) try the following options:

Change the rock size.
Change the face area and the depth of the rockbed.

Repeat the calculation if necessary.

System-related rockbed design

A thermally coupled rockbed radiates heat to the room. Therefore, the U-value and the area of the radiant surfaces have to provide the desired rate of heat release. A common variation is a flat, shallow rockbed under the floor of the heated space. Alternatively, the rockbed may have a vertical flow and a long and narrow containment situated between two uninsulated walls. If a thermally coupled rockbed is coupled with open-loop distribution, the U-value and area of the containment surfaces will be designed to distribute only a part of the heat.

An open-loop distribution over a conventional air heating system (application E) will make use of a relatively small rockbed containing smaller stones to achieve a higher pressure drop for an even distribution within the rockbed. Closed-loop systems (applications B and C) are more sensitive to pressure drop and therefore should have larger stones.

A rockbed coupled to a room with direct gain (application F) differs greatly from the rest. The storage temperature varies together with the room air temperature. If the air temperature variation in the room has to be small, the volume of the rockbed can be unusually large.

CONSTRUCTION

Layout/location of the rockbed

If a rockbed is in a central position in the building, the rockbed's unintentional heat losses are still kept within the building. The air collector, the rockbed, and the technical equipment should be located close together to avoid long ducts and hence duct heat losses.

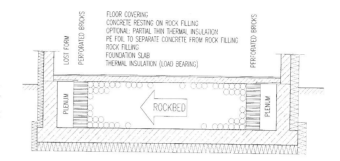

Figure IV.7.5. Thermally coupled horizontal rockbed (applications B, D and F)

The containment

The floor above a rockbed with horizontal airflow (Figure IV.7.5) is typically constructed of a layer of approximately 10 cm of reinforced concrete, poured directly onto a polyethylene film on top of the rocks to prevent the wet concrete from running into the stones. This method ensures that there are no shortcuts in the air stream of the rockbed. A thin layer of insulation may be added to the floor to delay the heat release and provide a lower floor surface temperature. The insulation may cover either the whole surface or the warmer part of it. Remember that there is a horizontal thermal stratification from one side of the floor to the other. The thermal insulation to the ground consists of bearing material (e.g. Roofmate) under the concrete slab. Often, too little insulation is used here (a minimum of 20 cm of polystyrene is needed, given the extensive contact area and the cold earth (in central Europe at a temperature of 10°C over the year). The height of the stone filling matches the necessary depth for frost protection of the foundations. This also influences

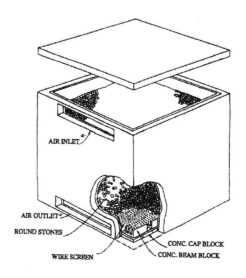

Figure IV.7.6. Stand-alone, lightweight containment (applications D and E) (Illustration courtesy of Solaron Corp.)

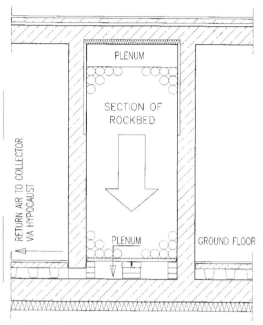

Figure IV.7.7. The construction of an integrated vertical uninsulated rockbed (application B)

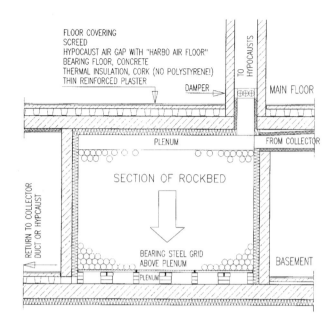

Figure IV.7.8. The construction of an integrated vertical insulated rockbed (application C)

the face area of the rockbed. In regions where frost normally requires deep foundations, horizontal perimeter insulation can reduce the depth of foundations.

A stand-alone rockbed (Figure IV.7.6) is also possible, i.e. a lightweight wooden construction. The sides and top can be insulated with mineral wool, the floor with foamglass on which concrete beam blocks used for the plenum are placed below the stone filling. The importance of the U-value of the insulation is apparent when the heat loss is calculated using the container surface area, average rockbed temperature and the ambient temperature. Care has to be taken to seal the joints for air-tightness. The top is closed after it has been filled with the rocks.

Figure IV.7.7 shows the cross section of a long uninsulated rockbed, designed to distribute radiant heat. The containment is a part of the building structure. The concrete walls give stability against horizontal pressure from the rock filling. Depending on the temperature level and the area of the walls, the insulation level of concrete is sometimes lower than desired. Then, a partial insulation of cork, mounted in the concrete, is placed around the upper warmest part of the rockbed. The cork insulation has been given a thin layer of reinforced plaster. Masonry walls often have a more suitable U-value, but have to be reinforced against the horizontal pressure of the rock filling. A bearing steel grid of the type shown in Fig. IV.7.8 creates the lower plenum. The insulation under the foundation slab has the same thickness as that of the rest of the floor, because the bottom of the rockbed is at the cool end of the thermal stratification. A hypocaust is used instead of a duct for the air returning to the air collector.

The insulated rockbed shown in Figure IV.7.8 is part of a system with closed-loop heat distribution to hypocausts. The distribution loop has been designed for natural convection.

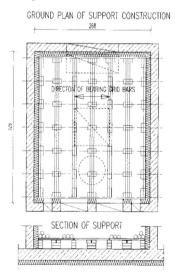
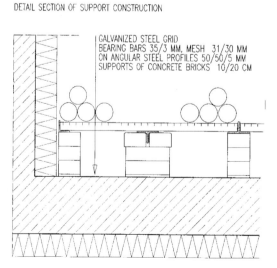

Figure IV.7.9. Support above the lower air gap of a vertical rockbed (applications A, C and E)

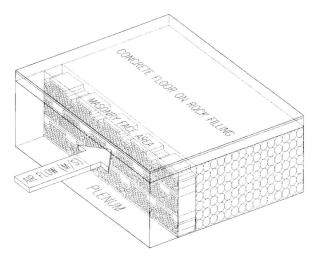

Figure IV.7.10. Side construction at the face area of a horizontal rockbed (applications B, D and F)

The galvanized bearing steel grid shown in Figure IV.7.9 forms the lower air gap of a vertical-flow rockbed. The supports are concrete bricks. Angular steel profiles are assembled as girders to carry the steel grid, which is of the type commonly used in industrial buildings and to support vehicles.

Figure IV.7.10 shows a masonry wall separating the stone filling from the plenum at the face area that can be made of any bricks with enough holes. In this case they were 25 cm wide tile bricks designed as filling elements for bearing floors.

Building matters

Rock material
As noted earlier, the stones should be round, such as found in rivers and sedimentary deposits. Ideally, the stones should be of a uniform size. Usually quarries offer material with stone diameters varying within a factor of two. This range is acceptable, but it is advisable to personally inspect the material before ordering. Smaller material is more readily available than stones with larger diameters. Using larger stones often requires a special order. Order 'washed' material. This only means that sand has been removed, so the stones have to be washed again on the building site.

Cleaning and filling
This involves manual labour and washing the material over a sieve. Be sure to remove sand and dust. Take time to plan this procedure thoroughly, because planning will save time and effort. Pipes for thermal sensors should be mounted before filling and should reach about 30 cm into the containment. Dropping the stones directly into the containment may cause them to split, so use a wooden slide. Filling the stones while they are still wet is most practicable, but this can only be done if the container is built of moisture-resistant materials. The rockbed must be allowed to dry. Depending on the building schedule, you may wish to close the rockbed and connect it to the rest of the system. Drying can then be accelerated by the use of a provisional fan drawing air from the collector to the rockbed and letting the air exit after passing through the rockbed. *At no time should moist air be blown in the opposite direction from the rockbed through the collector.* Use a provisional filter at the inlet to avoid dust.

Building schedule
Choose the best time in the building schedule to fill the rockbed, not too early and not too late. Even though this may seem obvious, filling at the wrong time is a common mistake. The earliest the rockbed should be filled is at a time when rain may no longer enter the building. If the stones are wet when they are put into the containment, time should be allowed for the stones to dry. The rockbed should be sealed before dusty work such as mounting plasterboards, cutting tiles or insulating with mineral fibre is carried out.

Sealing
Air-tightness is most easily achieved with a containment of poured concrete. The filling hole in the roof is closed with help of lost forms. Lightweight constructions need more care as sealant material is used.

Operation and maintenance

Rockbeds require no maintenance, providing that no mistakes have been made during construction. Problems occurring after construction are difficult to correct.

Filters should be used in the duct system to avoid the accumulation of dust in the rockbed. Open systems (applications A, D, E, and F) should have a filters both at the inlet and at the outlet of the rockbed. Closed systems (applications B and C) usually require only one filter after the rockbed and before the air collector. The filters should be inspected at regular intervals and, if necessary, changed. The inspection intervals are longest for closed-loop systems and may be up to five years. An alternative is to use electronic air cleaners. These have the advantage of not adding any pressure drop to the system. However, this option increases electricity consumption.

Condensation
There is usually no danger of condensation in rockbeds. The normal operating temperature does not allow the air to get near saturation point. This is true for both closed- and open-loop applications using air collectors. An exception is when the controls in an open-loop system malfunction and a fan is blowing when it should not, either causing the rockbed to undercool or letting humid summer night air to enter a colder rockbed. Condensation in the rockbed may under these circumstances cause mould and unhygienic air, but this is easily avoided by ensuring that the system operates correctly.

Economy and construction
The cost of a rockbed can be reduced by several means:

- Use the building for two or more walls of the containment.
- Position the rockbed so that the length of ducts remains as short as possible.
- Use the depth of foundations required for frost protection for horizontal rockbeds.

REFERENCE

1. Balcomb JD *et al.* (1983) *Passive Solar Design Handbook*, Vol. 3. Jones, RW (ed.). Boulder, American Solar Energy Society, Boulder, CO. (Available from Publications Office: 110 W. 34th Street, New York 10001.)

FURTHER READING

Abbud GO, Löf G, Hittle DC (1995) 'Simulation of Solar Air Heating at Constant Temperature'. *Solar Energy* Vol. 54, pp. 75–83.

Choudhury C, Chauhan PM, Garg HP (1995) 'Economic Design of a Rock Bed Storage Device for Storing Solar Thermal Energy'. *Solar Energy* Vol. 55, pp. 29–37.

Duffie J A, Beckman W A (1991) *Solar Engineering of Thermal Process*. Wiley-Interscience, New York. [German original: *Sonnenenergie – Thermische Prozesse*. (1976). U. Pfriemer Verlag, Munich.]

Erhorn H, Franke A, Gertis K, Kiesl K, Rath J. (1987) *Thermisches Verahlten luftdurchströmter Kiesspeicher*. WB 10/87 des Fraunhofer-Instituts für Bauphysik, POB 800469, D-70504 Stuttgart.

Ong KS (1995) 'Thermal Performance of Solar Air Heaters: Mathematical Model and Solution Procedure'. *Solar Energy* Vol. 55, pp. 93–109.

Chapter author: Sture Larsen

IV.8 Phase change materials (pcm) for heat storage

INTRODUCTION

Description and principles

Phase change materials (pcm) have the ability to store a certain amount of heat without a temperature change, i.e. at the temperature where they change between a liquid and solid. This is of course also the case with water, but a pcm can be chemically composed so that it changes state at a useful temperature, for example 30°C. Thus pcm can store more heat in a given volume within a limited temperature interval than materials that store only sensible heat. Figure IV.8.1 shows this advantage for two salt hydrates in comparison with rock and water.

Generally a pcm consists of a material or a combination of materials that go through a chemical or physical process that requires heat. The most common process is melting/solidifying, for example a salt hydrate. For solar air systems two salt hydrates are widely used:

- $Na_2SO_4.10H_2O$ (Glauber's salt) with a melting point of 32°C and a latent heat of 100 kWh/m^3;
- $CaCl_2.6H_2O$ with a melting point of 29°C and a latent heat of about 78 kWh/m^3.

Theoretically the process of changing from the solid to the liquid phase and vice-versa (phase change) occurs at a constant temperature (dotted line in Figure IV.8.2). The total heat storage capacity consists of the specific heat needed to heat the material up to the melting point in the solid phase, the actual phase-change energy and the specific heat of the liquid phase when the material is heated up above the melting point. Practically, the heat must be transported between the periphery and the centre of the material. Therefore the process does not occur simultaneously within the whole material and a behaviour of the material like the solid line in Figure IV.8.2 is the result.

Variations

For practical applications, a pcm cannot be used in its pure form; additives are needed for stabilization and to avoid subcooling. The properties given above (especially the latent heat) are strongly influenced by these. The melting temperature can also be customized by additives. The properties of commercial products are stated in the manufacturer's information.

Equally as important as the pcm itself is the container that holds it. Products have come and gone with diverse geometries ranging from cylinders, balls, cans and more complicated forms. The containers are made of an equally wide palette of materials including diverse metals and thermoplastics (Figure VI.8.3).

Applications

Phase change materials can be used in different kinds of solar air systems including System 6 with air and water combined as the heat transport medium. The most appropriate solar air systems are 2, 4 and 5. System 1 is also possible, if heat storage is necessary. Pcms can be

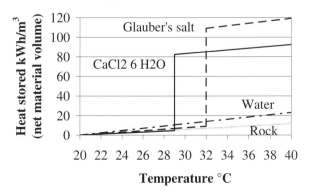

Figure IV.8.1. Comparison of stored heat in pcm and sensible materials

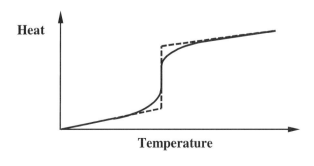

Figure IV.8.2. Theoretical and real heat–temperature curve for a phase change material

Figure IV.8.3. Examples of pcm containers (source: manufacturer's brochures)

applied to all building types where these systems are applicable.

A common configuration is placing pcm elements in a bin. Solar-heated air is passed through the void space between the elements for the charging process. Heat is transferred through the surface of the elements to their centre. When pcm spheres are used, this is very analogous to a rockbed, but thanks to the much greater heat capacity, the pcm bin can be much smaller than a rockbed. Solutions with interior walls with just one layer of pcm elements can also be found.

Characteristics

Subcooling
During the cooling process a pcm sometimes remains in the liquid phase rather than crystallizing at temperatures below the melting point. Crystallization then occurs spontaneously at times and under circumstances that have not been planned. The temperature curves in Figure IV.8.9 shows such behaviour.

This subcooling effect can be avoided by using additives and generally does not occur today with modern products. Its occurrence, if detected, may be an indicator of 'untight' containers. As long as it is not too dominant, subcooling can be tolerated.

Stability
Phase change materials must keep their properties over the system life. Some products have been tested by undergoing a certain number of melting/crystallization cycles. For instance, a product with $CaCl_2$ was exposed to 1,000 cycles and no degradation of its properties was found.

Diffusion of water through the container walls is a common physical process that degrades the properties of pcm materials. Changing the water content of the salt hydrate influences its thermal–physical properties. Diffusion tight containers are essential.

Toxicity/disposal
Phase change materials may be toxic, but the most frequently used materials, Glauber's salt and $CaCl_2.6H_2O$, are non-toxic.[1]

Corrosion
Salt hydrates corrode metals.[1] Leaking containers are therefore a potential source of corrosion of metal components or concrete reinforcement. Accessibility for inspection over the pcm lifetime has to be provided.

DESIGN (SIZING)

Sizing

Sizing a pcm storage consists of the following steps:

- Determine the heat demand of the building.
- Determine the desired fraction of the demand that is to be met by the solar system.
- Determine the collector area.
- Determine the desired time period for heat storage.

These design steps lead to the storage size and are described in more detail in the related system chapters (Chapters II.2, II.4 and II.6).

Requirements

Melting temperature
A high melting temperature of the pcm decreases the collector efficiency, but the overall system performance depends on the load, storage and collector match. The pcm melting temperature is important for passive discharge configurations, where the heat release is driven by the temperature difference between the storage and the room(s) to be heated.

General characterization of products
The design process determines a certain storage requirement, expressed in MJ or kWh, which may be

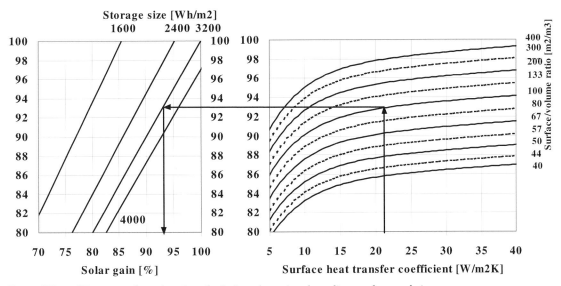

Figure IV.8.4. Nomogram for estimation of relative solar gains, depending on the pcm design

varied within a certain range. This can be achieved by different combinations of products, e.g. a bin of ball-type containers, different packages of cans, rows of cylinders etc. These different arrangements can be compared considering two properties:

- *The surface/volume ratio.* A higher ratio yields a good heat exchange. Therefore, small elements are better than large ones. The heat transfer of large elements can be improved by increasing the heat transfer area, for example with fins. The elements should be stacked so that the collector air comes in contact with the whole container surface.
- *The surface heat transfer coefficient.* This is influenced by the air velocity, the geometry of the pcm elements and their arrangement, which also influences the air velocity. This has to be calculated according to the classical heat transfer equations.[2–4] Higher velocities and higher turbulence increase the heat transfer coefficient. Increasing the air velocity not only improves the heat transfer coefficient, but also unfortunately increases the pressure drop, which then necessitates more fan power.

Figure IV.8.4 is a nomogram for estimating the relative solar gains as a function of the pcm storage size, the surface/volume ratio and the surface heat transfer coefficient, derived from simulations.[1] The absolute solar gains are not given, because they depend on further parameters like collector performance and climate. This nomogram is intended as a help for choosing the best arrangement of pcm elements.

In Figure IV.8.4, the pcm storage size is expressed in Wh/m² of collector area. Only the latent heat is accounted for, sensible gains are not considered. The gross volume depends on the pcm container geometry and arrangement. The numbers are derived from a cylinder-type container and would lead to a gross volume of 27% more than the net volume.

Even flow and pressure drop
The containers must be arranged so that the air can pass evenly through the whole bin, reaching all containers without an excessive pressure drop. The pressure drop depends strongly on the form and arrangement of containers and must be computed using classical equations, which are given in the fluid mechanics literature.

Passive discharge
Passive discharge of pcm storage can be envisaged. This may be by means of natural convection, by conduction or by a combination of the two. The design of the configuration of passive discharge storage is more critical than that of fan-driven discharge. The rate of passive discharge must not be so great as to prevent melting of the pcm by charging. Figure IV.8.5(a) gives

a) One-layer construction:

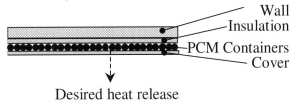

b) Multilayer construction:

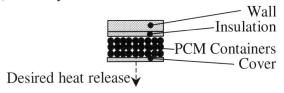

Figure IV.8.5. Example pcm constructions

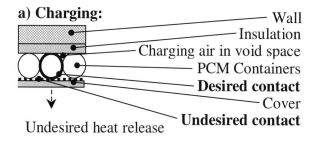

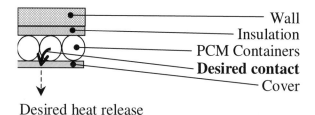

Figure IV.8.6. Heat transfer of a pcm in charging and discharging modes for the different parts of the construction

an example of a construction with one layer of pcm elements (of any form).

For charging, good contact between air and ideally the whole pcm element surface is important. Contact between the charging solar air and the bin cover is not desirable, but unavoidable (Figure IV.8.6(a)). It is a source of direct heat release to the room, bypassing the pcm. This effect can be reduced with a compact multilayer bin (Figure IV.8.5(b)).

For passive discharge, the construction must be optimized such that the heat transfer from the elements through the cover to the room occurs when the room needs heat (Figure IV.8.6(b)). The multilayer storage mentioned above may not be suitable for this purpose (Figure IV.8.5(b)).

Design procedure

1. From the system design procedure determine the following requirements:
   - storage size (MJ or kWh);
   - duration of storage;
   - air flow rate for charging.
2. Specify the pcm product characteristics as described above. These include:
   - the melting temperature;
   - the container geometry (form and size).
3. Determine the number of elements necessary to provide the required storage capacity.
4. Design an arrangement of the elements.
5. Calculate the surface/volume ratio for the chosen product and arrangement.
6. Calculate the air velocity, using the air volume flow rate for the overall system and the pcm arrangement. (The definition of the velocity depends on the equation used for the calculation of the surface heat transfer coefficient, which in turn depends on the arrangement.)
7. Calculate the surface heat transfer coefficient with the appropriate equation for the arrangement.
8. Determine the relative solar gain using the nomogram in Figure IV.8.4.
9. Calculate the pressure drop with the appropriate equation for the arrangement and check if it is admissible.
10. Check the discharge of the arrangement against the required duration.

Repeat steps 4 to 10 for different arrangements and repeat steps 2 to 10 for different products.

For a more detailed design process a variety of computer models or routines have been developed.[1,5–7]

CONSTRUCTION

General

Spherical pcm storage containers can be built exactly the same way as a rockbed storage. Similar constructions are possible with other container geometries, e.g. with cylinders arranged parallel to the air flow. This is the case as long as the heat discharge is active

Because of the smaller size of pcm storage, it can more easily be integrated into the structure of the building, internal walls or ceilings. Such solutions will probably be very thin in one dimension, e.g. with just one or a few layers of pcm elements.

Example construction

A successfully tested construction, which can be added to an internal wall, is shown in Figure IV.8.7. It consists of four layers of pcm cylinders with a diameter of 38 mm and a length of 1 m. Two element lengths are always allowed, giving a total length of 2 m, which leaves enough space for inlet and outlet plenums above and below within the normal room height. The construction is designed for a passive heat release in one direction and could be put on one face of an interior wall.

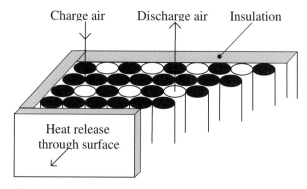

Figure IV.8.7. Example of pcm storage construction

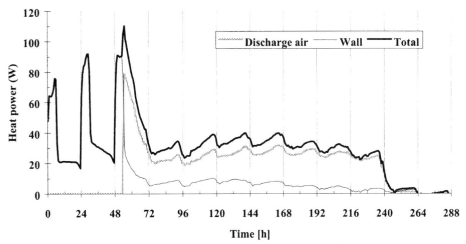

Figure IV.8.8. Measured results for heat-release power as a function of time

Sixty-eight cylinders were packed in a box 0.16 by 0.44 by 2 m. 10 empty tubes for discharge were distributed between the cylinders. The salt hydrate in the cylinders is Glauber's salt with a melting temperature of 31°C. The total latent storage capacity is 5,100 Wh or 18,360 kJ. This could be scaled up to any width, keeping the same height and depth.

Charging is done by circulating the solar air from top to bottom through the void space between the cylinders and parallel to the cylinders. Charging in the opposite direction (bottom–top) was also tested and turned out to be possible, but required higher supply air temperatures, probably because of mixing effects due to internal circulation. With a volume flow rate of 250 m^3/h and a remaining space between the cylinders of 0.19 m^2 cross section area, the velocity in the void space was about 3.7 m/s. This resulted in a pressure drop, including the inlet and outlet chambers, of about 90 Pa.

Passive discharge occurs in two ways: the layer next to the cover separating the unit from the room to be heated releases its heat through radiation and natural convection to the cover and from there to the room. The layers behind are not sufficiently connected to the cover. Therefore, empty tubes were added in the second and fourth layers, in order to increase the heat release from these rows. They release their heat to the empty tubes rather than to the cover. In the empty tubes, the room air can circulate by natural convection. With this arrangement, the heat release can be controlled by opening or closing the discharge tubes.

Performance

Figures IV.8.8 and IV.8.9 show measured results from a test facility with this arrangement. A typical three-day charging period with charging intervals of 6 h (sunny winter days) was performed, followed by a period of discharge.

During the charging period, the discharge tubes were closed; for discharge they were half open. From Figure IV.8.8 it can clearly be seen that the discharge

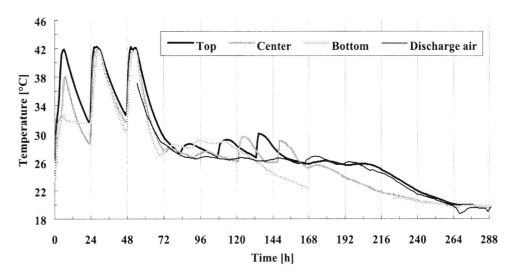

Figure IV.8.9. Measured and simulated results for two temperatures (top and bottom) on the pcm cylinder surfaces as a function of time

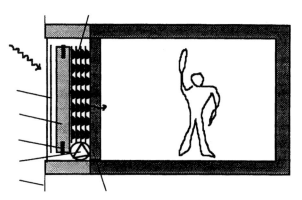

Figure IV.8.10. Compact facade-mounted collector–pcm module with passive discharge (from Manz et al.[5])

period lasts eight days from the end of the last charging interval. The portion of discharge power through the wall is very high during the charging intervals owing to the high charging air temperature. It decreases rapidly after the charging periods and becomes small compared with the convection part through the discharge tubes, despite the fact that these were half closed. The discharge power fluctuates because the room air temperature was not constant.

The temperatures in Figure IV.8.9 show stratification. The top part of the device is discharged after eight days, the centre part after seven days and the bottom after five days. The temperatures remain flat until the pcm is fully crystallized, except for some randomly appearing peaks, which are the result of subcooling effects.

The duration of the discharge process could be reduced to three to five days by opening the discharge tubes more.

Integrated design

An example of an integrated design is shown in Figure IV.8.10.[5] The system of collector, air circuit and pcm storage is combined into one module, which can be mounted on a facade, i.e. as a retrofit measure.

SUPPLIERS

- Cristopia Energy Systems, 8 Chemin du moulin de la clue, F-06140 Vence, France. Tel. +33 493 58 40 00; Fax: +33 493 24 29 38.

- Climator AB, :t Helenagatan 10, -541 30 Skövde, Sweden, Tel. +46 500-48 23 50; Fax +46 500-48 57 22; Email: rolf@climator.com.

REFERENCES

1. Zweifel G, Koschenz M, Tscherry J (1996) *Optimierung der Luftkollektoranlage mit Latentspeicher in Schüpfen – Schlussbericht*. Forschungsstelle Solararchitektur, ETH Hönggerberg, CH-8093 Zurich.
2. Whitaker S (1977) *Fundamental Principles of Heat Transfer*. Pergamon Press, New York.
3. *VDI-Wärmeatlas, Berechnungsblätter für den Wärmeübegang* (19??) 7th expanded edition. Verein Deutscher Ingenieure; VDI-Verlag GmbH, Düsseldorf.
4. Eckert E (1959) *Einführung in den Wärme- und Stoffaustausch*. Springer-Verlag, Berlin/Göttingen/Heidelberg.
5. Manz H et al. (1992) *'Paroi Chauffante Solaire' - Untersuchung eines fassadenintegrierten Luftkollektor-Latentwärmespeicher-Systems zur Nutzung der Sonnenenergie für die Raumheizung*. EMPA, ZEN, CH-8600 Dübendorf.
6. Egolf PW et al. (1994) *Latentwärmespeicher für die Sonnenenergienutzung: Lade- und Entladevorgänge*. NEFF-Projekt 515. EMPA, ZEN, CH-8600 Dübendorf.
7. Egolf PW et al. (1996) *Latentwärmespeicher mit kleinen Kapseln für Luftsysteme*. NEFF-Projekt 593. EMPA, ZEN, CH-8600 Dübendorf.

FURTHER READING

Zweifel G (1989) *Messprojekt Schüpfen, Schlussbericht, BEW/EMPA*. EMPA, Abt. 175, CH-8600 Dübendorf.

Lane GA (1985) *Solar Heat Storage: Latent Heat Materials*, Volumes 1 and 2. CRC Press, Florida

Schuler M (1992) *Technology Simulation Set – Phase Change Materials*; Working document, IEA Solar Task 13. Institut für Thermodynamik und Wärmetechnik, Prof. Dr.-Ing. E. Hahne, Universität Stuttgart, Pfaffenwaldring 6, D-7000 Stuttgart 80.

Keller L (1989) *Stockage latent: Etat de la technique*. Swiss Federal Office of Energy, Bureau d'Etudes Keller-Burnier, Ch. Du Renolly, CH-1175 Lavigny.

Chapter author: Gerhard Zweifel

# V. Accessories

V.1 Fans

INTRODUCTION

Description

Fans are flow machines designed to convey a certain air volume and to increase the pressure in order to overcome the resistance of the system. They should work with the best possible efficiency and at lowest possible noise level.

Variations

Fans can be classified according to the air flow direction through the fan. Major types are axial, radial and mixed-flow fans. Other possible criteria are the location of the drive or the use of the fan. Usually fans are driven by an electric motor positioned within or outside the air flow.

Axial fan
The air enters and leaves the fan axially. The main modules of an axial ventilator (Figure V.1.1) are the hub with the blades, the casing and the drive. There are several components designed to increase the efficiency of an axial blower, such as an inlet nozzle, stator, diffuser and moveable blades. The air flow of an axial fan can be controlled by:

- dampers on the extraction or pressure side of the fan;
- variation of the blade angle;
- variation of the rotational speed.

If energy savings are considered, the best way to change the air flow rate of a fan is to vary the speed of rotation. A common position for the drive is the hub of the fan.

The motor is cooled by the air flow and all the heat from the motor can be used, while the open cross section is reduced. Typical operating conditions of standard axial fans are high volume rates and low delivery pressure, e.g. ventilation of a sunspace or exhaust-air extraction from rooms.

Radial fan
The air enters axially and leaves radially. These fans have usually a spiral casing with single or double inlet (Figure V.1.2). Some special casings allow the use of a radial fan in a round duct. There are different shapes of blades, which can be forward-curved, radial or backward-curved. It is possible to have the motor external to withstand high flow temperatures and to avoid heat gains due to the motor. A radial fan typically achieves a lower volume rate and a higher pressure increase than an axial fan. Radial fans are typically used in ducts with a higher flow resistance, e.g. ducts with air heating, cooling and filter devices.

Cross-flow fan
The air flow runs through the impeller transversally. The volume increases proportionally to the impeller width. Cross-flow fans are suitable for blowing into narrow ducts or grooves with a broad width but with

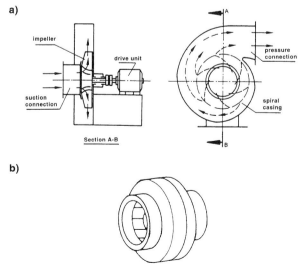

Figure V.1.2. Radial fan: (a) axial inlet and radial outlet; (b) axial inlet and outlet

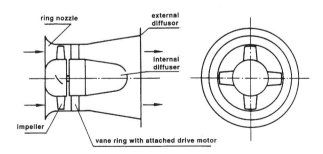

Figure V.1.1. Axial fan

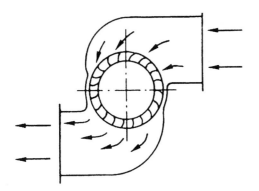

Figure V.1.3. Cross-flow fan

typically low volume rate and efficiency. An advantage of cross-flow fans is their low noise emission, on account of which they are used in applications where the fan is placed near the user, e.g. an underfloor convector.

Electric power demand

The motor output of the fan can be calculated as follows:[1]

$$p_M = \frac{\dot{V}\Delta p_t}{\eta_L \eta_M},$$

where $\dot{V}$ is the volume flow rate (m³/s), Δp_t is the total pressure increase (Pa), η_L is the fan efficiency factor and η_M the motor efficiency factor.

The total pressure increase Δp_t consists of the static pressure difference Δp_{St} and the dynamic pressure p_d occurring at the outlet of the fan. The static pressure difference is the pressure loss in the system (pipe friction, formed parts, aggregates). Table V.1.1 shows various efficiency values of the fan and motor for different volume flow rates.

The efficiency is dependent on the type and size of the fan as well as on the working point. Attention must be paid to the fact that radial fans with backward-curved or forward-curved blades have the highest power consumption when they are blowing free.

DESIGN

General selection criteria

Pressure losses in the system
Specifying a fan requires knowledge of the system characteristic to calculate pressure losses due to ducts, elbows, branches, flaps, collectors, heat exchangers, storages, etc.

In general, the air flow circulating inside the system is determined according to the collector surface area.[2] To determine the flow velocity, consider the following:

High velocity of flow	Low velocity of flow
small duct diameter	low pressure loss
low space requirement	low-noise flow
low mounting costs	low fan output
low system costs	low power costs
but:	but:
high pressure loss	large duct diameter
heavy flow-noise	considerable space requirement
high fan output	expensive mounting
high power costs	expensive systems.

In practice, a compromise must be reached between 'affordable' system costs and associated future operating costs (saved energy and running costs for electricity). The flow velocities compiled in Table V.1.2 may be used as reference values.

Generally, the air flow inside the duct is turbulent. Here, the following equation applies:

$$\bar{c} = 0.8 \ldots 0.9 c_{max},$$

where $\bar{c}$ is the mean flow velocity and c_{max} is the maximum flow velocity.

Pressure losses in straight tubes
For pressure losses inside straight circular pipes the equation is:[1]

$$\Delta p = \lambda \frac{l}{d} \frac{\rho_m}{2} \bar{c}^2,$$

where λ is the pipe coefficient of friction, l is the length of the pipe, d is the pipe diameter, ρ_m is the mean density between the first and last sections of pipe and $\bar{c}$ is the mean flow velocity.

In the case of rectangular ducts, the hydraulic diameter d_H should be used:

$$d_H = \frac{4A}{U},$$

Table V.1.1. Fan and motor efficiency at different volume flow rates

Volume flow rate (m³/h)	Efficiency		
	Fan	Motor	Total
Up to 300	0.40–0.50	0.80	0.32–0.40
300–1000	0.60–0.70	0.80	0.48–0.56
1000–5000	0.70–0.80	0.80	0.56–0.64
5000–100,000	Up to 0.85	0.82	Up to 0.70

Table V.1.2. Compilation of reference values for air flow rates and flow velocities for different system sizes

Application	Volume flow rate (m³/h)	Flow velocity (m/s)
Small-scale systems in housing construction	Up to 500	≤ 4
Medium-sized systems	Up to 2000	< 5
Large-scale systems	Up to 10,000	< 8

where A is the cross-sectional area and U the perimeter.

The pressure difference per unit length is referred to as pressure gradient R:

$$R = \frac{\lambda}{d} \frac{\rho_m}{2} \bar{c}^2.$$

This, for the pressure loss one obtains:

$$\Delta p = Rl.$$

The pressure gradient R is determined by the roughness ε of the internal duct surfaces. In Table V.1.3 some values for ventilating pipes and air ducts are given.

Based on the diagram in Figure V.1.4, pressure losses per unit length for an air flow of $\rho = 1.2$ kg/m³ can be computed for straight tubes with different roughness coefficients according to the flow rate.[1]

Table V.1.3. Roughness values of several ventilating pipes and air ducts[1]

Material	Roughness ε (mm)
PVC piping	0.01
Sheet-metal ducts, folded	0.15
Sheet-metal ducts, smooth	0.50
Concrete ducts, rough	1.00–3.00
Masonry ducts, rough	3.00–5.00
Flexible pipes	0.20–3.00

Pressure loss due to single resistances
A pressure loss due to elbows, branchings, flaps etc. is expressed by the following equation:

$$\Delta p = \xi \frac{\rho_m}{2} \bar{c}^2,$$

where ξ is the friction coefficient. The ξ values have mainly been determined in test series. In Figure V.1.5 various friction coefficients are given.[1]

Pressure loss due to air collectors
Usually, the air flows through air collectors inside small, rectangular air ducts arranged in parallel. Owing to these parallel air flows, the pressure loss is obtained by computing the pressure loss of one single duct according to the above calculation.

Depending on the type of collector and the flow velocity, the values are between 2 and 10 Pa per metre of collector length. Additionally, there is a pressure loss at the collector inlet and at the collector outlet. In Figure V.1.6 pressure losses are given for a collector with a sectional area of flow of 0.05 m²/m and for various pipe connections.[2]

Pressure loss due to rockbed storage
According to Chapter IV.7, the pressure loss caused by a rockbed storage depends on the length of the storage,

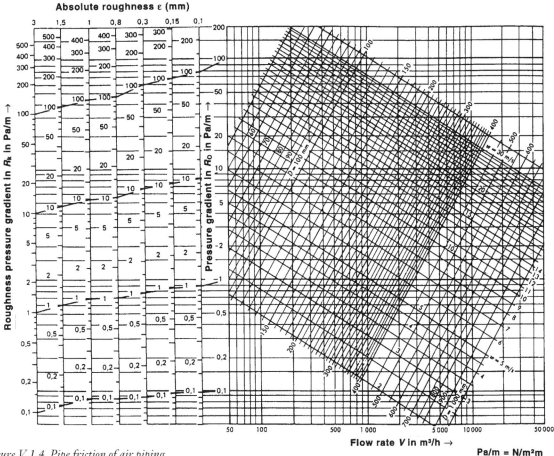

Figure V.1.4. Pipe friction of air piping

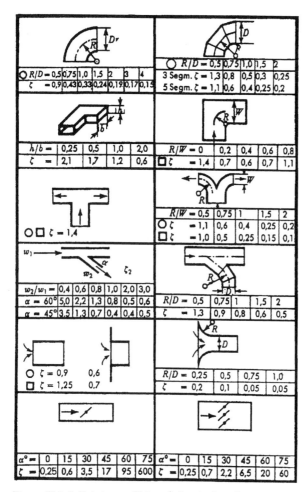

Figure V.1.5. Friction coefficients ξ for single resistances

the size of the pebbles and the air flow velocity. For a rockbed of 1.5 m in length, with an air flow velocity of 0.1 m/s and a pebble size of 20 mm, the pressure loss will be 35 Pa. With a storage length of 1.85 m, the same pebble size of 20 mm and a flow velocity of 0.15 m/s the pressure loss will be 100 Pa.

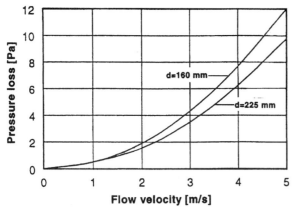

Figure V.1.6. Pressure losses at the collector inlet and the collector outlet for different flow velocities inside the connecting pipes

Table V.1.4. Pressure losses due to devices

Device	Pressure loss (Pa)
Heat exchanger	70–100
Filter, clean	40–60
Filter, polluted	200–300
Silencer	40–80

Pressure loss due to devices
When devices are installed inside the system, further pressure losses will result. In Table V.1.4 some pressure losses are compiled for several of these devices. These values are to be used for approximate calculations only. Precise data have to be taken from the respective product specifications.

Total pressure losses
Once all the pressure losses of the system have been calculated, the system characteristic may be established. It forms the basis for selecting the appropriate ventilating fan. The pressure drops by the square of the velocity:

$$\Delta p = \left(\sum \lambda \frac{l}{d} + \sum \xi \right) \frac{\rho_m}{2} \bar{c}^2.$$

Selecting the appropriate type of fan
The fan may be selected on account of various criteria, such as:[3]

- operating costs;
- sound level;
- space requirement and mounting conditions;
- characteristic profile (steep, flat, steady);
- price.

Selection according to acoustic criteria
The noise emitted by the blower is transferred to the adjoining duct and thus also into adjacent spaces. Part of the noise is emitted into the surrounding space and part of it is transmitted to the floors and ceilings as structure-borne sound. In principle, noise emissions must be kept as low as possible at the sound source itself, i.e low-noise fans and motors are to be preferred. If this is not possible, measures for sound insulation and sound attenuation have to be taken to prevent further sound propagation. There are many factors that have an influence on fan noise, for instance:

- number of blades;
- shape of blades;
- volume flow rate;
- pressure difference;
- circumferential velocity.

The sound power level L_W of fans differing in construction and efficiency can be determined using a comparatively simple formula, namely:[1]

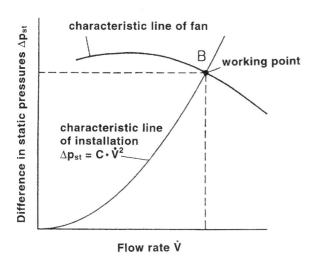

Figure V.1.7. Working point of system and fan

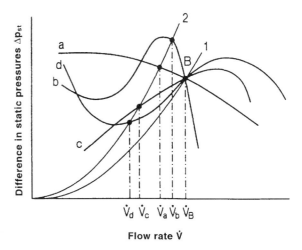

Figure V.1.8. Working point at different fan characteristics: a radial fan with backward-curved blades; b axial fan in steady range; c radial fan with forward-curved blades; d axial fan in non-steady range; 1 designed system characteristic; 2 system characteristic with minor deviation

$$L_W = L_{WS} = 10\log \dot{V} + 20\log \Delta p_t,$$

where L_W is the specific sound power, $L_{WS} = 37 \pm 4$ dB applies approximately for all fans. $\dot{V}$ is the flow rate (m³/s) and Δp_t is the total pressure difference (Pa).

If the fan does not reach its optimum working point, the sound pressure level may well increase by 5 dB. Ventilating fans have an acoustic minimum close to their maximum efficiency. This does not apply for radial fans with forward-curved blades.

Fan operating conditions
As shown in Figure V.1.7, the working point B of a fan is the point of intersection between the system characteristic and the fan characteristic. At this point, there is a balance between the system pressure and the fan feed pressure at a specified flow rate. If the calculated pressure drop is not maintained, the working point will change. The air flow rate decreases with higher resistances, while it increases with lower resistances. If the fan characteristic shows a steep curve close to the calculated working point, there will be only a small change in the air flow conveyed. A flat fan characteristic, however, may signify major differences in the volume flow rate.[4]

In order to guarantee a steady working point to handle minor resistance fluctuations, it is desirable that system and fan characteristics should meet at a rather large angle. As shown in Figure V.1.8, the volume flow rate V is only slightly modified in the case of the declining fan characteristics *a* and *b*, and has a low resistance fluctuation between system characteristics 1 and 2. The ascending fan characteristics *c* and *d*, however, show significant deviations from the desired flow rate $\dot{V}_B$.

If the total pressure increase is to be maintained at a constant level, even with minor changes in resistance, a horizontal fan characteristic should be preferred. A fan with a steep characteristic is recommended, however, if only minimal changes of flow rate are allowed in case of an overall pressure fluctuation.

Parallel operation
Within one system it is possible to operate an arbitrary number of fans in parallel.[5] Parallel operation of several fans is generally used if very high volume flow rates are required relative to the pressure increase, or if the volume flow rate is to be controlled by switching a fan on or off. With this kind of flow-rate control, however, it should be noted that the working point with parallel operation is different than with only one fan. The characteristic line results from the addition of the volume flow

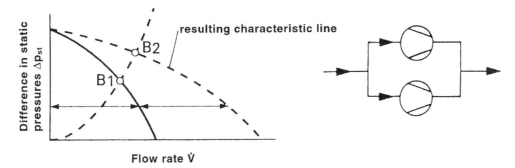

Figure V.1.9. Resulting characteristic for two fans operated in parallel

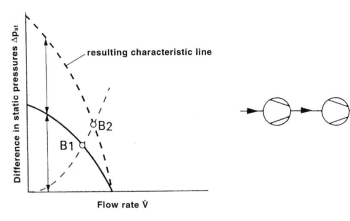

Figure V.1.10. Resulting characteristic for two equal fans operated in series

rates at same delivery pressure, as shown in Figure V.1.9. The figure shows that the volume flow rate of the two fans operated in parallel is smaller than the added volume flow rates of the two fans operating separately.

Series connection
In order to increase the pressure, two or more fans can be connected in series. The characteristic line theoretically results in the addition of the total pressures at the same volume flow rate, as shown in Figure V.1.10. In practice, however, the achieved pressure increases are smaller as a result of the friction losses occurring between the fans and because the air does not flow optimally to the second fan.

In most systems the fans are replaced by several smaller fans, because of the lack of space for large fans. This is disadvantageous, as smaller fans are less efficient than larger ones and have a higher power consumption.

Fan control
The air flow or the required pressure increase is adjusted to the respective requirements by way of different control mechanisms, such as:

- throttling;
- change of rotational speed;
- parallel operation;
- series connection;
- on/off control;
- variable-inlet guide vanes;
- change of rotor blades (axial fans)
- bypass control.

For solar air systems the first two control strategies are generally the most important.

Throttling
Using a throttling flap or a damper is the easiest and least expensive way to vary the volume flow rate. Therefore, this is the control used most often, even though it has the highest energy consumption. The fan can be throttled on its suck or pressure side. The throttling causes a pressure loss in the throttling device, which is converted to heat and noise. This leads to a high loss of energy.

Control by changing the speed of rotation
The control having the lowest losses is the speed control. If the speed of rotation of a fan is changed, the throttling curve changes according to the relations:

$$\dot{V} \propto n$$
$$\Delta p_t \propto n^2$$
$$P_L \propto n^3,$$

where $\dot{V}$ is the volume flow rate, Δp_t is the total pressure increase, P_L is the impeller power and n is the speed of rotation.

The last relation given above is relatively inaccurate as it does not take into account the dependence of the flow efficiencies and the variability of the mechanical efficiency.

The most important electronic speed controls are:

- phase-angle control;
- voltage control;
- frequency converter.

For fans conveying volume flow rates of up to 500 m³/h, speed control via phase-angle control is the most economic way.[6] With phase-angle control the fan is switched on only for a limited time during a running period. The compact phase-angle control module can be used in front of any standard fan.

Voltage control is usually done with an adjustable transformer. Transformers, however, are considerably larger than phase-angle control modules.

Frequency converters present another way to control the speed of rotation. However, they are more expensive than the other two control strategies and are mainly used with larger fans. An advantage of the speed control is the significant noise reduction. The sound level can be reduced down to:

$$\Delta L = 50 \log\left(\frac{n}{n_0}\right) \text{ (dB)},$$

where n_0 is the rated speed.

Fans with EC motors
Fans with single-phase AC motors work most efficiently at their rated speed. The speed control reduces the efficiency of the fan. A significant advancement in the development of fans is fans with EC motors (EC = Electronic Commutation).

The EC motor is comparable to the DC shunt motor, except that the magnetic field is produced by permanent magnets in the rotor. The commutation is done electronically and therefore without wear. Depending on the layout and the applications, EC motors can be connected and operated from a DC power supply or – via a separate rectification unit – from an AC mains supply. The electronics of the EC motor can be designed to offer certain additional functions, such as speed control, closed-loop speed control and alarm functions. EC motors have a high efficiency and are maintenance-free.

Power supply from solar cells
Photovoltaic (PV) cells offer an ideal energy supply for fans. The electricity produced by the solar cells is not only supplied at the same time but also increases automatically with an increasing demand. As solar cells produce DC, the choice of DC motors is obvious. For solar air systems, PV modules with a voltage of 12 V, a power of 50 W_p and a size of 0.5 m² have proved to be suitable; this corresponds to a maximum power at an insulation of 1000 W/m² and a solar cell temperature of 25°C. This power, however, can only be supplied to the consumer if the solar module is operated at the point of maximum power (MPP). In Figure V.1.11 these points are depicted for different radiation intensities.

The power decreases proportionally to the radiation intensity. It also decreases with increasing module temperature, namely by about 0.5% per K.

Connecting n solar modules, voltage sources, in series produces n times the voltage of one module. By connecting n solar modules in parallel, a current source that has n times the current strength of one module, can be produced. By connecting several modules in parallel, each with modules connected in series, solar generators of any size can be produced (see Figure V.1.12).

The size of the solar generator required by the solar air system results from the formula:

$$p_M = \frac{\dot{V}\Delta p_t}{\eta_L \eta_M},$$

where the variables are as defined previously.

A fan connected to a solar generator dimensioned from this equation supplies the desired air volume flow rate $\dot{V}_0$ only with a solar radiation I_0 of 1000 W/m². With a small solar radiation I_N, the air volume flow rate $\dot{V}_N$ decreases approximately according to the following formula:

$$\dot{V}_N = \dot{V}_0 \sqrt[3]{\frac{I_n}{I_0}},$$

where $\dot{V}_0$ is the air volume flow rate at a solar radiation of 1000 W/m², I_0 is the solar radiation of 1000 W/m², $\dot{V}_N$ is the air volume flow rate with arbitrary solar radiation and I_N arbitrary solar radiation. In general, the smaller air volume flow rate occurring with decreased solar radiation is not of importance, and most

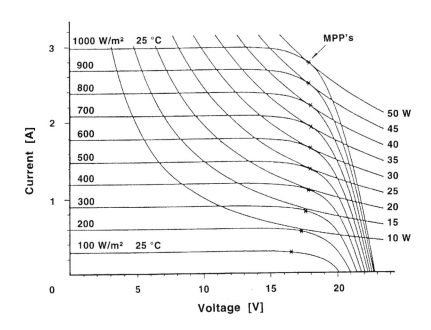

Figure V.1.11. Characteristic lines of a solar module at different insulations and a solar cell temperature of 25°C

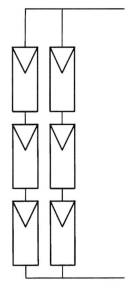

Figure V.1.12. Solar generator consisting of six solar modules. Three modules per run are connected in series and two runs are connected in parallel

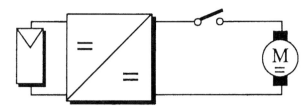

Figure V.1.13. Solar power supply of a DC motor with matching transformer ensuring operation near the maximum power point[7]

often it is even desirable, because the air has more time in the collector to be heated. This leads to a higher efficiency of the overall system.

Between solar generator and fan a matching transformer has to be installed, as shown in Figure V.1.13.

The matching transformer is set such that the power of the solar generator at constant input voltage, i.e. near the maximum power point (MPP), is adapted to the resistance of the fan motor, where a power-dependent output voltage occurs.

Air collectors with integrated solar modules are available, but the solar cells may also be installed outside the collectors. Integrating them near the air inlet of the air collector has the advantage that the solar cells are cooled and, at the same time, the energy thus gained can be supplied to the air flow. The example below demonstrates how to calculate the power required by a solar module of a small solar air system.

Example

collector area:	5 m²
volume air flow rate/collector area:	60 m³/m²h
assumed fan power:	0.3 W per m³/h

$$p_m = 5\,\text{m}^2 \times 60 \frac{\text{m}^2}{\text{m}^2\text{h}} \times 0.3 \frac{\text{Wh}}{\text{m}^3} = 90\,\text{W}.$$

The required fan power of 90 W can be produced by two 12 V solar modules, each having a power of 50 W_p. By connecting the two modules in parallel, a voltage of 12 V is produced at the inlet of the matching transformer, together with a power of 100 W, which is sufficient to operate a DC fan of 12 V and 100 W. The area required for the solar module is 1 m². For smaller systems of up to about 15 m² the 50 W_p modules can be connected in parallel and thus operate a 12 V fan. For larger systems, the modules have to be connected in series. The voltage produced then amounts to a multiple of 12 V.

Special selection criteria

For solar air systems different types of fans can be used. The selection of the fan mainly depends on the volume flow rate $\dot{V}$ and the pressure difference Δp in the installation system. Further selection criteria are investment costs, running-power costs, required installation space, sound level, etc. Table V.1.5 gives an overview of fan types that are suitable for solar air systems. In addition, the volume flow rates and pressures, as well as the efficiencies and the advantages and disadvantages are depicted.

In general, both types of radial fan with forward- (trummel rotor) and with backward-curved rotor blades are suitable. Radial fans with forward-curved rotor blades are recommended for volume flow rates of up to 1000 m³/h. For higher values radial fans with backward-curved rotor blades should be used; these are about 30% more expensive but have a significantly higher efficiency, although the fan noise is slightly higher. Both types of fans are available as internal rotors and with a flange-mounted motor. The motor of the internal rotor is situated within the air flow, so that its waste heat can be used.

Inline circular duct fans can be used for volume flow rates of up to 1500 m³/h and for maximum pressure differences of up to 400 Pa, i.e. only for smaller systems. This type of fan also requires less space and installation work than the other types described above. The waste heat of the motor can be used.

Inline rectangular duct fans have the same properties as inline circular duct fans but transport a larger volume flow of up to 10,000 m³/h.

Axial fans are only suitable for smaller systems, because they can only overcome small pressure differences.

The fan types described can be operated with either AC or DC. Within the power spectrum of less than 1 kW, DC motors generally achieve a significantly better efficiency than AC motors.[7] Whether the motor is commutated electronically or mechanically is not important. However, electronically commutated motors (EC motors) are maintenance-free as there are no brush contacts and collectors to wear out. The use of DC motors is recommended for systems where the power to operate the fan is generated by solar cells.

CONSTRUCTION

Calculation example for fan dimensioning

This section illustrates the dimensioning of a fan to be used in a solar air system and is based on the project 'Wannseebahn Rowhouse in Berlin'. Figure V.1.14 shows a diagram of the solar air system. The murocaust wall, which actually extends over two storeys, is presented in the diagram only on one storey to simplify the diagram.

Data

The data for the example are given in Table V.1.6.

Figure V.1.15 shows the system, including the components that are relevant for the calculation of pressure losses.

Table V.1.5. *Range of use of the different fan types suitable for solar air systems*

Type of fan	Air flow rate (m^3/h)	Pressure range (Pa)	Efficiency (%)	Advantages	Disadvantages
Radial fan with forward-curved rotor blades	50–50,000	100–500	40–60	Low investment costs Suitable for lower pressures	Higher power consumption
Radial fan with backward-curved rotor blades	1000–50,000	200–1500	80–85	Lower power consumption	Higher investment costs Not usable at lower pressures
Inline circular duct fan	150–1500	50–400	< 40	Easy to install Waste heat of motor can be used Only little space required	Higher power consumption
Axial fan	200–10,000	10–400	< 50	Easy to install Waste heat of motor can be used Only little space required	Higher power consumption Only usable at lower pressures

Table V.1.6. *Data for the Wannseebahn rowhouse example*

Air collectors	16 m^2, three rows in parallel, each with three collectors connected in series
Air flow rate	600 m^3/h
Hypocaust area	12.4 m^2
Connecting ducts between collector and hypocaust wall	200 mm diameter

Table V.1.7 shows the pressure losses of the various components calculated by means of the equations given in the section General selection criteria. The total pressure loss amounts to 330 Pa.

The pressure distribution depicted in Figure V.1.16 shows that the murocaust is operated at an overpressure, whereas the air collector is operated at an underpresssure. The large pressure loss in the duct section E–F is caused by the branches and elbows.

If the total pressure losses are plotted as a function of the air volume flow rate $\dot V$, the characteristic line depicted in Figure V.1.17 results. It is the task of the planner to select a fan such that the characteristic line of the fan crosses that of the system at the desired point. The characteristic line of the fan selected for this solar air system meets these requirements perfectly. The

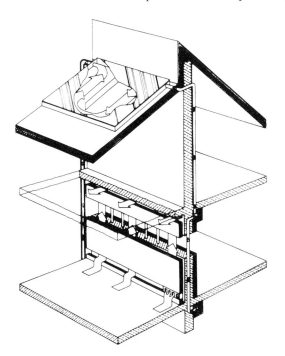

Figure V.1.14. *Diagram of the solar air system applied in the Wannseebahn rowhouse in Berlin*

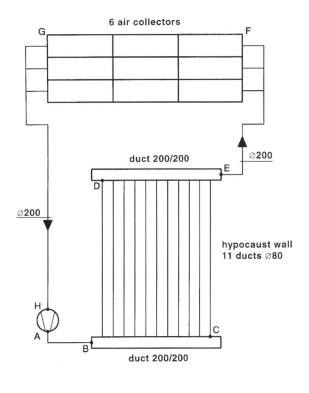

Figure V.1.15. *Diagram of the system, showing the components relevant for the pressure loss calculation*

Table V.1.7. Pressure losses and pressure distributions in the various systems components

Connection/point	Resistance	Pressure (Pa)	Pressure distribution (Pa) $\left(\sum \lambda \frac{l}{d} + \sum \xi\right)\frac{\rho_m}{2}\bar{c}^2$
A	–	–	165
A–B	Elbows; duct distance	18	
B			147
B–C	Rectangular pipe duct, branches	22	
C			125
C–D	Hypocaust wall	38	
D			87
D–E	Rectangular pipe duct, branches	22	
E			65
E–F	Elbows; duct distance	107	
F			–42
F–G	Collector	32	
G			–73
G–H	Elbows; duct distance	92	
H			–165
Total pressure loss		330	–

transported volume flow rate amounts to 600 m³/h and the pressure increase caused by the fan comes to 330 Pa.

Recommendations for fan installation

The position of the fan in the pipework

Every time a solar air system is planned, the question where to install the fan arises. For System 1, where the outdoor air is preheated in the collector before being supplied to the living space, the only position to install the fan is between collector and living space. With the other systems the fan can be installed such that either positive or negative pressure occurs in the collector. Although an air flow is created with both installation variants, there are several advantages in installing the fan such that a negative pressure occurs in the collector:

- Leakages in the collector component do not lead to heat losses.
- If leakages occur in the closed air system, the air in the air flow cannot absorb moisture because the outdoor air, which possibly continues to flow into the collector, has a lower absolute air humidity than the indoor air.
- If the fan is installed on the pressure side of the collector, in most collector connecting branches the air passes through the sheet bar of the collector, which leads to a flow resistance that is seven times as high as the rest of the system. This is why the total pressure loss is much smaller if the fan is installed at the extraction side of the collector.

If space is available, it is thus advantageous for all solar air systems for the fan to be installed such that the

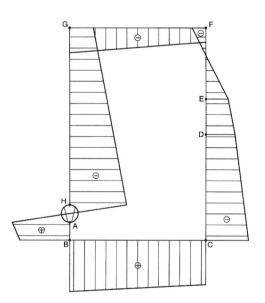

Figure V.1.16. Pressure distribution in the system

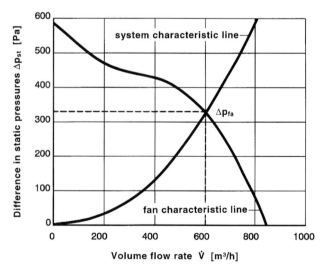

Figure V.1.17. Characteristic lines of the system and the fan of the solar air system applied in the Wannseebahn rowhouse, Berlin

collector is under negative pressure. It is not recommended that several smaller fans are connected in series, because smaller fans are generally less efficient.

The position of the fan motor
The fan motor should be installed such that it is located within the air flow; this allows use of the waste heat of the motor. In general, fan motors should only be continuously operated with an ambient temperature of up to 40°C, according to fan product information sheets. During the operation of a solar air system, however, air temperatures of up to 60°C may occur, even though only for short periods. In practice, it became evident that these generally short periods do not damage the fan motors. Moreover, there are fans available on the market that are equipped with thermal circuit breakers that are connected to the motor winding and switch off the motor if the temperature becomes too high. The motor switches on automatically when the temperature drops. If requested by the client, the fans can be provided with a better insulation, which allows them to operate continuously at an ambient temperature of up to 80°C. Permissible ambient temperatures should be discussed with the fan manufacturer during the design phase.

With the often applied radial fans, where the air flow is reversed by 90°, standard motors are generally located outside the air flow. It is not therefore possible to use the waste heat of this type of fan.

REFERENCES

1. Recknagel H, Sprenger E, Schramek E-R (1997/98) *Taschenbuch für Heizung und Klimatechnik*. Oldenburg-Verlag, Munich and Vienna.
2. Planning documents Firma Grammer, Amberg.
3. (1988)*Arbeitskreis der Dozenten für Klimatechnik: Handbuch der Klimatechnik*. Vol.3. *Bauelemente*. C. F. Müller-Verlag, Karlsruhe.
4. Schlender F, Klingenberg G (1996) *Ventilatoren im Einsatz: Anwendungen in Geräten und Anlagen*. VDI-Verlag, Düsseldorf.
5. Bohl W (1983) *Ventilatoren. Berechnung, Konstruktion, Versuch, Betrieb*. Vogel-Buchverlag, Würzburg.
6. Linder H, Brauer H, Lehmann C (1989) *Taschenbuch der Elektrotechnik und Elektronik*. Verlag Harri Deutsch, Thun and Frankfurt/Main.
7. Ladener H (1996) *Solare Stromversorgung. Grundlagen, Planung, Anwendung*. Ökobuch Verlag, Staufen bei Freiburg.

Chapter author: Johann Reiss

V.2 Air-to-water heat exchangers

INTRODUCTION

Description

An air-to-water heat exchanger is an arrangement of surfaces separating an air stream and a water stream in such a way that it maximizes heat transfer between the two streams. The most common air-to-water heat exchanger is the fin-tube heat exchanger, which is placed directly in the air ductwork and on the water side is connected to the pipes of the hydraulic system, see Figure V.2.1. Fins on the outside of the tubes in which the water runs increase the heat transfer from the air, thereby improving the performance.

Principles

Heat exchangers can be characterized as counter-flow, parallel-flow or cross-flow types. Optimum heat transfer is obtained with the counter-flow heat exchanger. The air-to-water heat exchanger is typically designed as a number of connected cross-flow heat exchangers. When these are connected in series a counter-flow effect is obtained; when they are connected in parallel cross-flow characterization is obtained.

Variations

The fin-tube heat exchanger can be varied in size, the dimensions and distance apart of the fins and combinations of construction materials. The tubes and the fins can be made of copper, aluminium, steel and different alloys of these metals. Even plastic can be used.[1] Copper tubes and aluminium fins are the standard type. The materials have to be carefully selected to avoid corrosion and galvanic action in the hydraulic system. Using large gaps between the fin tubes reduces the pressure drop if this is a problem.

Applications

Air-to-water heat exchangers can be used for systems especially designed for water heating (System 6 in this book) or for domestic hot-water heating by air systems designed mainly for heating. Making use of surplus heat occurring during the summer greatly improves the economics of a system.

Characteristics

The product of the overall heat transfer coefficient U and the heat exchanger surface A determines the overall heat transfer capability of the heat exchanger. If the product UA were infinite and the capacity flows on both sides equal, the outlet temperature of the water would equal the inlet temperature of the air.

The thermal ratio of effectiveness ε represents the performance of the actual heat exchanger compared to the ideal of 1.0. The thermal ratio of effectiveness is defined as the relationship between the temperature rise of the water and the maximum possible rise, from water inlet to air inlet temperature:

$$\varepsilon = \frac{(T_{\text{water, outlet}} - T_{\text{water, inlet}})(\dot{m}C)_{\text{water}}}{(T_{\text{air, inlet}} - T_{\text{water, inlet}})(\dot{m}C)_{\text{min}}} \cdot \quad (V.2.1)$$

The efficiency of heat exchangers is often expressed as a function of NTU (Number of heat Transfer Units). For an ideal counter-flow heat exchanger the expression is:

$$\varepsilon = \frac{\text{NTU}}{\text{NTU}+1}, \quad (V.2.2)$$

where

$$\text{NTU} = \frac{UA}{(\dot{m}C)_{\text{min}}} \quad (V.2.3)$$

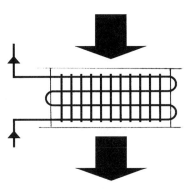

Figure V.2.1. Typical air-to-water heat exchanger

and $(\dot{m}C)_{min}$ is the smaller of the two heat capacity flows.

As the heat transfer coefficient depends on the air velocity along the surface, then both the numerator and denominator of equation (V.2.3) depend on $(\dot{m}C)_{min}$ if the air capacity flow is the smaller of the two flows, which it generally is. Therefore the capacity flow on the air side is the critical and governing parameter for the effectiveness of the heat exchanger.

The heat transfer rate across the heat exchanger is given in simple terms as:

$$\dot{Q} = UA\Delta T, \qquad (V.2.4)$$

which applies when the two capacity flows are identical. The reader is referred to textbooks for further information.[2]

Expressed in terms of capacity flows and temperature rises, the heat transfer is:

$$(\dot{m}C)_{air}(T_{air,\,inlet} - T_{air,\,outlet}) = (\dot{m}C)_{water}(T_{water,\,outlet} - T_{water,\,inlet}). \qquad (V.2.5)$$

DESIGN (SIZING, SELECTION)

The parameters to be considered when designing an air-to-water heat exchanger are:

- effectiveness, ε;
- pressure drop, ΔP;
- cost.

Design procedure

Often an iterative process is required because the optimum solution varies from case to case. If domestic hot-water heating (DHW) is an additional use of a large heating system (see Chapter III.1), the solar collector output is high compared to the DHW load. Accordingly the heat exchanger does not need to be very efficient and should be designed for a low pressure drop and low cost. If, on the other hand, the collector output is generally smaller than the DHW load, the heat exchanger should be more efficient. At the same time, a higher pressure drop and higher cost can be accepted. The right heat exchanger is found in an optimization process involving efficiency, pressure drop and cost. It is also important to identify which temperature levels can be reached with a reasonable efficiency for a given collector and which temperature levels are at all useful for the application.

When using manufacturers' catalogues, be aware that the heat exchangers have been designed for different operating conditions (typically cooling systems) than occur in solar air heating systems. This leads to a tendency to select heat exchangers that are too small. The manufacturers' guidelines can, therefore, not be used directly for the selection or optimization of heat exchangers for solar air heating systems.

A manufacturer uses computer programs to design a heat exchanger for a particular application. The following specifications for a typical operational situation are required:

- on the air side, at least three of the following:
  - flow rate;
  - inlet temperature and relative humidity;
  - outlet temperature and relative humidity;
  - yield;
- on the liquid side, at least two out of the following:
  - water flow rates;
  - inlet temperature;
  - outlet temperature.

On the basis of the information available, the manufacturer suggests a design and quotes a price. The consultant now has to decide whether the temperature efficiency, the pressure drop and the cost are acceptable. If not, the iteration process begins.

The temperature efficiency can be inserted in equation (V.2.1) to produce a graph like that in Figure V.2.2, which can be used to test whether the obtained water temperatures match the application.

Hydraulic aspects

An equal distribution of the air flow is required to ensure optimal performance of the heat exchanger. If the heat exchanger is placed in the duct system immediately after the fan, there is a risk that the air flow will be a jet of air that is not evenly distributed over the surface of the heat exchanger. Thus, a location before the fan is generally recommended. An exception would be if the duct between the fan and the heat exchanger bends. Also, funnels before and after the heat exchanger improve the uniformity of the air flow.

The pressure drop over the heat exchanger is often the single largest pressure drop within the air solar system. The efficiency of an air-to-water heat exchanger in-

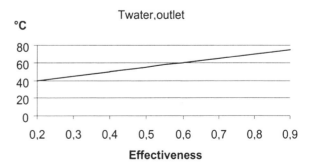

Figure V.2.2. The temperature of the water leaving the heat exchanger as a function of the heat exchanger effectiveness; $T_{air,\,inlet} = 80°C$, $T_{water,\,inlet} = 30°C$

Figure V.2.3. A conventional air-to-water heat exchanger

creases with the number of rows of tubes and with the total area of the fins. However, this also increases the pressure drop across the air side of the heat exchanger. As the overall performance of a solar air system depends on the electricity used for the fan, it is worth while to investigate different heat-exchanger configurations to find the one with the least pressure drop. A reduction of the pressure drop is achieved by enlarging the gap between the fins.

Generally the efficiency of the heat exchanger in the summer operation of a large system is not critical, but the pressure drop is.

CONSTRUCTION

Figure V.2.3 shows a photograph of a conventional air-to-water heat-exchanger.

Location, installation, weight and size

Location and installation
The installation of an air-to-water heat exchanger requires additional space in the standard ductwork. Funnels and duct bends require extra space.

Materials
Beware of corrosion risks and choose appropriate materials with the manufacturer.

Maintenance
- *Freeze protection*. The heat exchanger will be damaged if the liquid in the tubes freezes. Two methods for freeze protection are:
  - Preventing cold air from back-thermosiphoning into the heat exchanger. It is not advisable to trust a simple backward gravity damper. A motorized damper with spring return can be used in combination with a freeze-protection thermostat that switches on the secondary (liquid-side) circuit pump to heat the heat exchanger momentarily.
  - The use of a freeze-protection liquid (e.g. glycol). This has to be a closed circuit, which involves the use of a secondary heat exchanger. However, if a leak occurs in the circuit between the two heat exchangers, a limited volume of liquid is spilt. A leakage in an open circuit that contains tap water can flood a building.
- *Service requirements*. The lamellae (fins) must be accessible for cleaning, particularly in open systems. Increasing the distances between fins eases cleaning. A dust deposit of 1 mm decreases the heat transfer between the air and the lamellae by as much as 50%.

Rules of thumb
The capacity flow rate on the two sides of the heat exchanger should be of the same order of magnitude. The liquid capacity flow could be chosen to be up to twice the air capacity flow in order to keep the temperature rise to a minimum.

Check list
- If varying fan speeds are used, calculate the efficiency and the pressure drop at varying air speeds.
- Locate the heat exchanger close to the collector to minimize pressure drops and heat losses in the duct system.
- Remember to freeze-protect the heat exchanger.

COMPONENTS

Components are available from the following suppliers:

- Ttcoil as, Blokken 17, Dk 3460 Birkerød, Denmark. Tel. +45 45 82 58 88; Fax +45 45 82 54 44;
- GEA Luftkühler GmbH, Südstrasse 48, D-4690 Herne 2, Germany. Tel. +49 2325 469 150; Fax +49 2325 469 127.

REFERENCES

1. Jensen SØ (1991) *Luft-til-vand varmevekslere til tagrumssolfangere*. Report No. 218, Institute for Energy and Buildings, Technical University of Denmark, 2800 Lyngby, Denmark.
2. Bayley FJ, Owen JM, Turner AB (1972) *Heat Transfer*. Thomas Nelson and Sons, London.

Chapter author: Ove Mørck

V.3 Controls

INTRODUCTION

The control system is the communication device between the three main components of a solar air system:

- the collector;
- the air transport system;
- the storage/release system.

These must interface correctly to obtain the maximum solar contribution.

Experience with large domestic hot water systems has shown that the choice of a controller, and its installation and adjustment, can cause differences in gained energy of several hundred kWh/m^{2}a of collector area. The control system is therefore a very important component.

Description

Principles

The control of a system is a process with one or more input parameters that influence other parameters according to the relevant system rules. In a solar system the controller senses the temperatures of components as input, compares these values with each other or with set values and accordingly generates output values. Values influencing the status of a component such as the solar radiation or the time of day can also be used as input signals.

Open- and closed-loop controllers have to be distinguished. A closed-loop controller senses not only the input values but also the system that they control. Based on the input values the open-loop controller can switch on/off the fans, dampers etc. The closed-loop controller is able to modify the speed of a fan, or adjust a damper to another position to keep a temperature constant. Controllers available for this purpose are:

- open-loop controller:
  - differential thermostats;
  - thermostats;
  - solar-cell switches;
  - microprocessor controls;

- closed-loop controller:
  - analogue controllers;
  - 'intelligent devices';
  - microprocessor controls.

A self-regulating system is a fan with a DC motor driven by a solar PV panel. The air flow is then directly influenced by the solar intensity and only an on/off control is necessary (depending on the minimum or maximum set temperature).

Description and application of devices

Thermostats and differential thermostats
Differential thermostats are the key component in control of active solar heating systems. They sense the temperatures in two components, compare them and, depending on the temperature difference, a relay is switched to control the operation of a fan or damper.

The temperature difference for the operation has to be adapted to the system. To avoid too much switching, a hysteresis or dead band for switching off is used. Instead of hysteresis, a delay device can be used.

For systems such as space heating systems, which always work with the same set temperature, a thermostat is used. In this case the sensed temperature is compared to a fixed value, which is called the set temperature or set point.

Compared to mechanical thermostats, electronic devices have the following advantages:

- set points and dead band adjustable (by the user);
- higher accuracy;
- no limitations for the location of the thermostat;
- reliability.

As a result, mechanical thermostats such as capillary tubes (remote bulbs), bimetal switches (snap discs) or two-stage thermostats are no longer recommended. Thermostats and differential thermostats use a simple electronic logic circuit (a compactor) and thermistor or resistance temperature sensors. Thermistors are electrical resistors that change their resistance with temperature and have negative temperature coefficients (NTC-sensors), while resistance temperature sensors have positive temperature coefficients. Thermostats are specified by the temperature range in which they operate, their temperature dead-band or differential, whether they function as a heating or cooling thermostat, or both, and the amount of power they can control directly.

Instead of thermistors, solar cells can be used to switch a system. In this case a threshold value of insulation has to be evaluated by calculating system or component efficiency. Thermostats have the disadvantage of a fixed set point. Because the set point has to be chosen for all operating conditions the ON temperature or ON irradiation has to be higher. This leads to lower performance. However, thermostats and solar-cell switches are useful in ventilation systems.

Manual on/off switches
A manual switch is necessary in a control system to be able to override all the other controls and turn the power off directly. Some differential thermostats already include manual switches. A switch should be readily accessible so that it is easy to turn off the system, when required.

Timers
Timers can be used as switches in solar air systems when there is a time-dependent energy demand, as in ventilation systems. They may also be combined with other controllers, such as a thermostat. Timers in most systems are necessary to control the backup heating.

Delay devices
A device is used to delay the operation of a fan in order to guarantee that the motor-driven dampers are open when the fan starts. In the case of long ducts with a large capacity, the delay also allows heat stored in the mass of the system to be recovered by delaying the shutdown.

Analogue controllers
Controllers with analogue output (closed-loop) are used in more sophisticated systems, for example systems with integrated backup heating or rockbed storage. They operate according to the difference between the set temperature and the measured temperature. They are used to keep a temperature or another value constant, for example to obtain a uniform outlet temperature from the collector; in this case the speed of the fan is changed by the controller. Additional devices for this purpose are necessary, such as a frequency converter for fans or power electronics to regulate the energy flux. In systems with an integrated rockbed, for example, a constant inlet temperature will improve the storage performance.

Microprocessor controls
In more complex systems, such as systems with collectors of different orientations, storage of different sizes, and heating priorities, a simple thermostat is not sufficient. The control system is then set up by connecting different thermostats, which is done using electric relays to build a logical network for switching on or off. Multiple thermostats increase the costs of the wiring and of the switchboard. Microprocessor control is a more elegant and economical solution. If a collector has to heat different storages to different temperature levels, the controller needs to determine which storage to heat first. The control system therefore has to calculate whether the collector is able to reach the temperature level of the system with the highest priority. This cannot be done by connecting thermostats.

Another advantage of a microprocessor controller is that the control system can be programmed, which means that changes in the solar or backup heating system can be controlled with the hardware without changing the wiring. Other features such as a timer or date-dependent schedules can be easily programmed. These controllers, called DDC (direct digital control), also have additional advantages, e.g. for calculation. They can calculate energy gained, efficiency etc. and values can be monitored. The features of analogous outputs are integrated by software. Another advantage is the possibility to display and to record temperatures and the status of each component, so these systems can be used for monitoring system performance. Open- and closed-loop controllers are available.

Microprocessor controllers are the only controllers that can include multiple heating sources, maximizing the solar use. They can be programmed to handle new sophisticated control concepts that use predictive, adaptive or stochastic control strategies.

Adaptive controllers are available that allow heating systems to find the optimum start time for heating up in the morning and the appropriate control parameters. However, these controllers have been developed for this special purpose and cannot be used in solar systems. Adaptive control could be used to optimize a whole system with the different efficiencies and losses taken into account. However, for small systems the additional costs would not be covered by the added gains.

Predictive controllers integrate a weather forecast into their control strategies. Only one system at the moment is available for floor heating systems.[1] The big advantage of these intelligent systems is that they can combine complex correlations of stochastic weather and the energy demand of a building, considering efficiencies and overall performance. For thermal storage with radiant discharge, these controllers might improve efficiency by avoiding overheating from the auxiliary heating and increasing the useful solar gains. The problem here is that each system would need to be customized and the costs would increase.

Sensors
The accuracy of a controller depends on the sensors providing input, their sensitivity and linearity, their reaction time and their placement in the system. The sensitivity of a sensor depends on the materials used, which change their electrical resistance with temperature. The controller measures this resistance and can convert it to a temperature.

DESIGN – SELECTION OF A CONTROL DEVICE

General selection criteria

The selection of a control system can be made following a cost/benefit analysis. With sophisticated microprocessor control any system can be controlled, but in most cases a simple differential thermostat gives the same solar contribution. Apart from the cost of a microprocessor system, it is more error-prone owing to its programming and installation complexity. The accuracy of sophisticated simulation has shown that, when a collector is sunlit, for each degree a thermostat switches too late, the performance decreases by 2–3%. Electrical thermostats can improve system performance by 10–20% over mechanical thermostatically controlled systems.

Steps to select a control device

1. Analysis of the system:
   - number of components having different thermal status, such as collectors with different orientations, storages etc.;
   - modes of operation, such as summer/winter operation, dhw mode, charge, discharge etc.;
   - additional requirements, such as a timer, schedules etc.;
   - integration of backup control.
2. Defining of rules of operation:
   - What should happen? When it should happen? How long should it happen?
     - for each operation mode;
     - for each component.

For each rule a switch is necessary. This can be a thermostat, a controller, a timer, a relay etc. The types of rules determine the type of the control device needed, as shown in Table V.3.1.

Additional considerations are:

- accuracy (< 1 K);
- usable sensors (NTC, Pt-100, Pt-1000, semiconductors);
- range of temperatures (0 to 80°C);
- range of adjustable ΔT (1 to 15 K);
- range of adjustable hysteresis (1 to 10 K);
- ambient conditions for installation (–5 to +40°C; relative humidity 0 to 90%);
- maximum power that can be switched (300 VA); additional features like:
  - display;
  - maximum/minimum switch;
  - freeze protection;
  - electric consumption;
  - error messages for, for example, broken sensors;
- costs.

Table V.3.1. Control devices needed for each type of rule

Rule	Device
ON/OFF without conditions	Manual switch
Condition IF > or IF NOT <	Thermostat or differential thermostat
Logical condition AND or NOT	Connecting input and output of two control devices (relay)
Logical condition OR	A component can be switched by different control devices (relay)
Components with priority	Microprocessor control
Schedules	Timer
Constant conditions $T = T_{set}$	Controller with analogue output

APPLICATION

Control parameters and environmental impact

Differential thermostats are the most common controllers. The adjustment of the ΔT and the hysteresis are critical.

The minimum ΔT can be calculated by comparing the energy flux from the system with the power of all the components in the system being switched on for that operation. To achieve a real reduction of the CO_2, the CO_2 emissions to generate the electricity consumed by the fans and controllers of the solar system must be less than the CO_2 emission of fossil fuel equivalent to the solar savings. The CO_2 emissions from producing electricity are much higher than for heat production. A typical factor for electricity generation is three (one unit of electricity requires three units of primary energy). The CO_2 values of different processes can be taken from GEMIS:[2]

$$\Delta T_1 = (P/mc_p) \times CO_2 \text{ factor,}$$

where P is the power, m the mass flow and c_p the specific heat of the fluid. To the minimum ΔT_1, the temperature losses in the ducts have to be added if the losses are to the ambient. Additionally, the hysteresis, which is useful in achieving operational stability, has to be added. Typically, the hysteresis should be smaller than $\Delta T_1/2$. If a delay is used, the hysteresis is 0. For the calculation of delay times see Chapter II.6.

Selection criteria for sensors

The choice of the sensor therefore depends on the chosen controller. However, different sensors having the same resistance but different shapes, sizes and lengths can be chosen. The requirements for sensors are:

- The capacity (size) should be as low as possible. The mass of the sensor has to be heated or cooled and delays the system.

- The shape of the sensor should fit the place of installation so as to have a high heat transfer from the fluid to the sensor.
- The resistance of the wire has to be taken into account or has to be compensated if the resistance of the sensor is low (platinum sensors of 100 ohms – Pt-100 – for example) and the wires to the sensors compared if they are not of the same length.
- Sensors made of semiconductor materials are sensitive to overvoltage (lightning) and have to be protected.
- The wires to the sensors are often exposed to solar radiation (UV) and to high temperatures (collector stagnation temperature). The insulating material of the wire must withstand this harsh environment.

Placement of temperature sensors

The placement of the sensors is important for the performance of the system in sensing the relevant temperature:

- In air collectors the sensor should be placed close to the outlet in the air flow, but should not touch the absorber.
- In air channels an inhomogeneous temperature distribution often occurs, so the temperature sensor should be placed at the position of the mean temperature.
- Sensors should not be placed in areas of reverse flow (in bends of ducts).
- Irradiation sensors should be placed close to the collector and have the same tilt and azimuth angle.
- Sensors in storages like hypocausts or rockbeds should sense the air outlet temperature, but still be installed inside the storage to sense the right temperature when the system is switched off. An additional sensor for maximum temperature is helpful in the case of hypocausts to avoid overheating. This should be mounted in the mass of the hypocaust.
- Light-mass and fast-responding murocausts or hypocausts can be controlled by the room temperature. In the case of heavy-mass hypocausts overheating may occur.
- Sensors in double envelope systems are often placed in the middle of the double envelope (half height).

The placement of sensors often can be optimized by dynamic simulation.

Accuracy of sensors

For the accuracy of sensors, national standards such as DIN or NBS are available. The typical accuracy of sensors used in control systems is ± 1 K within the operation range. For an accuracy better than 1K for the whole operation range of solar systems, the sensors have to be specified as a fraction of the DIN or NBS value (e.g. 1/3 of DIN 43760). If the sensors are specified to a national standard, a calibration is not necessary.

The accuracy of the sensor should be of the same order as the accuracy of the controller.

Operation

For solar air systems a standard use cannot be defined and standard controllers with factory-aligned parameters are unknown. A *commissioning* test is therefore necessary. A control system with an integrated display showing temperature and status is an advantage for such tests. The following checks should be made:

- adjustment of calculated control parameters;
- placement of the sensors, as well as the material used for wiring;
- switching of the controller by doing an on/off check of the fans and the position of the dampers;
- the function of switched devices like motor-driven dampers and fans;

If there is no sun, the initiation check of the system and the function of the components have to be checked by simulating a status of operation by heating the sensors etc.

Maintenance

Most control systems are fail-safe and do not need any calibration or adjustment. Their displays often give information about errors like broken sensors etc. A check by the user is then sufficient. However, an inspection of the system functions from time to time should be carried out for the switched devices like dampers and fans because they have mechanical parts that may fail.

Do's and don'ts

- Keep the costs of the control system in sensible relationship to those of the solar air system.
- Check if an analogue controller should be used, for example to adjust the fan speed (system with rockbed storage and air-to-water heat exchanger).
- Keep the system as simple as possible. Surplus gains by complex control strategies for additional system functions raise costs and often cause operational problems.
- Relays inside controllers should switch a contactor and not the fan directly to extend the lifetime of the relay (impedance).
- Provide for the replacement of sensors, which like any component can fail some day (especially in a hypocaust or rockbed).
- Do not install cabling to sensors close to power lines, which can falsify the sensed temperature by electrical induction.

Table V.3.2. Necessary devices

	Device	Operation
Heating demand	Thermostat	Release of Rules 2–4
Collector heating mode	Differential thermostat An integrated minimum switch allows the integration of Rule 1	Open damper to the room; fan on
Discharge mode	Differential thermostat	Discharge storage fan on, damper to room open If charge mode off – Relay
Backup heating	Relay	Backup heating on Released if Rules 2 and 3 do not hold
Charge mode	Differential thermostat with maximum thermostat	Charge fan on, open damper to storage Released if Rule 1 does not hold If discharge of storage is off and charge mode is on – Relay Optional: constant outlet temperature Closed-loop controller and frequency inverter for the fan
Dhw mode	Differential thermostat	Collector fan on, damper to heat exchanger open, pump to storage on Released by Rule 4 (maximum T) or manual switch in summer
Summer mode	Manual switch	Switching all other controllers off; other operation see Charge mode and dhw mode

- Use protection tubes for cables installed outside. Silicon covering for sensor cables is resistant to most environmental influences but not to biting by martens or other animals!

EXAMPLE

Select the control of a solar system with rockbed storage.

System analysis

The purpose is to heat the room if there is an energy demand:

$T_{room} < T_{set}$ (Rule 1. Heating mode).

Where should the energy come from? Available components are the collector, the storage and the backup heating system. To maximize the solar contribution the collector should have first priority, storage second priority and finally backup:

collector mode: $T_{coll} > T_{room}$ (Rule 2).

If Rule 2 is does not hold (relay), the storage should be discharged:

discharge mode: $T_{storage} > T_{room}$ (Rule 3).

If Rule 3 does not hold (relay), the backup heating should be used. When Rules 1–3 hold, the backup must be blocked.

If Rule 1 does not hold, the storage should be charged:

charge mode: $T_{coll} > T_{storage}$ (Rule 4).

An additional condition is that the rockbed should only be charged to a maximum T.

If: $T_{storage} = T_{max}$ (Rule 5. Hot-water mode (dhw)):

dhw mode: $T_{coll} > T_{dhw}$ (Rule 6).

In summer, only hot water should be produced without charging the rockbed first (Rule 7).

To avoid too much switching between components, the set temperatures for backup heating and solar heating should be slightly different.

Mode summary
- charge;
- discharge;
- collector heating;
- summer mode;
- backup.

Four components have a thermal influence:

- collector;
- storage;
- room;
- domestic hot water.

Necessary devices are summarized in Table V.3.2. Calculation of ΔT on/off has to be done for each controller.

The problem can be solved either by using the mentioned number of differential thermostats, thermostats and relays or by using a programmable controller. The solution using a programmable controller is more elegant but more expensive.

For a constant charge temperature that increases the system performance, an analogue controller and a frequency converter for the fan are necessary.

REFERENCES

1. Ferguson AMN (1990) *Predictive Thermal Control of Building Systems*. Thesis 876, Ecole Polytechnique Federale de Lausanne, Dept. of Physics, CH-1001 Lausanne, Switzerland.
2. Rausch L, Fritsche ER (1998) *GEMIS Gesamt Emissions Modell integrierter Systeme*. Ökoinstitut, Bunsenstrasse 14, D-64293 Darmstadt, Germany. See also http://www.oeko.de/service/gemis/gmsfiles.htm.

Chapter author: Joachim Morhenne

V.4 Fire protection

INTRODUCTION AND TYPOLOGY

Fire protection is of the same importance for solar air heating systems as for ventilation systems.

- *Important! Local fire regulations must always be regarded. The local fire department/authorities must always be informed and consulted!*

Depending on the building type, air-system type, the risk of spreading of fire/smoke/gas and local fire regulations, a fire protection system should be installed. A system can include the following components:

- smoke detectors that will record the occurrence of smoke in the air system;
- a control/alarm system connected to the smoke detection system; the alarm system will report that smoke/fire has occurred and the control system will switch off fans, close shutters etc.;
- installations to actively stop the spreading of fire, such as automatic shutters;
- installations to evacuate smoke, such as fans, shutters and vents;
- fire- and heat-proof gaskets and sealing compounds;
- fire-insulation materials for ducts etc.

STRATEGY

Fire protection strategy

The purpose of a fire protection system is to prevent the spreading of fire, smoke and gas through the air system.

In order to achieve a safe solar air system, the following strategy is be recommended:

- Local fire regulations concerning both buildings and ventilation systems must be followed.
- The building must be divided into relevant fire cells according to local regulations.
- Passages between fire cells must be air-tight and built to the same fire protection standard as the fire cell itself.
- Smoke detection, fire alarms and shutter systems should be installed when found necessary according to local regulations.

- Combustible materials that can emit poisonous gas should be avoided.
- High-performance, high-temperature air collectors can achieve high stagnation temperatures (100–200 °C) and must be installed without any direct contact with materials that are easily inflammable (e.g. wood in close proximity to sparks).

Open/closed systems

Open-loop systems function in a very similar way to a conventional mechanical pre-heating ventilation system from a fire protection point of view. It is important that regulations for ventilation systems are regarded and followed in order to prevent smoke and fire from spreading from one room to another through the solar air system. In this type of system outside air enters the building and circulates. There is therefore the risk that the system may spread smoke and fire. The solar air heating system should thus adhere to the local fire regulations for ventilation systems. The materials used in the collector should comply with the fire regulations for materials in facades.

Closed-loop systems do not connect one room to another. Thus, there is less risk of fire and gas spreading through the air system. It is, however, important that the air system is separated from the fire cells to the same standard as the fire cells are separated from each other.

Detectors

Detectors placed in a duct should not be located in the following positions:

- close to a fan;
- close to a bend;
- close to a connection or a junction;
- close to a change of dimension;
- close to a heat exchanger or a shutter.

The air inlet of the detector should be placed as close as possible to the centre point of the duct (where the air velocity is high).

Smoke detectors in ducts can either be placed in a central spot or decentralized, i.e. spread out through the system. A decentralized system will need more detectors, but give an earlier alarm.

It is important to know that detectors in a duct system will only function when the system is running. The degree of safety will increase if detectors in ducts are combined with other detectors in the building.

Maintenance and supervision

Since a solar air system operates at high temperatures, there is always a risk of false alarms, which can be quite annoying. It is recommended that the fan-motor reset functions be placed where they are easy to reach and where the indication lights are clearly visible.

It is also important that all smoke/fire detectors have indicator lamps. Otherwise, it may be hard to determine which detector to replace or clean when there is a false alarm or a service alarm.

SELECTION

Selection of system – general advice

Smoke detectors, alarms and control units are mostly manufactured and marketed as complete systems. The local dealer and the manufacturer will tell you which system will be the right choice for your solar air system. As well as the local authority/fire department, it is recommended that you also contact your insurance company for advice.

Optical detectors

In optical detectors, a beam of light passes through the unit. When smoke enters the unit, the light is reflected to a photocell, causing an alarm signal. Optical detectors are used when the smoke contains fewer and larger particles.

Heat detectors

These detectors react to temperatures or temperature differences and are mostly recommended for industrial applications.

Shutters

Fire-shutters are used to maintain the fire-cell sections in case of fire in the building. The shutters will, when closed, prevent fire or/and smoke from entering one fire cell from another.

Most automatic shutters are closed automatically when a constant electric current is interrupted and in some cases also when the temperature in the ducts exceeds a set value.

There are different types/qualities of shutters according to their function and degree of air-tightness under different pressures. Selection should be made according to the local fire and building regulations.

Shutters are manufactured for circular and rectangular duct sections, which make them easy to install in most air systems.

Control/alarm units

The control unit will supervise all connected detectors, fans, shutters and alarms automatically. An advanced control unit will also test the function of the entire system at preset intervals.

In the case of fire/smoke the control unit should (minimum standard):

- Switch off all fans in the solar air system.
- Close all automatic shutters.
- Indicate that something has happened.

A high quality control unit will also:

- Give a service alarm, indicating that smoke detectors are out of order and need to be cleaned/replaced.

All control units should be equipped with outputs for external alarm units to allow connection to a telephone line or a central fire alarm. It is also important that the control unit can operate on both 240 V and 12 V supplies.

Chapter author: Christer Nordström

Appendices

Appendix A: Glossary

Closed-loop system: A system in which solar-collector air circulates in a closed channel between the collector and heat storage or radiating surface. Room air does not enter the system, as is the case with an open-loop system.

COP: The coefficient of performance for a system is the energy output divided by the energy input, i.e. for a heat pump this would be kWh heat output per kWh electricity input (3–4 is a realistic range). For a solar air system the COP is kWh hot air output per kWh electricity consumed by the fan/s (15–30 is possible according to monitored projects).

Design temperature: The coldest ambient temperature that, per the local building code, a building must cope with in guaranteeing that a required room temperature (i.e. 20°C) is achieved. The design temperature under night conditions (no solar) is a basis for sizing the capacity of the heating plant for a given building design.

Heating degree days: A unit that characterizes how cold a climate is. There is unfortunately no consensus how this should be calculated.

- When a single base number is cited, i.e. 18.3°C in the USA, then heating degree days are the sum of (18.3°C – $T_{ambient}$) for all days where $T_{ambient}$ < 18.3°C.
- When two numbers are cited, i.e. 20/12 in central Europe, then the heating degrees are the sum of (20°C – $T_{ambient}$) for all days where the average $T_{ambient}$ ≤ 12°C in months where the monthly average temperature ≤ 12°C (to exclude summer months).

Hybrid system: Solar air systems are often referred to as hybrid systems vs 'passive systems' because they use a fan to assist in moving the air from the collector to the point of use. The system is hybrid in that it makes use of both solar energy and electricity to power the fan.

Hypocaust: An old Roman construction in which smoke from a wood fire was circulated through a space beneath a massive floor to provide radiant heating. In this book, hypocausts refer to massive floors with channels through which solar-heated air passes, thereby warming the floor mass and providing a delayed release of radiant heat to the room.

Murocaust: Similar to a hypocaust, except that a massive wall with air channels is used and the air flow is vertical instead of horizontal.

Thermosiphon: Flow induced by a warm fluid or gas being less dense than a cool medium and hence rising. In a thermosiphon solar air system, solar-warmed air in the collector rises and so can be self-transported to either a room or storage.

Shoebox: A box geometry representing a building, simplified for the purpose of analysing specific features or systems. Since these features are the issue of study, the geometry is kept simple to save computation time and not confuse the output.

UNITS

1 joule (J) = 1 watt second (Ws)
1 K = 1°C temperature difference as used here
1 m = 39.37 in or 3.28 ft
1 m² = 10.76 ft²
1 litre = 1.06 quarts (US)
1 kg = 2.2 lbs
$ = US dollars unless otherwise noted
€1 =1 Euro = $1.10 as used here

Table A.1. Conversion table between joules, kWh and Btu

J	kWh	Btu
1	2.78×10^{-7}	9.48×10^{-4}
3.6×10^6	1	3.41×10^3
1.055×10^3	2.93×10^{-4}	1

Table A.2. Conversion table for annual energy consumption

MJ/m²a	kWh/m²a	Btu/ft²a
1	0.278	88.1
3.6	1	317
0.0114	3.16×10^{-3}	1

Table A.3. Conversion table for pressure units

	Pa	mbar	mm water	atmospheres	inch water	mm Hg
Pa	1	10^{-2}	0.102	9.869×10^{-6}	4.02×10^{-3}	7.5×10^{-3}
mbar	100	1	10.197	9.869×10^{-4}	0.402	0.75
mm water	9.807	9.807×10^{-2}	1	9.678×10^{-5}	3.937×10^{-2}	0.0736
atmosphere	1.013×10^{5}	1013	1.0332×10^{4}	1	406.77	760
inch water	249.1	2.491	25.4	2.453×10^{3}	1	1.869
mm Hg	133.3	1.333	13.59	1.316×10^{-3}	0.535	1

Table A.4. Heating values for some fuels

Energy source	MJ/kg	kWh/kg
Heating oil, extra light	42.7	11.9
Heating oil, heavy	40.2	11.2
Hard coal, coke	29.3	8.1
Soft coal/brown coal	20	5.6
Wood	15.5	4.3
Softwood (360 kg/m^3)		
Hardwood (500 kg/m^3)		
Wood chips (250 kg/m^3)		
	MJ/m^3	kWh/m^3
Natural gas (Zurich)	37.6	
Propane, Butane	46	12.8

Prepared by Robert Hastings

Appendix B: Properties of materials

	ρ (kg/m³)	λ(kJ/hm²K)	c(kJ/kgK)
Natural stones and natural soil			
Granite	2800	12.6	1.0
Marble	2800	12.6	1.0
Mussel shell lime	2600	8.28	1.0
Sand gravel	1800–2000	2.52	1.0
Pumice gravel	1000	0.68	1.0
Concrete			
Common concrete	2400	7.56	1.0
Steam hard gas concrete	500	0.58	1.0
	800	0.83	1.0
Lightweight concrete with natural pumice	500	0.54	1.0
	800	0.86	1.0
	1200	1.58	1.0
Lightweight concrete with swell clay	500	0.65	1.0
	800	0.94	1.0
	1200	1.66	1.0
Plaster, clay			
Lime mortar	1800	3.13	1.0
Lime-cement mortar	1800	3.13	1.0
Cement mortar	2000	5.04	1.0
Gypsum mortar	1400	2.52	1.0
Gypsum plaster	1200	1.26	1.0
Cement clay	2000	5.04	1.0
Under floors	1400	1.69	1.0
Industrial floors	2300	2.52	1.0
Poured asphalt	2300	3.24	1.0
Wood			
Spruce, pine	600	0.47	2–2.4
Beech	700	0.72	2–2.4
Oak	800	0.72	2–2.4
Plywood	800	0.54	1.2
Particle board	200	0.16	1.0
	300	0.20	1.0
Hardboard, high density	1000	0.61	1.38
Masonry			
Clinker	1800	2.92	1.0
	2200	4.32	1.0
Clinker hollow	1400	2.09	1.0
	2000	3.46	1.0
Lightweight clinker hollow	900	1.51	1.0
	1000	1.40	1.0
Limestone	1000	1.8	1.0
	1600	2.84	1.0
	2200	4.68	1.0
Gas concrete block	500	0.79	1.0
	800	1.04	1.0

	ρ (kg/m³)	λ(kJ/hm²K)	c(kJ/kgK)
Hollow block 2–4k	500	1.04	1.0
	900	1.58	1.0
	1400	2.63	1.0
Hollow block 2k	500	1.04	0.3
	900	1.98	0.3
	1400	3.24	0.3
Hollow block 3k	500	1.04	0.365
	900	1.98	0.365
	1400	3.24	0.365
Stone massive	500	1.15	1.0
	800	1.44	1.0
	1100	1.94	1.0
	1500	3.56	1.0
Block massive	500	1.04	1.0
	900	1.55	1.0
	1200	2.27	1.0
	1500	3.56	1.0
Block massive pumice	400	0.72	1.0
	700	1.01	1.0
Block massive swell clay	300	0.72	1.0
	600	1.12	1.0
Hollow block	1500	3.31	1.0
	1800	4.68	0.365
Insulation			
Wooden fibre panels	400	0.33	0.25
Foamed plastic	20	0.54	0.15
Polyurethane 020	40	0.07	2.09
Polyurethane 035	40	0.13	2.09
Polystyrene 025	15	0.09	1.25
Polystyrene 040	30	0.14	1.25
Foam glass 045	100	0.16	1.0
Foam glass 060	250	0.22	1.0
Cork 050	100	0.18	1.8
Mineral wool 035	80	0.13	0.9
Mineral wool 050	80	0.18	0.9
Construction board			
Asbestos cement	2000	2.09	1.0
Gas concrete	500	0.79	1.0
	800	1.04	1.0
Wall board	800	1.04	1.0
	1400	2.09	1.0
Wall board gypsum	600	1.04	1.0
	900	1.48	1.0
	1200	2.09	1.0
Plaster card	900	0.76	1.0
Other materials, metals			
Cork granulated	200	0.18	1.0
Linoleum	1000	0.61	1.0
Cork linoleum	700	0.29	1.0
Bitumen	1100	0.61	1.0
Poly-vinyl-chloride	1500	0.83	1.0

		ρ (kg/m^3)	λ (kJ/hm^{2}K)	c (kJ/kgK)
Swell clay		400	0.58	1.0
Glass		2500	2.88	1.0
Tile		2000	3.60	1.0
Rubber		1100	0.72	6.0
Steel		7800	54	1.8
Copper		8300	1340	4.19
Aluminium		2700	720	3.43
Water		1000	–	4.182
Air layer				
Perpendicular	5 mm	1.2	0.043	1.0
	10 mm		0.065	
	20 mm		0.115	
	40 mm		0.221	
Horizontal				
(ground to top)	10 mm	1.2	0.072	1.0
	20 mm		0.137	
	50 mm		0.307	
Horizontal				
(top to ground)	10 mm	1.2	0.061	1.0
	20 mm		0.110	
	50 mm		0.243	

Prepared by Aninja Grilc from data provided by Alex Knirsch

Appendix C: Meteorological data

This Appendix provides:

- a survey of meteorological data that can be used to run computer simulations for other climates than those displayed in the book or to apply the theory presented in Appendix D;
- diagrams for selected sites showing the relative shift of solar energy input into a solar system as a function of different collector orientations.

The simulation tool Trnsair is supplied with a few reference climates, which are presumed to be representative for northern climates as shown in Table C.1.

In order to relate any given building location to one of the reference climates, a solar similarity index (SSI) – comparable, but not identical with the index used in IEA Task 8[1] – was computed. This solar similarity index is a measure of the solar contribution to the energy balance of a system for the given climate. Its exact meaning and practical applications are explained in more detail in Appendix D. Here, the index is defined as the ratio of horizontal radiation summed up during the heating season (GH) to the total of heating degree days (DD):

$$\text{SSI} = \frac{\text{GH}}{\text{DD}}. \tag{C.1}$$

Because North America and Europe do not have a common standard for computing heating degree days, the number of heating degree days used here is computed from monthly means of daily averaged outdoor temperatures $T_{\text{O,month}}$ (with $T_{\text{O,month}} = 12°\text{C}$ as border temperature for heating):

Table C.1. Example of reference climates

Climate, description	Example region	Example cities
Cold/Sunny	Continental North America	Denver
Cold/Overcast	North/Scandinavia	Stockholm
Temperate/Overcast	Middle Europe	Copenhagen, Zurich
Temperate/Sunny	Pre-Alpine	Brig
Mild/Sunny	Mediterranean	Rome

$$\text{DD} = \sum_{\substack{\text{all months of} \\ \text{heating season}}} \sum_{\substack{\text{number of} \\ \text{days per month}}} (20°\text{C} - T_{\text{O, month}}). \tag{C.2}$$

Similarly, the seasonal solar radiation GH was summed up for months with a mean outdoor temperature less than or equal to 12°C from daily values GH_d:

$$\text{DD} = \sum_{\substack{\text{all days of} \\ \text{heating season}}} \text{GH}_\text{d}. \tag{C.3}$$

Table C.2 summarizes the characteristic data of the reference climates delivered by Trnsair. The following publications were used as data sources to compile the relevant weather data (DD, GH and SSI) for altogether 87 additional stations in Europe and 56 locations in North America:

- Müller:[2]   Europe – outdoor air temperatures,
- *Atlas über die Sonnenstrahlung Europas*:[3] Europe – solar radiation,
- Duffie and Beckman:[4]   North America – air temperatures and solar radiation.

These data are presented as tables sorted by solar similarity index (Table C.3: Europe; Table C.5: North

Table C.2. Characteristic data for various reference climates delivered by Trnsair

Location	Stockholm	Copenhagen	Zurich	Brig	Rome	Denver
Latitude (°)	59.1	54.4	47.2	46.3	42.1	39.5
Longitude (°)	−17.6	−12.4	−8.3	−8.0	−12.3	104.5
Heating season	October–May	October–May	October–April	October–April	November–March	October–April
Degree days (Kd)	4454	3814	3370	3463	1610	3562
Global radiation (kWh/m²)	359	453	380	520	375	802
Solar Similarity Index (kWh/m²Kd)	0.081	0.119	0.113	0.150	0.233	0.225

America) and by heating degree days (Table C.4: Europe; Table C.6: North America). In these tables, positions of reference climates are typeset in bold. This allows easy comparison of any given location with the reference climates in terms of solar similarity index (SSI) or heating degree days (DD). It is worth noting that the meteorological data of the reference stations taken from the tables are not always identical with those of the weather files from Trnsair, which used different data sources. There may be slight differences (generally not greater than 10%).

Table C.3. European stations and reference stations (in bold type) listed by solar similarity index

Solar Similarity Index (kWh/(m²Kd))	Station	Country	Latitude	Horizontal solar radiation during heating season (kWh/m²)	Heating degree days (Kd)
0.078	Potsdam	Germany	52°23′	276	3508
0.089	Klagenfurt	Austria	46°39′	348	3877
0.092	Berlin	Germany	52°28′	316	3449
0.100	Bergen	Norway	60°12	394	3906
0.103	Milan	Italy	45°26′	230	2249
0.103	De Bilt	Netherlands	52°06′	323	3113
0.103	**Stockholm**	**Sweden**	**59°21′**	**439**	**4306**
0.106	London	Great Britain	51°28′	290	2732
0.108	Essen	Germany	51°24′	340	3138
0.108	Lille	France	50°44′	329	3018
0.108	Salzburg	Austria	47°48′	400	3653
0.108	Bolzano	Italy	46°28′	263	2405
0.108	Trier	Germany	49°45′	356	3248
0.111	Frankfurt am Main	Germany	50°07′	366	3309
0.111	**Zurich**	**Switzerland**	**47°27′**	**386**	**3488**
0.111	Karlstad	Sweden	59°22′	526	4688
0.114	Kew	Great Britain	51°28′	310	2732
0.114	Vienna	Austria	48°15′	378	3339
0.117	Visby	Sweden	57°39′	478	4055
0.117	Udine	Italy	46°02′	254	2164
0.119	Bologna	Italy	44°32′	262	2176
0.119	Strasbourg	France	48°33′	384	3236
0.122	Nancy	France	48°42′	385	3160
0.122	Venice	Italy	45°30′	256	2098
0.122	Hamburg	Germany	53°38′	447	3667
0.122	Braunlage	Germany	51°43′	550	4545
0.125	**Copenhagen**	**Denmark**	**55°41′**	**463**	**3709**
0.128	Turin	Italy	45°04′	299	2322
0.131	Reims	France	49°18′	392	2994
0.133	Dresden	Germany	51°07′	466	3528
0.133	Trieste	Italy	45°39′	256	1904
0.133	Lerwick	Great Britain	60°09′	556	4158
0.136	Dijon	France	47°16′	411	3039
0.139	Auxerre	France	47°48′	399	2869
0.139	Innsbruck	Austria	47°16′	499	3566
0.144	Eskaldemuir	Great Britain	55°19′	603	4194
0.144	Weihenstephan	Germany	48°24′	587	4029
0.144	Cambridge	Great Britain	52°12′	460	3157
0.147	Paris	France	48°58′	419	2830
0.147	Norderney	Germany	53°43′	508	3483
0.147	Kirkwall	Great Britain	58°57′	584	3936
0.151	**Brig**	**Switzerland**	**46°11′**	**523**	**3471**
0.156	Vlissingen	Netherlands	51°28′	503	3202
0.156	Lyons	France	45°43′	444	2866
0.156	Tours	France	47°25′	418	2694
0.156	Clermont-Ferrand	France	45°48′	448	2879
0.156	den Helder	Netherlands	52°58′	513	3304
0.156	Ancona	Italy	43°37′	279	1785
0.158	Birr	Ireland	53°05′	497	3115
0.161	Nantes	France	47°10′	371	2316

Table C.3. (cont.)

Solar Similarity Index (kWh/(m²Kd))	Station	Country	Latitude	Horizontal solar radiation during heating season (kWh/m²)	Heating degree days (Kd)
0.161	Rennes	France	48°04'	419	2600
0.164	Kilkenny	Ireland	52°40'	522	3158
0.164	Limoges	France	45°49'	471	2858
0.172	Pisa	Italy	43°40'	303	1756
0.172	Lugano	Switzerland	46°00	443	2574
0.181	Pescara	Italy	42°28'	264	1457
0.186	Toulouse	France	43°38'	428	2295
0.186	Le Puy-en-Velay	France	45°03'	647	3454
0.186	Genova	Italy	44°25'	290	1557
0.192	Madrid	Spain	40°27'	370	1936
0.192	Bordeaux	France	44°50'	426	2226
0.192	Valentia	Ireland	51°56'	511	2682
0.197	Brest	France	48°27'	547	2767
0.197	Plymouth	Great Britain	50°21'	544	2772
0.200	Montélimar	France	44°35'	459	2296
0.211	**Rome**	**Italy**	**41°54'**	**348**	**1641**
0.219	Zaragoza	Spain	41°40'	385	1746
0.219	Salamanca	Spain	40°57'	524	2405
0.219	Nice	France	43°40'	354	1616
0.219	Barcelona	Spain	41°25'	196	895
0.222	Bastia	France	42°33'	349	1576
0.236	Tortosa	Spain	40°49'	214	905
0.236	Braganca	Portugal	41°49'	630	2678
0.239	Coimbra	Portugal	40°12'	210	878
0.244	Naples	Italy	40°51'	286	1162
0.247	**Denver**	**USA**	**39°75'**	**848**	**3447**
0.261	Alghero	Italy	40°38'	299	1145
0.267	Porto	Portugal	41°08'	320	1206
0.269	Lisbon	Portugal	38°43'	211	784
0.269	Cagliari	Italy	39°13'	309	1145
0.275	Palma de Mallorca	Spain	39°33'	229	833
0.278	Seville	Spain	37°24'	156	561
0.283	Valencia	Spain	39°29'	230	814
0.286	Messina	Italy	38°12'	144	504
0.292	Pantelleria	Italy	36°49'	151	514
0.311	Murcia	Spain	37°59'	255	823
0.317	Trapani	Italy	38°01'	156	493
0.372	Almería	Spain	36°51'	182	487

Table C.4. European stations and reference stations (in bold type) listed by heating degree days

Heating degree days (Kd)	Station	Country	Latitude	Horizontal solar radiation during heating season (kWh/m²)	Solar Similarity Index (kWh/(m²Kd))
487	Almería	Spain	36°51'	182	0.372
493	Trapani	Italy	38°01'	156	0.317
504	Messina	Italy	38°12'	144	0.286
514	Pantelleria	Italy	36°49'	151	0.292
561	Seville	Spain	37°24'	156	0.278
784	Lisbon	Portugal	38°43'	211	0.269
814	Valencia	Spain	39°29'	230	0.283
823	Murcia	Spain	37°59'	255	0.311
833	Palma de Mallorca	Spain	39°33'	229	0.275
878	Coimbra	Portugal	40°12'	210	0.239
895	Barcelona	Spain	41°25'	196	0.219
905	Tortosa	Spain	40°49'	214	0.236
1145	Alghero	Italy	40°38'	299	0.261
1145	Cagliari	Italy	39°13'	309	0.269
1162	Naples	Italy	40°51'	286	0.244
1206	Porto	Portugal	41°08'	320	0.267
1457	Pescara	Italy	42°28'	264	0.181
1557	Genova	Italy	44°25'	290	0.186
1576	Bastia	France	42°33'	349	0.222
1616	Nice	France	43°40'	354	0.219

Table C.4. (cont.)

Heating degree days (Kd)	Station	Country	Latitude	Horizontal solar radiation during heating season (kWh/m²)	Solar Similarity Index (kWh/(m²Kd))
1641	**Rome**	**Italy**	**41°54'**	**348**	**0.211**
1746	Zaragoza	Spain	41°40'	385	0.219
1756	Pisa	Italy	43°40'	303	0.172
1785	Ancona	Italy	43°37'	279	0.156
1904	Trieste	Italy	45°39'	256	0.133
1936	Madrid	Spain	40°27'	370	0.192
2098	Venice	Italy	45°30'	256	0.122
2164	Udine	Italy	46°02'	254	0.117
2176	Bologna	Italy	44°32'	262	0.119
2226	Bordeaux	France	44°50'	426	0.192
2249	Milan	Italy	45°26'	230	0.103
2295	Toulouse	France	43°38'	428	0.186
2296	Montélimar	France	44°35'	459	0.200
2316	Nantes	France	47°10'	371	0.161
2322	Turin	Italy	45°04'	299	0.128
2405	Salamanca	Spain	40°57'	524	0.219
2405	Bolzano	Italy	46°28'	263	0.108
2574	Lugano	Switzerland	46°00	443	0.172
2600	Rennes	France	48°04'	419	0.161
2678	Braganca	Portugal	41°49'	630	0.236
2682	Valencia	Ireland	51°56'	511	0.192
2694	Tours	France	47°25'	418	0.156
2732	London	Great Britain	51°28'	290	0.106
2732	Kew	Great Britain	51°28'	310	0.114
2767	Brest	France	48°27'	547	0.197
2772	Plymouth	Great Britain	50°21'	544	0.197
2830	Paris	France	48°58'	419	0.147
2858	Limoges	France	45°49'	471	0.164
2866	Lyons	France	45°43'	444	0.156
2869	Auxerre	France	47°48'	399	0.139
2879	Clermont-Ferrand	France	45°48'	448	0.156
2994	Reims	France	49°18'	392	0.131
3018	Lille	France	50°44'	329	0.108
3039	Dijon	France	47°16'	411	0.136
3113	De Bilt	Netherlands	52°06'	323	0.103
3115	Birr	Ireland	53°05'	497	0.158
3138	Essen	Germany	51°24'	340	0.108
3157	Cambridge	Great Britain	52°12'	460	0.144
3158	Kilkenny	Irland	52°40'	522	0.164
3160	Nancy	France	48°42'	385	0.122
3202	Vlissingen	Netherlands	51°28'	503	0.156
3236	Strasbourg	France	48°33'	384	0.119
3248	Trier	Germany	49°45'	356	0.108
3304	den Helder	Netherlands	52°58'	513	0.156
3309	Frankfurt am Main	Germany	50°07'	366	0.111
3339	Vienna	Austria	48°15'	378	0.114
3447	**Denver**	**USA**	**39°75'**	**848**	**0.247**
3449	Berlin	Germany	52°28'	316	0.092
3454	Le Puy-en-Velay	France	45°03'	647	0.186
3471	**Brig**	**Switzerland**	**46°11'**	**523**	**0.151**
3483	Norderney	Germany	53°43'	508	0.147
3488	**Zurich**	**Switzerland**	**47°27'**	**386**	**0.111**
3508	Potsdam	Germany	52°23'	276	0.078
3528	Dresden	Germany	51°07'	466	0.133
3566	Innsbruck	Austria	47°16'	499	0.139
3653	Salzburg	Austria	47°48'	400	0.108
3667	Hamburg	Germany	53°38'	447	0.122
3709	**Copenhagen**	**Denmark**	**55°41'**	**463**	**0.125**
3877	Klagenfurt	Austria	46°39'	348	0.089
3906	Bergen	Norway	60°12	394	0.100
3936	Kirkwall	Great Britain	58°57'	584	0.147
4029	Weihenstephan	Germany	48°24'	587	0.144
4055	Visby	Sweden	57°39'	478	0.117
4158	Lerwick	Great Britain	60°09'	556	0.133
4194	Eskaldemuir	Great Britain	55°19'	603	0.144
4306	**Stockholm**	**Sweden**	**59°21'**	**439**	**0.103**
4545	Braunlage	Germany	51°43'	550	0.122
4688	Karlstad	Sweden	59°22'	526	0.111

Table C.5. *North American stations and reference stations (in bold type) listed by solar similarity index*

Solar Similarity Index (kWh/(m²Kd))	Station	State	Country	Latitude	Horizontal solar radiation during heating season (kWh/m²)	Heating degree days (Kd)
0.103	**Stockholm**		**Sweden**	**59°21′**	**439**	**4306**
0.111	**Zurich**		**Switzerland**	**47°27′**	**386**	**3488**
0.114	Burlington	Vermont	USA	44°70′	496	4352
0.119	International Falls	Minnesota	USA	48°57′	734	6142
0.119	Concord	New Hampshire	USA	43°20′	488	4080
0.119	Churchill	Manitoba	Canada	58°75′	1179	9846
0.125	**Copenhagen**		**Denmark**	**55°41′**	**463**	**3709**
0.125	Minot	North Dakota	USA	48°22′	671	5326
0.125	Bethel	Alaska	USA	60°78′	989	7948
0.131	Edmonton	Alberta	Canada	53°57′	764	5871
0.133	Parkersburg	West Virginia	USA	39°27′	332	2503
0.133	Detroit	Michigan	USA	42°23′	495	3686
0.136	Hartford	Connecticut	USA	41°93′	498	3687
0.136	Lexington	Kentucky	USA	38°03′	339	2506
0.139	Chicago	Illinois	USA	41°98′	456	3257
0.139	Moncton	New Brunswick	Canada	46°12′	849	6170
0.139	Pierre	South Dakota	USA	44°38′	598	4324
0.139	Montreal	Quebec	Canada	45°50′	665	4783
0.142	Missoula	Montana	USA	46°92′	626	4455
0.144	Evansville	Indiana	USA	38°05′	354	2475
0.144	Ottawa	Northwest Territories	Canada	45°45′	729	5021
0.144	Green Bay	Wisconsin	USA	44°48′	678	4689
0.147	Portland	Maine	USA	43°65′	635	4300
0.147	St. Johns	Newfoundland	Canada	47°52′	782	5328
0.150	Blue Hill	Massachusetts	USA	42°22′	559	3716
0.151	**Brig**		**Switzerland**	**46°11′**	**523**	**3471**
0.153	Ames	Iowa	USA	42°03′	619	4022
0.156	New York	New York	USA	40°77′	433	2775
0.156	Columbus	Ohio	USA	40°00′	509	3295
0.156	Manhattan	Kansas	USA	39°20′	415	2688
0.158	Portland	Oregon	USA	45°60′	423	2694
0.164	Lakehurst	New Jersey	USA	40°03′	476	2925
0.169	Philadelphia	Pennsylvania	USA	39°88′	473	2805
0.181	Grand Island	Nebraska	USA	40°97′	666	3687
0.186	Richland	Washington	USA	46°28′	541	2906
0.192	Pullman	Washington	USA	46°73′	743	3881
0.194	Ashville	North Carolina	USA	35°43′	428	2203
0.194	Vancouver	British Columbia	Canada	48°98′	639	3262
0.194	Mount Weather	Virginia	USA	39°07′	651	3356
0.194	Chattanooga	Tennessee	USA	35°03′	382	1962
0.197	Patuxent River	Maryland	USA	39°17′	517	2624
0.208	Newport	Rhode Island	USA	41°48′	676	3263
0.211	**Rome**		**Italy**	**41°54′**	**348**	**1641**
0.214	Oklahoma City	Oklahoma	USA	35°40′	464	2173
0.217	Salt Lake City	Utah	USA	40°77′	727	3358
0.219	Pocatello	Idaho	USA	42°92′	889	4063
0.222	Cheyenne	Wyoming	USA	41°15′	936	4211
0.222	Little Rock	Arkansas	USA	34°73′	434	1963
0.236	Jackson	Mississippi	USA	32°32′	234	993
0.236	Ely	Nevada	USA	39°28′	1049	4426
0.242	Montgomery	Alabama	USA	32°30′	233	962
0.247	Atlanta	Georgia	USA	33°65′	433	1750
0.247	**Denver**	**Colorado**	**USA**	**39°75′**	**848**	**3447**
0.269	Houston	Texas	USA	29°97′	76	279
0.269	Farmington	New Mexico	USA	36°75′	898	3330
0.289	Columbia	South Carolina	USA	33°95′	455	1571
0.300	New Orleans	Louisiana	USA	29°98′	158	527
0.325	Phoenix	Arizona	USA	33°43′	191	589
0.364	Pasadena	California	USA	34°15′	91	248
0 / 0	Miami	Florida	USA	25°78′	0	0
0 / 0	Honolulu	Hawaii	USA	21°30′	0	0

Table C.6. North American stations and reference stations (in bold type) listed by heating degree days

Heating degree days (Kd)	Station	State	Country	Latitude	Horizontal solar radiation during heating season (kWh/m^2)	Solar Similarity Index (kWh/(m^{2}Kd))
0	Miami	Florida	USA	25°78'	0	0 / 0
0	Honolulu	Hawaii	USA	21°30'	0	0 / 0
248	Pasadena	California	USA	34°15'	91	0.364
279	Houston	Texas	USA	29°97'	76	0.269
527	New Orleans	Louisiana	USA	29°98'	158	0.300
589	Phoenix	Arizona	USA	33°43'	191	0.325
751	Pensacola	Florida	USA	30°47'	275	0.367
962	Montgomery	Alabama	USA	32°30'	233	0.242
993	Jackson	Mississippi	USA	32°32'	234	0.236
1571	Columbia	South Carolina	USA	33°95'	455	0.289
1641	**Rome**		**Italy**	**41°54'**	**348**	**0.211**
1750	Atlanta	Georgia	USA	33°65'	433	0.247
1962	Chattanooga	Tennessee	USA	35°03'	382	0.194
1963	Little Rock	Arkansas	USA	34°73'	434	0.222
2173	Oklahoma City	Oklahoma	USA	35°40'	464	0.214
2203	Ashville	North Carolina	USA	35°43'	428	0.194
2475	Evansville	Indiana	USA	38°05'	354	0.144
2503	Parkersburg	West Virginia	USA	39°27'	332	0.133
2506	Lexington	Kentucky	USA	38°03'	339	0.136
2624	Patuxent River	Maryland	USA	39°17'	517	0.197
2688	Manhattan	Kansas	USA	39°20'	415	0.156
2694	Portland	Oregon	USA	45°60'	423	0.158
2775	New York	New York	USA	40°77'	433	0.156
2805	Philadelphia	Pennsylvania	USA	39°88'	473	0.169
2906	Richland	Washington	USA	46°28'	541	0.186
2925	Lakehurst	New Jersey	USA	40°03'	476	0.164
3257	Chicago	Illinois	USA	41°98'	456	0.139
3262	Vancouver	British Columbia	Canada	48°98'	639	0.194
3263	Newport	Rhode Island	USA	41°48'	676	0.208
3295	Columbus	Ohio	USA	40°00'	509	0.156
3330	Farmington	New Mexico	USA	36°75'	898	0.269
3356	Mount Weather	Virginia	USA	39°07'	651	0.194
3358	Salt Lake City	Utah	USA	40°77'	727	0.217
3447	**Denver**	**Colorado**	**USA**	**39°75'**	**848**	**0.247**
3471	**Brig**		**Switzerland**	**46°11'**	**523**	**0.151**
3488	**Zurich**		**Switzerland**	**47°27'**	**386**	**0.111**
3686	Detroit	Michigan	USA	42°23'	495	0.133
3687	Hartford	Connecticut	USA	41°93'	498	0.136
3687	Grand Island	Nebraska	USA	40°97'	666	0.181
3709	**Copenhagen**		**Denmark**	**55°41'**	**463**	**0.125**
3716	Blue Hill	Massachusetts	USA	42°22'	559	0.150
3881	Pullman	Washington	USA	46°73'	743	0.192
4022	Ames	Iowa	USA	42°03'	619	0.153
4063	Pocatello	Idaho	USA	42°92'	889	0.219
4080	Concord	New Hampshire	USA	43°20'	488	0.119
4211	Cheyenne	Wyoming	USA	41°15'	936	0.222
4300	Portland	Maine	USA	43°65'	635	0.147
4306	**Stockholm**		**Sweden**	**59°21'**	**439**	**0.103**
4324	Pierre	South Dakota	USA	44°38'	598	0.139
4352	Burlington	Vermont	USA	44°70'	496	0.114
4426	Ely	Nevada	USA	39°28'	1049	0.236
4455	Missoula	Montana	USA	46°92'	626	0.142
4689	Green Bay	Wisconsin	USA	44°48'	678	0.144
4783	Montreal	Quebec	Canada	45°50'	665	0.139
5021	Ottawa	Northwest Territories	Canada	45°45'	729	0.144
5326	Minot	North Dakota	USA	48°22'	671	0.125
5328	St. Johns	Newfoundland	Canada	47°52'	782	0.147
5871	Edmonton	Alberta	Canada	53°57'	764	0.131
6142	International Falls	Minnesota	USA	48°57'	734	0.119
6170	Moncton	New Brunswick	Canada	46°12'	849	0.139
7948	Bethel	Alaska	USA	60°78'	989	0.125
9846	Churchill	Manitoba	Canada	58°75'	1179	0.119

In addition to these tables, maps of Europe and North America have been included in this Appendix, showing geographical isolines of SSI (Figure C.1 for Europe, Figure C.3 for North America) and DD (Figure C.2 for Europe, Figure C.4 for North America). In this way, regions for which the indices (DD, SSI) are similar to those of the reference cases may be identified. Isolines of solar radiation are not presented, because they are well known and accessible from other publications such as *Atlas über die Sonnenstrahlung Europas*.[3]

All the maps are constructed from the data contained in the corresponding tables. They were generated by using Axum 5.0,[5] a software package for technical graphics and data analysis. This software creates isolines of any physical quantity, which are given station-wise in a two-dimensional array by numerical interpolation between the stations. Thus, influences on the parameters DD or SSI originating from the geographical altitude of a location are not considered explicitly, but are only taken into account in an implicit way, because stations at higher altitudes may contribute to the interpolation process by having greater amounts of solar intensity or heating degree days than those at lower altitudes. Therefore, in mountain regions, where there is also a lower

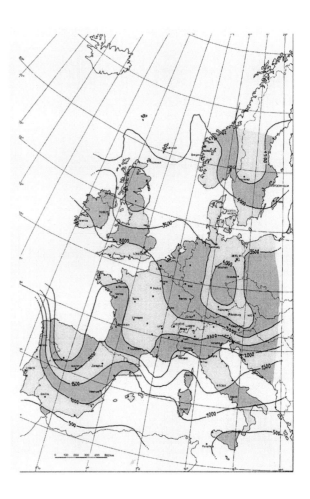

Figure C.2. *Heating degree days for Western and Central Europe (units Kd)*

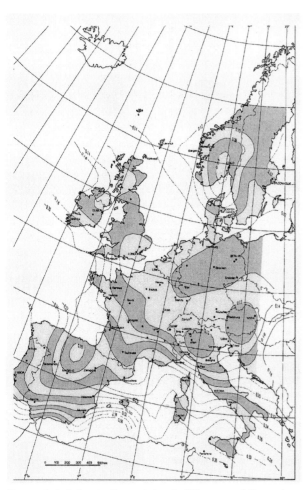

Figure C.1. *Solar similarity index for Western and Central Europe (units kWh/(m²Kd))*

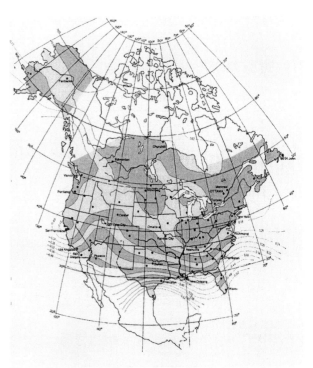

Figure C.3. *Solar similarity index for North America (units kWh/(m²Kd))*

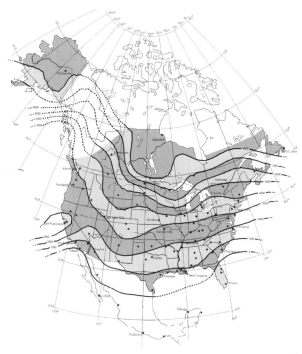

Figure C.4. Heating degree days for North America (units Kd)

density of meteorological stations (e.g. the Alps and the Pyrenees) the isoline presentations of DD and SSI may be questionable.

Another way to improve access to meteorological information is to use the database Meteonorm.[6] Version 3.0 (1997) of this software includes solar radiation data and outdoor temperatures from 626 meteorological stations all over the world. Hourly time series of global radiation and air temperature can be generated by stochastic modelling. In one of the next versions of Meteonorm the data format of these time series will be made compatible with the data reader of Trnsair. Table C.7 displays the characteristic properties of 21 locations, for which such hourly time series have been produced from the Meteonorm database for further investigation. Again, the stations with the same names as the reference climates delivered by Trnsair are set in bold and their values for DD, GH and SSI usually do not coincide completely with those of Table C.1.

Global solar radiation GH is given in all the tables as the total sum incident on horizontal surfaces during the heating season. This is basic information, but not the value that is usually received by a solar air collector with its building-dependent or system-specific orientation. To bridge this knowledge gap, diagrams have been developed for various locations (Figures C.5 to C.10 for Stockholm, Copenhagen, Zurich, Brig, Rome and Denver), in which global radiation – summed over the heating season – is presented as a function of collector orientation. The abscissa is the azimuth and the ordinate shows the elevation angle of the collecting target. To obtain a presentation that is independent of absolute figures, maximum values have been marked as 100% and several isolines of percentage (97.5, 95, 92.5, 90, 80, 70, 60, 50) have been constructed. Each line encloses an area that receives a higher percentage of incident solar energy than the isoline indicates. Obviously, the abscissa is such an isoline, as an elevation angle of zero means a horizontal surface. The isoline value of the abscissa varies slightly with the reference station (about 62% for Rome, 65% for Brig, 68% for Denver, 70% for Stockholm, 72% for Copenhagen and 77% for Zurich). The results of such calculations depend – as well as on location – on both the model used to separate direct from global radiation and the model describing the distribution of sky radiosity. Here, the widely acknowledged model of Perez *et al.*[7] was used to compute Figures C.5 to C.10.

Table C.7. Characteristic climatic data for various locations obtained by stochastic modelling with Meteonorm; reference locations in bold type

Location	Latitude (°)	Longitude (°)	Heating season	Degree days (Kd)	Global radiation (kWh/m^2)	Solar Similarity Index (kWh/(m^{2}Kd))
Oslo	59.5	–10.4	October–May	4113	441.5	0.107
Stockholm	**59.1**	**–17.6**	**September–May**	**4661**	**514.0**	**0.110**
Aberdeen	57.1	2.1	September–May	3702	485.3	0.131
Copenhagen	**55.4**	**–12.3**	**October–May**	**3699**	**458.6**	**0.124**
Hamburg	53.3	–9.4	October–May	3379	448.4	0.133
Kassel	51.2	–9.3	October–April	3373	327.6	0.097
Trier	49.5	–6.4	October–April	3261	351.1	0.108
Munich	48.1	–11.2	October–May	3773	570.8	0.151
Freiburg	48.0	–7.5	October–April	3091	384.0	0.124
Zurich	**47.2**	**–8.3**	**October–April**	**3237**	**371.7**	**0.115**
Innsbruck	47.2	–11.2	October–April	3461	458.1	0.132
Brig	**46.2**	**–7.6**	**October–April**	**3471**	**522.6**	**0.151**
Geneva	46.1	–6.1	October–April	2997	419.5	0.140
Turin	45.1	–11.4	November–April	2565	429.6	0.168
Genova	44.3	–8.6	December–March	1241	242.8	0.196
Carpentras	44.0	–5.1	November–April	2216	481.7	0.217
Pisa	43.4	–10.2	November–March	1724	298.2	0.173
Montpellier	43.4	–3.6	November–March	1790	331.2	0.185
Marignane	43.3	–5.1	November–March	1706	348.6	0.204
Rome	**42.1**	**–12.3**	**November–March**	**1539**	**356.1**	**0.231**
Denver	**39.5**	**104.5**	**October–April**	**3478**	**732.0**	**0.210**

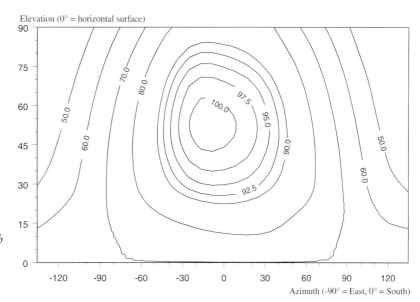

Figure C.5. Isolines of dimensionless solar irradiance for Stockholm on a collector with various orientations during the heating season; optimum tilt is 50° ± 10° at an azimuth of 0° ± 10°

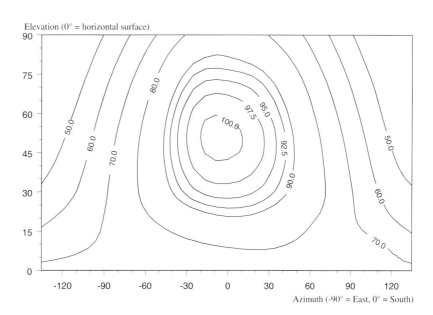

Figure C.6. Isolines of dimensionless solar irradiance for Copenhagen on a collector with various orientations during the heating season; optimum tilt is 50° ± 10° at an azimuth of 0° ± 10°

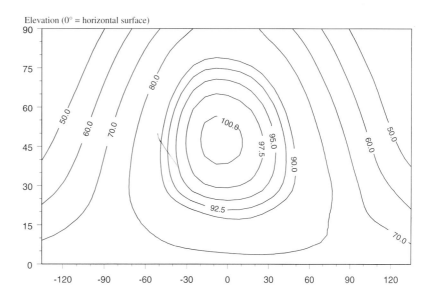

Figure C.7. Isolines of dimensionless solar irradiance for Zurich on a collector with various orientations during the heating season; optimum tilt is 50° ± 10° at an azimuth of 0° ± 10°

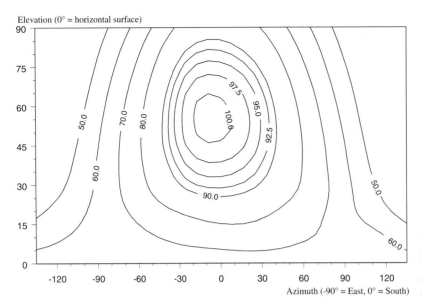

Figure C.8. Isolines of dimensionless solar irradiance for Brig on a collector with various orientations during the heating season; optimum tilt is $50° \pm 10°$ at an azimuth of $0° \pm 10°$

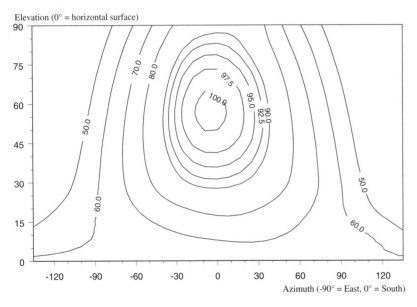

Figure C.9. Isolines of dimensionless solar irradiance for Rome on a collector with various orientations during the heating season; optimum tilt is $50° \pm 10°$ at an azimuth of $0° \pm 10°$

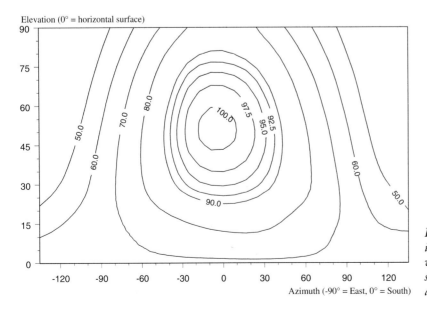

Figure C.10. Isolines of dimensionless solar irradiance for Denver on a collector with various orientations during the heating season; optimum tilt is $50° \pm 10°$ at an azimuth of $0° \pm 10°$

For all locations, the optimum orientation of solar collectors for energy gains during the specific heating period is not defined in a very precise way. The 100% array has some spatial extension and is always situated facing true south (± 10°) with an elevation angle of about 50°, again with allowed deviations of ± 10°. Vertical facades directed predominantly towards the south collect between 82% and 88% of the maximum solar energy. Deviations in the azimuth angle by ± 30° reduce this value to about 80%.

From Figures C.5 to C.10, solar energy gains of arbitrarily oriented collectors can be estimated approximately from horizontal solar gains for every reference station. One should keep in mind, however, that solar air systems do not behave linearly with solar energy input. Utilizability of solar gains depends on time of energy supply; this means on the history of the temperatures in the system. Application of Figures C.5 to C.10 to predict energy performances of solar air systems are therefore recommended only for first quick guesses.

REFERENCES

1. *IEA Task 8: Design Guidelines: An International Summary* (1990) Report No. IEA SHAC T.8.C.3. The Superintendent of Documents, US Government Printing Office, Washington, DC 20402.
2. Müller MJ (1980) *Handbuch ausgewählter Klimastationen der Erde*. 2nd edn. Universität Trier, Forschungsstelle Bodenerosion, Mertesdorf (Ruwertal), Trier. Available from University of Trier, Fachbereich VI, Selbstverlag, 54286 Trier, Germany.
3. *Atlas über die Sonnenstrahlung Europas* (1984) Band 1: *Horizontale Flächen*. Kommission der Europäischen Gemeinschaften, Verlag TÜV Rheinland, Köln, Germany.
4. Duffie JA, Beckman WA (1980) *Solar Engineering of Thermal Processes*. John Wiley & Sons, New York.
5. Axum 5.0 for Windows. MathSoft Inc., 101 Main Street, Cambridge, Massachusetts 02142-1521, USA.
6. METEONORM, Global Meteorological Database for Solar Energy and Applied Climatology. Meteotest, Fabrikstrasse 14, CH-3012 Bern, Switzerland.
7. Perez R, Ineichen P, Maxwell E, Seals R, Zelenka A (1991) 'Dynamic models for hourly global-to-direct irradiance conversion', *Proceedings of the Biennial Solar World Congress of the International Solar Energy Society*, Denver, Colorado, USA. Vol. 1, Part II, pp. 951–956. ISES International Headquarters, Wiesentalstrasse 50, 79115 Freiburg, Germany.

Prepared by Frank Heidt

Appendix D: Climate impact on performance

This Appendix describes how the energy performance of buildings varies with climate. It addresses the important problem how the energy balance of a building changes if the building is moved to another climate. To help answer this question, the next section presents the theory and results of general investigations. This is followed by two examples of how the energy results for a given climate can be transferred to other climatic conditions. Moreover, these predictions have been compared with corresponding simulation results.

DEFINITIONS, THEORY AND RESULTS

The thermal balance of a building is established for an extended period, say for some months or for the heating season. Thus, heat stored or released from building mass does not significantly contribute to the long-term energy balance. This balance can be stated for three different conditions of the building:

(0) without any solar gains;
(S) with passive solar gains but without an active solar system;
(SS) with both passive and active gains.

The nomenclature in Table D.1 is used, where Q (in kWh) stands for the energy accumulated during the period considered and where the subscripts specify its type.

Heat losses and solar gains are dominated by the climatic parameters degree days (DD in K × d) and global horizontal solar radiation (GH in kWh/m^2). Again, these values are summed up for the considered period.

According to common practice total heat losses of a building in case (0) are written as:

Table D.1. Nomenclature

Q_I	Internal gains
$Q_{LOAD,0}$	Losses (transmission, infiltration or ventilation), case (0)
$Q_{LOAD,S}$	Losses, case (S)
$Q_{LOAD,SS}$	Losses, case (SS)
$Q_{AUX,0}$	Auxiliary heating, case (0)
$Q_{AUX,S}$	Auxiliary heating, case (S)
$Q_{AUX,SS}$	Auxiliary heating, case (SS)
$Q_{SOL,S}$	Passive solar gain, case (S)
$Q_{SOL,SS}$	Passive solar gain, case (SS)
Q_{SYS}	Active solar gain, case (SS)

$$Q_{LOAD,0} = A_H \times BLC \times DD \qquad (D.1)$$

where A_H is the heated floor area in m^2 and BLC is the building load coefficient in kWh/(m^{2}Kd).
In a similar way solar gains in cases (S) and (SS) may be expressed as:

$$Q_{SOL,S} = A_P \times SGC_S \times GH \qquad (D.2)$$

$$Q_{SOL,SS} = A_P \times SGC_{SS} \times GH \qquad (D.3)$$

$$Q_{SYS} = A_C \times SGC_{SYS} \times GH, \qquad (D.4)$$

where A_P is the aperture area for passive gains in m^2, A_C is the aperture area for active gains in m^2 and SGC is the solar gain coefficient (dimensionless), different for a mere passive (subscript S) or a combined passive/active system (subscript SS).
BLC and SGC are complex functions of building parameters and aperture area orientations and may even depend slightly on geographical position. They are not further discussed in detail.
The main and most important feature of equations (D.1) through (D.4) is the proportionality of $Q_{LOAD,0}$ to DD, and of $Q_{SOL,S}$ and $Q_{SOL,SS}$ + Q_{SYS} to GH. With the terms defined above, thermal energy balances for cases (0), (S) and (SS) result in:

$$(0)\ Q_{LOAD,0} = Q_I + Q_{AUX,0} \qquad (D.5)$$

$$(S)\ Q_{LOAD,S} = Q_I + Q_{AUX,S} + Q_{SOL,S} \qquad (D.6)$$

$$(SS)\ Q_{LOAD,SS} = Q_I + Q_{AUX,SS} + Q_{SOL,SS} + Q_{SYS} \qquad (D.7)$$

Obviously, not all of the solar energy offered to a building can be used for the reduction of auxiliary heating. Therefore, solar gains must be split up into usable and useless fractions. The usable fraction reduces the required auxiliary heating, whereas the useless fraction augments heating losses. $Q_{SOL,S}$ and $Q_{SOL,SS}$ + Q_{SYS} may be each presented by the identities:

$$Q_{SOL,S} = F_S \times Q_{SOL,S} + (1 - F_S) \times Q_{SOL,S} \qquad (D.8)$$
$$Q_{SOL,SS} + Q_{SYS} = F_{SS} \times (Q_{SOL,SS} + Q_{SYS})$$
$$+ (1 - F_{SS}) \times (Q_{SOL,SS} + Q_{SYS}), \qquad (D.9)$$

where the first terms on the right-hand sides describe the usable and the second terms the useless parts. The dimensionless factors F_S and F_{SS} are called utilization factors and are in the range between zero and one:

$$0 < F_S, F_{SS} \leq 1. \quad (D.10)$$

According to the definition of usability (= reduction of auxiliary heating) and uselessness (= increase of heat losses) one obtains:

$$Q_{AUX,S} = Q_{AUX,0} - F_S \times Q_{SOL,S} \quad (D.11)$$

(reduction of auxiliary heating due to useful passive solar gains);

$$Q_{LOAD,S} = Q_{LOAD,0} + (1 - F_S) \times Q_{SOL,S} \quad (D.12)$$

(increasing the heat losses due to useless passive solar gains);

$$Q_{AUX,SS} = Q_{AUX,0} - F_{SS} \times (Q_{SOL,SS} + Q_{SYS}) \quad (D.13)$$

(reduction of auxiliary heating due to passive and active solar gains);

$$Q_{LOAD,SS} = Q_{LOAD,0} + (1 - F_{SS}) \times (Q_{SOL,SS} + Q_{SYS}) \quad (D.14)$$

(increasing the heat losses due to useless passive and active solar gains).

In order to see how climate influences energy balances equation (D.7) is transformed into a dimensionless relation. Substitution of $Q_{LOAD,SS}$ in equation (D.7) from equation (D.14) and solving for $Q_{LOAD,0}$ yields:

$$Q_{LOAD,0} = Q_I + Q_{AUX,SS} + F_{SS} \times Q_{SOL,SS}.$$

Division by $Q_{LOAD,0}$ and substitution of $Q_{LOAD,0}$ and $Q_{SOL,SS}$, Q_{SYS} from equations (D.1), (D.3) and (D.4) yields:

$$1 = CON_{SS} + SOL_{SS}, \quad (D.15)$$

with

$$CON_{SS} = \frac{Q_{AUX,SS} + Q_I}{A_H \times BLC \times DD} \quad (D.15a)$$

$$SOL_{SS} = F_{SS} \times \frac{(A_P \times SGC_{SS} + A_C \times SGC_{SYS}) \times GH}{A_H \times BLC \times DD}. \quad (D.15b)$$

The dimensionless variables CON_{SS} and SOL_{SS} designate contributions to the dimensionless total thermal energy demand due to transmission and ventilation (without gains) that are realized by 'conventional' (CON) or by 'solar' (SOL) sources. 'Conventional' stands for auxiliary heating and internal gains. In other words, SOL_{SS} is the percentage or fraction of the total thermal energy demand without gains that is going to be covered by passive and active solar means.

In the dimensionless equations (D.15) the influence of climate on building energy balance is represented by the two meteorological parameters DD in (D.15a) and GH/DD in (D.15b). For buildings with almost no solar gains (this means buildings with negligibly small values of A_P, A_C or SGC) the variable GH/DD is of almost no importance because SOL_{SS} is close to zero. This implies that CON_{SS} has a value close to one and from equation (D.15a) it follows that:

$$Q_{AUX,SS} + Q_I \approx A_H \times BLC \times DD.$$

In this case, no solar contributions reduce the auxiliary heating demand $Q_{AUX,SS}$ and DD is the only relevant climatic parameter.

For buildings with noticeable solar contributions, however, SOL_{SS} is significantly greater than zero and CON_{SS} is smaller than one by the same value. Then the ratio GH/DD becomes a second climatic parameter, which is relevant for the description of the dimensionless thermal energy balance of the building. This ratio has dimensions $(kWh/m^2)/(K \times d)$ and measures, for a given building, the climate impact on the ratio of solar gains and total thermal losses. It is called the 'Solar Similarity Index' (SSI):

$$SSI = GH/DD. \quad (D.16)$$

The reason for this nomenclature can be explained as follows: If SSI is given as a fixed value, GH can still vary if DD varies accordingly. SOL_{SS}, however, remains essentially constant as the solar system (SGC, A_P, A_C) and the building itself (BLC, A_H) are not altered, if additionally it is assumed that F_S or F_{SS} are not – or at most insignificantly – modified when the climate (DD, GH) is changed. This hypothesis seems to be justified for cases where the solar–load ratio does not vary much.

From equation (D.15), constant values of SOL_{SS} result in constant values of CON_{SS}. This means that a constant SSI forces constant quantities of SOL and CON and hence a constant ratio of solar and conventional heating. This is exactly what solar similarity means.

If an existing building and solar system are moved from one site with an SSI value SSI(1) to another location with value SSI (2), according to equation (D.15b) the corresponding SOL quantity varies as:

$$SOL(2) = SOL(1) \times SSI(2)/SSI(1). \quad (D.17)$$

Here again the above-mentioned hypothesis of almost constant utilization factors has been used. Combining equations (D.17) and (D.15) results in the rule covering how CON quantities are transformed when SSI is altered:

$$CON(2) = 1 - SOL(1) \times SSI(2)/SSI(1). \quad (D.18)$$

According to equation (D.1), the total heat loss of a building without any solar gains, $Q_{LOAD,0}$, transforms with climate as:

$$Q_{LOAD,0}(2) = Q_{LOAD,0}(1) \times DD(2)/DD(1). \quad (D.19)$$

For the energetic and economic characterization of solar systems the term 'saved energy' (SE) is commonly used. The energy (SE_{SS}) saved by an active solar system is the difference between the auxiliary heating required without and with the system. With equations (D.11) and (D.13) it follows that:

$$SE_{SS} = Q_{AUX,S} - Q_{AUX,SS}$$
$$= F_{SS} \times (Q_{SOL,SS} + Q_{SYS}) - F_S \times Q_{SOL,S}. \quad (D.20)$$

Thus, SE_{SS} varies almost linearly with GH if utilization factors F_S and F_{SS} do not depend much on global radiation input GH = DD × SSI.

As a second, dimensionless characterization of solar systems, the term 'saved energy fraction' SF is also commonly used for an active solar system. It is defined by:

$$SF_{SS} = (Q_{AUX,S} - Q_{AUX,SS})/Q_{AUX,S}$$
$$= SE_{SS}/Q_{AUX,S}. \quad (D.21)$$

SF_{SS} indicates precisely the percentage of the existing auxiliary heating demand that can be saved if an active solar system is applied.

Both indicators for solar energy savings, the physical quantity 'saved energy' SE_{SS} and the dimensionless characteristic figure 'saved energy fraction' SF_{SS}, can be expressed as functions of the meteorological parameters DD and GH or the equivalent parameters DD and SSI. If the utilization factors F_S and F_{SS} are assumed to be constant, from equations (D.2), (D.3), (D.4) and (D.20) it follows that SE_{SS} is proportional to GH. Thus, with a a constant not influenced by the climate:

$$SE_{SS} = aGH = aSSI \times DD. \quad (D.22)$$

Under the same assumptions it follows from equation (D.21) by substitution from equations (D.22), (D.11), (D.5), (D.2) and (D.1) that SF_{SS} can be represented as

$$SF_{SS} = aSSI/(b - Q_I/DD - cSSI), \quad (D.23)$$

where b and c are two more constants that do not depend on climate. Coefficients a, b and c, however, mainly depend on the solar system and the building type and can be determined if the energy terms $Q_{AUX,SS}$, $Q_{AUX,S}$, $Q_{LOAD,0}$, $Q_{SOL,S}$ and Q_I are given for a reference site with known DD and GH. Once the values of a, b and c have been calculated, SE_{SS} and SF_{SS} can be obtained for climates other than that of the reference site from equations (D.22) and (D.23) by inserting the corresponding values of DD and GH or SSI = GH/DD, respectively.

With these results a stationary model for the thermal energy balance of solar buildings has been derived, which shows how the following quantities and characteristic numbers depend on climatic parameters:

- total building heat loss without solar gains, $Q_{LOAD,0}$;
- combined (passive and active) solar fraction SOL_{SS} needed to compensate for the total building heat loss;
- saved fraction of energy SF_{SS} in a solar building (combining active and passive solar gains) with reference to the passive building only;
- saved energy SE_{SS} in a solar building (combining active and passive solar gains) in comparison to the passive solar building.

Thus it is possible to predict all of these quantities for a new location with given DD and SSI, provided the corresponding values are known for a reference station.

The next step is to check the theoretical modelling by comparing its predictions with results of detailed simulations for several climates, buildings and types of solar air systems. This was done in March 1998; climates were generated with the Meteonorm software (see Appendix C) for 21 locations (Oslo, Stockholm, Aberdeen, Copenhagen, Hamburg, Kassel, Trier, Munich, Freiburg, Zurich, Innsbruck, Brig, Geneva, Turin, Genova, Carpentras, Pisa, Montpellier, Marignane, Rome and Denver). Building construction was characterized as solid (S) or light-weight (L) and its degree of thermal insulation as high (H), medium (M) or poor (P). The active solar system was varied between four and six collector modules (about 10 m² and 15 m² respectively). Table D.2 gives a survey of the cases investigated.

Investigations of solar air systems were confined to System 1 (solar heating of ventilation air), System 2 (Bara–Costantini loop = collector–room–collector) and System 4 (closed loop with radiant or forced discharge). Periods of time for which Trnsair simulations were performed were the heating seasons. As heating is supposed to be necessary only during those months in which the monthly averaged daily outdoor tem-

Table D.2. Characteristics of building types investigated

Case No.	1	2	3	4	5	6
Construction type	S	S	L	L	S	L
Thermal insulation	H	H	H	H	P	M
Number of collectors	6	4	6	4	6	6

Table D.3. Climatic characteristics of investigated locations

Location	Location set	Latitude (°)	Longitude (°)	Heating season	Degree days (Kd)	Global radiation (kWh/m²)	SSI (kWh/(m²Kd))
Kassel	I	51.2	−9.3	October–April	3373	327.6	0.097
Oslo	I	59.5	−10.4	October–May	4113	441.5	0.107
Trier	I	49.5	−6.4	October–April	3261	351.1	0.108
Stockholm	I	59.1	−17.6	September–May	4661	514.0	0.110
Zurich	I	47.2	−8.3	October–April	3237	371.7	0.115
Copenhagen	I	55.4	−12.3	October–May	3699	458.6	0.124
Freiburg	I	48.0	−7.5	October–April	3091	384.0	0.124
Aberdeen	I	57.1	2.1	September–May	3702	485.3	0.131
Innsbruck	II	47.2	−11.2	October–April	3461	458.1	0.132
Hamburg	I	53.3	−9.4	October–May	3379	448.4	0.133
Geneva	II	46.1	−6.1	October–April	2997	419.5	0.140
Brig	II	46.2	−7.6	October–April	3471	522.6	0.151
Munich	II	48.1	−11.2	October–May	3773	570.8	0.151
Turin	II	45.1	−11.4	November–April	2565	429.6	0.168
Pisa	III	43.4	−10.2	November–March	1724	298.2	0.173
Montpellier	III	43.4	−3.6	November–March	1790	331.2	0.185
Genova	III	44.3	−8.6	December–March	1241	242.8	0.196
Marignane	III	43.3	−5.1	November–March	1706	348.6	0.204
Denver	III	39.5	104.5	October–April	3478	732.0	0.210
Carpentras	III	44.0	−5.1	November–April	2216	481.7	0.217
Rome	III	42.1	−12.3	November–March	1539	356.1	0.231

perature falls below 12°C, different lengths of the heating season come out for different climates. Table D.3 summarizes the climatic conditions for the chosen locations, as obtained from the Meteonorm weather files.

The locations in Table D.3 are sorted by their SSI value. They extend over a broad range of SSI from less than 0.100 kWh/(m²Kd) to more than 0.230 kWh/(m²Kd). Therefore, locations have been grouped to three sets which represent (I) northern and central Europe (0.097 kWh/(m²Kd) ≤ SSI ≤ 0.133 kWh/(m²Kd)), (II) the pre-alpine region (0.132 kWh/(m²Kd) ≤ SSI ≤ 0.168 kWh/(m²Kd)) and (III) the Mediterranean area (0.173 kWh/(m²Kd) ≤ SSI ≤ 0.231 kWh/(m²Kd)), which includes for the sake of simplicity also the sunny and cold location of Denver. Copenhagen, Geneva and Genova have been identified as suitable reference stations for the respective groups.

The combination of all locations (21), building types (six) and solar air systems (three out of six system types) would give far too many cases to be presented in this context. As the emphasis of this appendix is on the prediction of climatic variability, the display of results is restricted to one solar air system (System 4: closed collector loop with radiant or forced discharge) and one building type (SH4: solid building construction with high standard of thermal insulation and 10 m² collector area). Figures D.1 to D.3 show the results of comparisons made between estimates from the theoretical model described above (short name: SSI-prediction) and values calculated from dynamic simulations with Trnsair (short name: Trnsair-simulation) for the dimensionless characteristic numbers SOL_{SS}, SF_{SS} and the saved energy SE_{SS} in kWh. The conclusion of these comparisons is that there is fair to good correspondence between SSI predictions and simulated values. The accuracy of

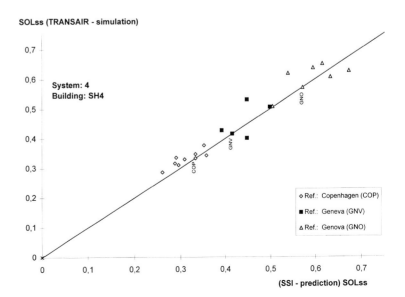

Figure D.1. Comparison of SSI-prediction with Trnsair-simulation of combined solar fraction SOL_{SS} of basic building heat demand

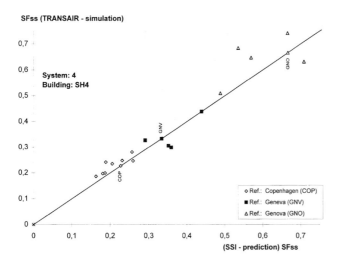

Figure D.2. Comparison of SSI-prediction with Trnsair-simulation of saved fraction of energy SF_{SS} in a combined solar building compared with a passive solar building

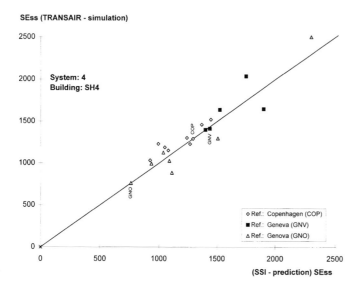

Figure D.3. Comparison of SSI-prediction with Trnsair-simulation of saved energy SE_{SS} (in kWh) in a combined solar building compared with a passive solar building

the estimated saved thermal energies is generally better than 250 kWh. For the whole heating season this uncertainty corresponds to a financial equivalent of less than US $10.

TWO ILLUSTRATIVE EXAMPLES OF PRACTICAL APPLICATIONS

A typical problem in practice is that one knows, maybe from measurements or from simulation results, the energy use for an actual building with its specific type of solar air system at a given location, and one would like to predict the corresponding values for another site, for which the climatic parameters (DD, GH or SSI) are given. The question cannot be restricted to $Q_{AUX,SS}$, the energy for auxiliary heating, alone, because this energy depends, according to equation (D.13), on other energies ($Q_{AUX,0}$, $Q_{SOL,SS}$, Q_{SYS}). If, additionally, saved energy SE_{SS} (saved by the active solar system) is of interest, $Q_{AUX,S}$ must also be given.

In order to show that the theoretical model described above also works for other solar air systems, two further examples have been calculated and displayed in Table D.4. Here, the energy quantities of systems may be known for Copenhagen (System 1: solar heating of ventilation air) and Zurich (System 2: Bara–Costantini loop = collector–room–collector) and are predicted for Freiburg (System 1) and Stockholm (System 2) as well.

The most interesting quantities for the considered destination sites are saved energy SE_{SS}, auxiliary heating demand $Q_{AUX,SS}$, energy delivered by the active solar system Q_{SYS}, passive solar gains $Q_{SOL,SS}$ and total heat loss without solar gains $Q_{LOAD,0}$. Formulae to estimate these energies are already known from the theoretcial section above and, therefore, will be referenced only in a short way.

Note, that internal gains Q_I are assumed from boundary conditions that depend on the lengths of the heating seasons for the buildings. They are based on an average internal heating power of 350 W (per 100 m² heated

Table D.4. *Summary of energy data for two reference sites (Copenhagen, System 1, and Zurich, System 2) and two locations of destination (System 1: Freiburg, System 2: Stockholm); predicted values are in bold type and simulation results in italic type*

Solar air system	Type 1			Type 2		
Reference climate/destination site	Copenhagen/Freiburg			Zurich/Stockholm		
DD (Kd)	3699/3091			3237/4661		
GH (kWh/m^2)	459/384			372/514		
SSI (kWh/(m^{2}Kd))	0.124/0.124			0.115/0.110		
Heating season	October–May/October–April			October–April/September–May		
Q_I (kWh)	2041/1780			1780/2293		
$Q_{LOAD,0}$ (kWh)	9673 /	**8083**	*8094*	8456 /	**12176**	*12211*
$Q_{SOL,SS}$ (kWh)	2237 /	**1871**	*1880*	1883 /	**2602**	*2780*
Q_{SYS} (kWh)	1334 /	**1116**	*1146*	445 /	**615**	*725*
$Q_{AUX,SS}$ (kWh)	4717 /	**3864**	*3818*	4734 /	**7201**	*7187*
SE_{SS} (kWh)	1008 /	**843**	*856*	296 /	**410**	*408*
F_S (dimensionless)	0.785 /	0.785	–	0.869 /	0.869	–
F_{SS} (dimensionless)	0.817 /	0.817	–	0.834 /	0.834	–

living area). The equations to determine the relevant quantities $Q_{LOAD,0}$, $Q_{SOL,SS}$, Q_{SYS}, $Q_{AUX,SS}$ and SE_{SS} for a new climate (2) from known quantities in a given climate (1) are therefore:

$$Q_{LOAD,0}(2) = Q_{LOAD,0}(1) \times DD(2)/DD(1) \quad (D.19)$$

$$Q_{SOL,SS}(2) = Q_{SOL,SS}(1) \times GH(2)/GH(1) \quad (D.20)$$

$$Q_{SYS}(2) = Q_{SYS}(1) \times GH(2)/GH(1) \quad (D.21)$$

$$SE_{SS}(2) = SE_{SS}(1) \times GH(2)/GH(1) \quad (D.22)$$

$$Q_{AUX,SS}(2) = Q_{LOAD,0}(2) - Q_I(2) - F_{SS}(1)$$
$$\times (Q_{SOL,SS}(2) + Q_{SYS}(2)), \quad (D.23)$$

where equation (D.23) is derived from equations (D.13) and (D.5) under the assumption $F_{SS}(2) = F_{SS}(1)$.

In Table D.4, values theoretically predicted from the equations above (in bold type) have been compared with data simulated with Trnsair (in italic type). In all cases a remarkable agreement in the results is observed.

Prepared by Frank Heidt

Appendix E: Economic evaluation

Very often the decision whether to use solar energy depends on economics. However, most decisions we make in life are not made based on mere economics. The purchase of clothing, furniture, cars etc. follows other rules. Quality, comfort or prestige associated with a product may dominate. Therefore, a decision to use solar energy for heating and domestic hot water may, for example, be based on a personal attitude towards reducing environmental impact, climate change or living with *natural* resources.

Two methods to analyse the economics of an investment are common: static and dynamic methods. Static economic analysis compares only yearly averaged savings with the corresponding costs and does not consider the influence of the time taken for payment in a correct manner. As solar systems integrated in buildings have long lifetimes (in the range between 15 and 25 years), dynamic economic analysis is preferable because it takes fully into account the importance of the time when payments occur.

Static analyses are commonly used because they require less mathematical effort than dynamic methods. As an example, simple static payback-time analysis (without consideration of interest) compares only initial investments I_0 with average annual revenues R due to energy savings in order to get the payback time n from:

$$n = \frac{I_0}{R}. \quad (E.1)$$

This (simple) payback time is just the number of years over which the sum of annual revenues from energy savings becomes equal to the amount of initial investment. The initial investment is the incremental cost of the solar system and does not involve costs that would arise in any other case for the corresponding building component. Annual revenues are the averaged energy cost avoided annually, which consists of the annual energy savings multiplied by the cost per energy unit. An initial (incremental) investment of €1,500, which leads to energy savings of 1.000 kWh with specific costs of €0.06/kWh, therefore, has a (simple) payback time of:

$$n = €1,500/(1000 \text{ kWh/a} \times €0.06/\text{kWh}) = 25 \text{ a}$$
(a = years).

This length of time is likely to approach the lifespan of a solar system. It will be compared below with the results of corresponding dynamic analyses.

If the dynamic nature of capital is considered, an initial investment is the origin of a time series of payments, as schematically illustrated in Figure E.1. Payments can be negative (expenditure, mainly due to initial investment, interest and cost of maintenance and operation) or positive (revenue, mainly due to saved energy costs). These are subjected to general inflation rates and perhaps different energy inflation rates.[1] For a proper balance of payments the time series of revenues and expenditures has to be discounted at a specific point in time.

The capital value C describes the amount of money that is obtained beyond a fixed (wished or payable) interest rate for capital or a loan. If related to the date of initial investment C becomes:

$$C(n,p) = \sum_{j=1}^{n} \frac{R_j - E_j}{(1+p)^j} + \frac{RV(n)}{(1+p)^n} - I_0, \quad (E.2)$$

where

- $C > 0$ means that the investment is economically viable for the given conditions.
- $C = 0$ means that the desired or required interest rate for the investment is just obtained.
- $C < 0$ means that the desired or required interest rate for the investment cannot be achieved for the given conditions.

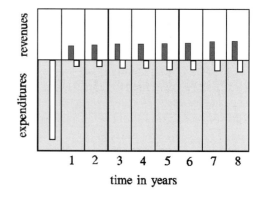

Figure E.1. Time series of payments (revenue and expenditure) as a consequence of an initial investment

If there are several technical system options, the one with maximum capital value C will be selected.

For a given initial investment I_0 the capital value C depends very much on the number n of years considered and on the interest rate p. The residual value $RV(n)$ after n years is often assumed to be zero.

Revenues R_j in the jth year depend on the solar system performance as well as on the inflation rate e for energy costs. Expenditures E_j in the jth year are costs of operation and maintenance of the solar system. These costs are usually assumed to be equal to a fixed percentage f of the initial investment I_0 and subject to the general inflation rate g. With yearly saved energy SE (in kWh) and specific energy costs CE (in €/kWh), under the assumption of invariable inflation rates e and g, R_j and E_j become:

$$R_j = \text{SE} \times \text{CE}(1+e)^j \quad (E.3)$$
$$E_j = fI_0(1+g)^j \quad (E.4)$$

In the following, several economic terms for analysing the capital value C are defined and the corresponding formulae are given. Illustrative examples are presented that show how to use these terms.

The evaluation of a capital value C can be done in different ways. One can calculate the following:

- payback time (n_*);
- internal interest rate (p_*);
- annuity (A).

Payback time

This is defined as that time span n_* (in years) since the date of investment, during which the initial investment costs I_0 including the expected or payable interests (at rate p) have been recovered. With vanishing residual value ($RV(n_*) = 0$) C yields:

$$C(n_*, p) = \sum_{j=1}^{n_*} \frac{R_j - E_j}{(1+p)^j} - I_0 = 0. \quad (E.5)$$

This is an implicit equation to determine n_* as the root of $C(n_*, p) = 0$. In practice, the solution of this equation has to be found by iterative calculations.

The shorter the payback time is, the more economic is the solar investment. In any case, the lifetime of a solar system must be longer than its payback period.

Internal interest rate

This is defined as that interest rate p_* (%), which after a given time period of n years reduces the capital value C to zero:

$$C(n, p_*) = \sum_{j=1}^{n} \frac{R_j - E_j}{(1+p_*)^j} - I_0 = 0. \quad (E.6)$$

Again this is an implicit equation to determine p_* as a root of $C(n, p_*) = 0$. To solve this equation, iterative numerical algorithms have to be applied.

The internal interest rate p_* of a solar investment should be as high as possible. In any case, the interest rate for capital or loan must be less than the internal interest rate p_*.

Annuity

This is defined as a constant amount A payable at the end of every n interest periods (years) for which the sum of payments is equal to the future amount of the initial capital value C after these n years:

$$A\frac{(1+p)^n - 1}{p} = C(1+p)^n.$$

With the capital interest rate p, annuity A is obtained directly from the above definition as:

$$A = \frac{p(1+p)^n}{(1+p)^n - 1}C. \quad (E.7)$$

Inserting C from equation (E.2) results in an explicit expression for the annuity A. The greater the annuity is, the better is the (solar) investment. In no case is annuity A allowed to be negative (equivalent to $C < 0$).

The annuity factor A/C is given by:

$$\frac{A}{C} = \frac{p(1+p)^n}{(1+p)^n - 1} \quad (E.8)$$

and this depends merely on interest rate p and the number of interest periods n. It is that fraction of the initial amount C that has to be paid at the end of each interest period in order to pay off the future amount of the capital value C (the accumulated gains or debts) after n years. This factor (the correct name of which is Equal-Payment Series Capital-Recovery Factor[2]) is listed in Table E.1 for $n = 6$ to 25 years and interest rates p between 3% and 8.5%, with steps of 0.5%.

For the special case where revenue minus expenditure is constant for every year, the annuity can easily be calculated from equations (E.7) and (E.2) and yields:

$$A = (R - E) - I_0 \frac{p(1+p)^n}{(1+p)^n - 1}. \quad (E.9)$$

In other words, the total annuity A is the difference between the surplus of revenues over expenditures ($R - E$) on the one hand and the annuity $A(I_0)$ for the initial investment I_0 on the other hand. To give an example, let $C = €1,500$, $p = 5\%$ and $n = 25$ years. From Table E.1 it can easily be determined that $A(I_0)$ for an initial amount I_0 is $A(I_0) = 0.0710 \times I_0 = 0.0710 \times €1,500 = €106.5$.

Table E.1. Annuity factors (Equal-Payment Series Capital-Recovery Factors) for different interest rates p and periods of interest n according to equation (E.8)

p (%) / n (years)	3	3.5	4	4.5	5	5.5	6	6.5	7	7.5	8	8.5
6	0.1846	0.1877	0.1908	0.1939	0.1970	0.2002	0.2034	0.2066	0.2098	0.2130	0.2163	0.2196
7	0.1605	0.1635	0.1666	0.1697	0.1728	0.1760	0.1791	0.1823	0.1856	0.1888	0.1921	0.1954
8	0.1425	0.1455	0.1485	0.1516	0.1547	0.1579	0.1610	0.1642	0.1675	0.1707	0.1740	0.1773
9	0.1284	0.1314	0.1345	0.1376	0.1407	0.1438	0.1470	0.1502	0.1535	0.1568	0.1601	0.1634
10	0.1172	0.1202	0.1233	0.1264	0.1295	0.1327	0.1359	0.1391	0.1424	0.1457	0.1490	0.1524
11	0.1081	0.1111	0.1141	0.1172	0.1204	0.1236	0.1268	0.1301	0.1334	0.1367	0.1401	0.1435
12	0.1005	0.1035	0.1066	0.1097	0.1128	0.1160	0.1193	0.1226	0.1259	0.1293	0.1327	0.1362
13	0.0940	0.0971	0.1001	0.1033	0.1065	0.1097	0.1130	0.1163	0.1197	0.1231	0.1265	0.1300
14	0.0885	0.0916	0.0947	0.0978	0.1010	0.1043	0.1076	0.1109	0.1143	0.1178	0.1213	0.1248
15	0.0838	0.0868	0.0899	0.0931	0.0963	0.0996	0.1030	0.1064	0.1098	0.1133	0.1168	0.1204
16	0.0796	0.0827	0.0858	0.0890	0.0923	0.0956	0.0990	0.1024	0.1059	0.1094	0.1130	0.1166
17	0.0760	0.0790	0.0822	0.0854	0.0887	0.0920	0.0954	0.0989	0.1024	0.1060	0.1096	0.1133
18	0.0727	0.0758	0.0790	0.0822	0.0855	0.0889	0.0924	0.0959	0.0994	0.1030	0.1067	0.1104
19	0.0698	0.0729	0.0761	0.0794	0.0827	0.0862	0.0896	0.0932	0.0968	0.1004	0.1041	0.1079
20	0.0672	0.0704	0.0736	0.0769	0.0802	0.0837	0.0872	0.0908	0.0944	0.0981	0.1019	0.1057
21	0.0649	0.0680	0.0713	0.0746	0.0780	0.0815	0.0850	0.0886	0.0923	0.0960	0.0998	0.1037
22	0.0627	0.0659	0.0692	0.0725	0.0760	0.0795	0.0830	0.0867	0.0904	0.0942	0.0980	0.1019
23	0.0608	0.0640	0.0673	0.0707	0.0741	0.0777	0.0813	0.0850	0.0887	0.0925	0.0964	0.1004
24	0.0590	0.0623	0.0656	0.0690	0.0725	0.0760	0.0797	0.0834	0.0872	0.0911	0.0950	0.0990
25	0.0574	0.0607	0.0640	0.0674	0.0710	0.0745	0.0782	0.0820	0.0858	0.0897	0.0937	0.0977

Therefore, in order to realize a positive total annuity A, the yearly surplus of revenue over expenditure $R - E$ must be greater than €106.5. If, however, n is equal to 20 years only, the annuity factor increases from 0.0710 to 0.0802 and the yearly surplus $R - E$ must exceed €120.5.

Of course, this case neglects the influences of yearly varying general inflation rates g_j, percentages of increasing or decreasing energy prices e_j and stochastically fluctuating saved energy SE_j. Nevertheless, as these fluctuations (g_j, e_j, SE_j) are hardly predictable with qualified accuracy, the model of a yearly constant surplus $R - E$ may be appropriate as a first guess at the real annuity A, which reflects the above-mentioned influences.

To illustrate the theoretical results of the economic evaluation methods presented, the conditions for a solar air system shown in Table E.2 are considered.

For the sake of simplicity, the residual value RV of the solar system is set equal to zero after n = 20 years. The values of interest for the economic evaluation are now:

- payback time $n_* = 26.9$ years;
- internal interest rate $p_* = 1.4\%$;
- annuity $A = -$€43.

Table E.2. Example conditions for a solar air system

Energy saved per year, SE	1000 kWh
Specific costs* for energy, CE	€0.06/kWh
Initial investment, I_0	€1500
Interest rate for capital, p	0.05 (5%)
General inflation rate, g	0.01 (1%)
Inflation rate for energy costs, e	0.07 (7%)
Fraction of initial investment for operation and maintenance, f	0.02 (2%)
Number of interest periods, n	20 years

*Specific thermal energy costs for district heating lie between €0.06/kWh and €0.09/kWh

The interpretation of these results is:

- Payback time n_* is almost seven years longer than the number of interest periods (= expected lifetime). Note that this value does not differ much from the example for simple payback time, presented following equation (E.1). This coincidence of results very much depends, however, on the assumptions about the various interest rates.
- Internal interest rate p_* amounts to only 1.4%, which is significantly less than the assumed interest rate for borrowing (5%).
- Annuity A is negative (–€43), which means that for a period of 20 years every year an additional amount of €43 has to be spent in order to yield a yearly energy benefit of 1000 kWh.

For this example, in the first year the real specific costs of saved energy are therefore:

$$CE = (€0.06 + €0.043)/kWh = €0.103/kWh.$$

This is one of many examples where solar energy applications are not yet competitive with the use of conventional energy. However, the payback time (26.9 years), internal interest rate (1.4%) and annuity (–€43) are not too far away from their break-even values (20 years, 5%, €0). If, for example, an initial government subsidy of €500 reduced the initial investment from €1,500 to €1,000, the results for the economic analysis would be changed to:

- payback time $n_* = 18.5$ years;
- internal interest rate $p_* = 5.9\%$;
- annuity $A = +$€8.

Thus, the solar system becomes economically competitive. Economic viability would also be guaranteed if specific costs for energy were higher: With CE = €0.09/kWh, almost the same economic parameters as before would be obtained without any governmental subsidy:

- payback time $n_* = 18.5$ years;
- internal interest rate $p_* = 5.9\%$;
- annuity $A = +€13$.

This demonstrates the very important influence of energy pricing and governmental supporting actions on the economic performance of solar energy systems.

For German-speaking users there exists a software product, OEKO-RAT,[3] which calculates all values of interest for an economic evaluation of solar and energy-saving systems (payback time, internal interest rate, annuity, required initial funding, required energy price, required energy price increase) under a user-friendly surface.

REFERENCES

1. Boer KW (1978) 'Payback of solar systems'. *Solar Energy* **20**, 225–232.
2. Stevens GT (1979) *Economical and Financial Analysis of Capital Investments*. John Wiley & Sons, New York, Chichester, Brisbane, Toronto.
3. OEKO-RAT. Building Physics and Solar Energy Department, University of Siegen, D-57068 Siegen, Germany. Price about €100.

Prepared by Frank Heidt

Appendix F: Collector testing

LABORATORY TESTING SOLAR AIR COLLECTORS

The testing requirements for solar *water* collectors have been intensively developed since about 1980 and, now, as well as an ISO (International Standardization Organization) standard there is also a new CEN (Comité Européen de Normalization) testing standard under development.

For testing solar air collectors no standard exists. Compared to water collectors, the measuring procedure for solar air collectors is more complex. Measuring air temperatures and air mass flows requires more effort than for water. Leakage, the air flow distribution inside the collector and the much lower heat transfer from the absorber to the heat-transfer medium complicate measurement.

A solar simulator – usually used for testing solar liquid collectors – was the source of radiation for testing solar air collectors within the work of IEA Task 19. Up to 55 simulator lamps create radiation that is very similar in wavelength and intensity to natural sunlight.

The new construction of an air duct and measuring system in combination with the existing climate cham-

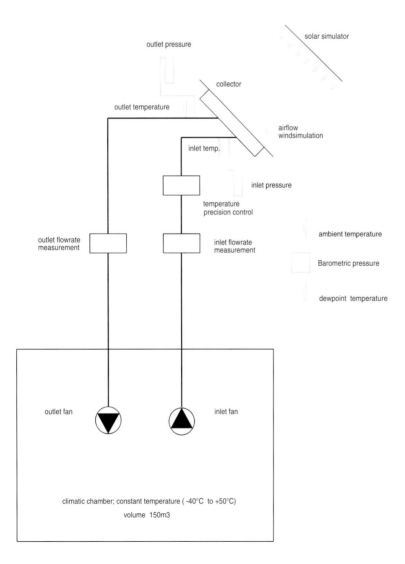

Figure F.1. Air collector testing facility

ber (volume 150 m³) met the requirements for carrying out a testing (Figure F.1).

Seven manufacturers from seven different countries provided collectors to be tested. Long-time proven products have been tested as well as promising prototypes.

General testing features

These include:

- an artificial radiation source, generating an insulation level of up to 1000 W/m² on the testing plane;
- an IR-shielding device (a cooled acrylic plate, simulating the cold sky temperature);
- surrounding air speed generated by two fans (wind simulation);
- a facility for measuring the global radiation;
- a temperature precision control on the stability of the inlet temperature;
- chillers for constant ambient temperature.

Measurement

Measurements were made of the following:

- inlet and outlet temperatures;
- surrounding air temperature;
- mass flow rate, using orifice plates in combination with high-precision pressure gauges;
- global radiation, using a pyranometer (highest standard);
- barometric pressure;
- surrounding air speed;
- dew point;
- static pressure at inlet and outlet.

In addition, there were devices for controlling and stabilizing the voltage supply for the lamps.

To test each system component (i.e. a collector), it is important to isolate it from the remaining system. A single component cannot be reliably assessed in combination with a variety of other components. By testing only one component at a time, it is easy to change the influencing parameters separately and analyse the effect of each parameter individually.

Why is collector-testing valuable?

The main goal of module testing in a solar laboratory is to assess the quality of the conversion of solar radiation to heat, not to predict yearly energy gains from a system.

Therefore, the tests were carried out under optimized, but realistic, conditions:

- constant inlet temperature by means of a climatic chamber and two controlled heating devices in the duct system near the collector inlet;
- constant mass flow rate;
- operation with leakage, thanks to the use of two fans (one at the inlet and one at the outlet); thereby the mean static pressure at the collector equals the ambient (barometric) pressure;
- constant solar irradiation by means of a solar simulator;
- variable, but for each measuring point constant, surrounding air speed (artificial wind);
- optimized air flow pattern inside (in order to have an even air flow from the inlet partition of the collector to the outlet; this was achieved, depending on the collector type, by metal sheets with small leakages mounted in front of the collector inlet and behind the outlet).

Measurements of mass flow rate and temperature can take place without influencing the performance of the collector.

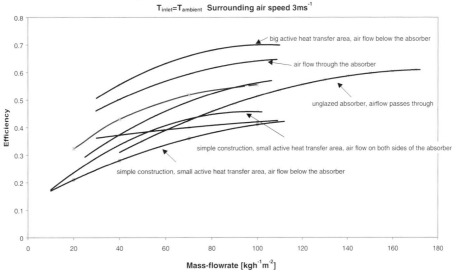

Figure F.2. Test results for different types of air collector

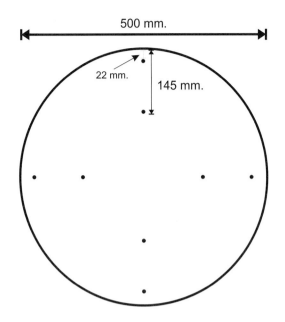

Figure F.3. Example of the locations of air temperature sensors for determination of the mean air temperature in a duct

The heat transfer coefficient (and therefore the performance) depends on the mass flow rate (rate of turbulence inside the air channel). All collectors were measured for at least three mass flow rates covering the typical range of operation.

The collectors that were intended to be operated with ambient air were also tested in a closed loop.

Pressure drop, leakage rate and dependence of surrounding air speed were also investigated. Test results are shown in Figure F.2.

Thermal performance of solar air collectors

Optical features (absorptance and emittance of the absorber, transmittance of the cover), materials used (absorber material, cover material, frame, insulation) and construction characteristics (effective heat transfer area, air flow principle) of the collector are of basic importance for the efficiency. However, the respective operation condition of the collector is decisive as well and the efficiency decreases with increasing temperature difference to ambient because of the increasing heat losses. The higher the mass flow rate and the lower the collector temperature, the lower the heat losses. However, for some applications minimum temperature is required, which necessitates lower mass flow rate.

IN-SITU DETERMINATION OF COLLECTOR EFFICIENCY

The efficiency of solar air collectors, in contrast to water collectors, is strongly influenced by the actual mass flow rate inside the collector. This is due to the often rather low heat transfer coefficient between the absorber and air. This heat transfer coefficient is highly dependent on the air speed. It is, therefore, often difficult/impossible to extrapolate from tests of small solar air collectors in test rigs to larger solar air collector arrays since the air speed will change.

A method for in-situ determination of the efficiency of solar air collectors has, therefore, been developed at the Solar Energy Center Denmark, DTI Energy.[1]

The main problems with in-situ determination of the efficiency of solar air collectors are:

- Accurate measurements must not disturb the flow pattern in the collector loop (may influence the efficiency of the collector).
- The efficiency has to be determined based on measurements carried out under dynamic conditions.

Measurement of temperatures and flow rates

Because a temperature profile often occurs across the sections of air ducts, it is necessary to establish a mean air temperature by measuring a number of air temperatures located at different positions within the same section of the duct (see Figure F.3). The temperature measurements are made using calibrated temperature sensors located in a well-defined grid in the duct. The sensors should be chosen in order to have only a small influence on the flow pattern in the duct (i.e. small sensors such as thermocouples connected to a thermopile).

Air flow measurements are made using either a calibrated pressure drop or a calibrated air speed. The calibrated pressure drop should preferably be an already existing pressure drop (e.g. an existing bending of the ducts). An orifice may also be used but introduces an extra pressure drop, which may change the flow rate noticeably. Pitot tubes and air-speed sensors can be used in straight duct.

The measurement accuracy of the temperature difference should be better than 0.1 K and for flow rates it should be better than 10%.

Identification

The collector efficiency is based on identification of the parameters within the standard efficiency equation for solar collectors. The method to identify parameters is an existing procedure – the Dynamic System Testing Procedure.[2] The procedure is shown graphically in Figure F.4. It has been used several years by the Solar Energy Center Denmark to determine the efficiency equations for liquid solar collectors.

Validation

The in-situ method has been applied and tested on two large Danish projects with solar air collectors: Havrevangen and Tjørnegade.

The solar air collector at Havrevangen is a traditional solar air collector with the air flowing behind the absorber and fins from the absorber down into the air stream – see Figure C.5. The absorber is made of aluminum. The absorber has a selective surface and the cover consists of a single layer of glass.

The solar air collector at Tjørnegade is a roof space collector and the southern part of the attic is utilized as a solar air collector. The cover consists of triple-walled ribbed sheets of UV stabilized polycarbonate. Behind the cover is a black felt mat acting as an absorber. The inlet air to the collector is blown into the space between the cover and the felt mat and sucked through the felt mat – see Figure F.5.

These two collectors have been selected for the tests, because they are very different and thereby represent a large span of collector types. The two collectors have one thing in common – the heat from the air stream is transferred to a liquid loop by an air to water heat exchanger. As it is easier to obtain accurate measurements in a stream of liquid, this made it possible to test the accuracy of the measurements in the air stream.

Within the area of interest the differences in the measured energy flows are below 2–3% between the air and liquid measurements. This is a surprisingly good agreement as it is difficult to obtain accurate measurements in air flows.

The accuracy of the determined efficiency equations for the two solar air collectors is within the range of the accuracy for the efficiency equations for liquid solar collectors.

Guidelines

Based on the measurements, guidelines on the measuring conditions have been reported – i.e. meteorological conditions measuring the duration of the measurements and the necessary accuracy of the measurement equipment.

The amount of solar radiation on the collector during the in-situ test should be minimum 120 MJ/m². Of this, at least 50% should be direct radiation and/or 50% of the measured irradiation should exceed 400 W/m².

An additional factor is the needed variability of conditions. The type and dimension of the solar air collector and its operation conditions influence the span of the variation of parameters. The values listed in Table F.1 should, therefore, be used for orientation only. T_m is the mean air temperature in the collector, T_a the ambient temperature, $(T_m - T_a)/G$ is the reduced temperature and G solar radiation on the collector.

CONCLUSION

This testing work fulfilled two primary goals:

- It was demonstrated that, with a careful measuring procedure, it is possible to achieve satisfactory accuracy of energy flow measurements in air ducts (comparable to the accuracy of energy flow measurements in liquid collectors).

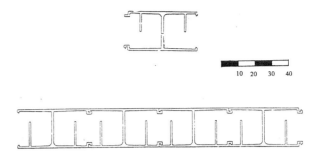

Figure F.4. Cross section of the absorber in the Havrevangen solar air collector

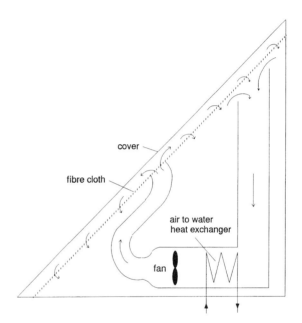

Figure F.5. The principle of the roof space collector at Tjørnegade.

- The method for in-situ characterization of air heating collectors was validated.

Testing procedures were demonstrated on two different types of solar air heating systems (Havrevangen and Tjørnegade), showing that the method is valid for a wide range of applications.

However, most important is that the method is able to accurately predict long-term energy yield (accuracy, better than 4.2% for the Havrevangen collector and 5.3% for the Tjørnegade collector). These figures are based on measured collector energy output for periods of at least one month. The testing method also allows system malfunctions to be diagnosed. This was demonstrated when the Havrevangen system

Table F.1. Recommended range of variation of influencing parameters (preferred values in brackets)

Influencing parameters	Recommended range of variation
$T_m - T_a$	10–50 (70) K
Reduced temperature	0.02–0.12 (0.2) Km²/W
Wind speed	1–5 m/s

had a periodic malfunction during the measuring period.[1] This feature of the method may be used to develop a tool for permanent monitoring of system performance and on-line detection of any system malfunctioning. Measured and simulated collector output may be continuously compared (using collector parameters identified *a priori*). Such a tool would be especially suitable for larger systems where a prompt warning of system malfunctioning would be cost-effective.

The method was accredited in 1998 and implemented in line with the already existing certified test methods at the Solar Energy Center Denmark at the Danish Technological Institute.

REFERENCES

1. Bosanac M, Jensen SØ (1997) *In-situ Solar Air Collector Array Test*. Solar Energy Center Denmark, DTI Energy, Gregersensvej, DK-2630 Taastrup, Denmark.
2. Spirkl W (1992) *Dynamic SDHW System Testing*, Program Manual. Section Physik der Ludwig-Maximilians Universität Munchen, Germany.

Prepared by Hubert Fechner

Appendix G: TRNSAIR – thermal simulation program for solar air systems

INTRODUCTION

Trnsair is a simulation program for solar air heating systems based on the well known dynamic simulation program TRNSYS. The nomograms in this book have been calculated by using TRNSAIR.

The program uses the algorithms of Trnsys in combination with a user-friendly interface. The program is available for Intel based platforms with operating systems Windows NT and Windows 95.

The program can simulate six generic solar air heating systems in combination with a two-zone building model. The user can modify the most relevant parameters of the building and the solar air system. Because of its user-friendly interface, the program can even be used by people who have little experience in using simulation programs. Program features are:

- full Trnsys simulation;
- changing of geometry data;
- five predefined typical reference buildings;
- predefined building usages (dwelling office etc.);
- user-defined insulation level;
- user-defined construction type (heavy or light);
- user-defined schedules of, for example, internal gains, occupation etc.;
- most predefined values can be changed;
- user-defined main orientation of the facade;
- a large database is included and this can be modified or extended;
- on-line display of simulation results, e.g. of room and air collector temperatures etc.

The default settings are very helpful for the user who wishes to simulate his or her own building projects. The program outputs are:

- heating demand;
- total building losses;
- solar gains;
- system performance.

BUILDING MODEL

The program can model a building with two thermal zones, each described by length, width and height and linked by one common wall.

As default, the Task 19 reference building is included. For this building the user can choose three different insulation levels, a light or heavy construction (low or high thermal mass) and office or dwelling building type. In addition, a large volume building is available; however, only System 1 or System 2 can be attached to this building type. The project initialization window is shown in Figure G.1.

This choice determines the types, orientations and areas of windows and walls. Once the building type has

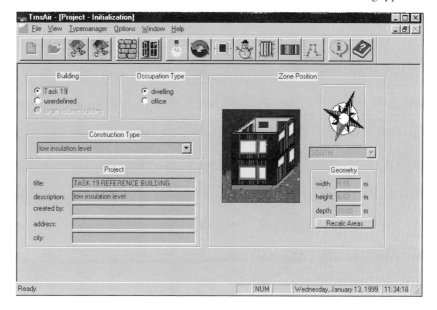

Figure G.1. Project initialization window of TRNSAIR

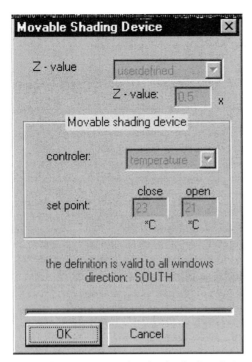

Figure G.2. Setting window for the shading device

ZONE DESCRIPTION

Changing the wall and window parameters can be done in the zone window, which is shown in Figure G.4. The regime data for heating, ventilation or infiltration can also be set in the zone window.

SYSTEM DESCRIPTION

After the building has been defined, it is possible to link it to one of the six different solar air systems (in practice, there are seven systems because System 5 has two different applications) – see Figure G.5. Each of these different systems is represented by a different logo. Further information about each system can be obtained by clicking the *Info* button.

In addition to a short description of the systems, a scheme of each system and a description of the information flow is provided with the program.

SIMULATION

Here the user can switch off the solar system so as to simulate, as a reference case, the building without any active system. The start and stop times (e.g. September till May) of the simulation can be set.

Because the system parameters differ from system to system, for each system an extra input window is provided, as shown in Figure G.6. With a parametric table simulation, an optimum for the system in question can be found.

The results of running a simulation may be displayed on the screen (Figure G.7) or can be exported as an ASCII table for import into any spreadsheet program (such as EXCEL). When the program run is finished, there will be output plots of the most relevant results (e.g. heating demand and system output).

been selected, data for regimes such as internal gains, heating schedules, occupation and ventilation schedules and set points for the shading device can be generated for each zone.

Most of the parameters in these default settings can be changed by the user and the database is extensible with user-defined wall types or regime data. It is also possible to define different values for each zone.

In order to model different control strategies, the schedules of heating, cooling, ventilation and the use of shading devices may be controlled by time or by using set points (e.g. room temperatures). See for example Figures G.2 and G.3.

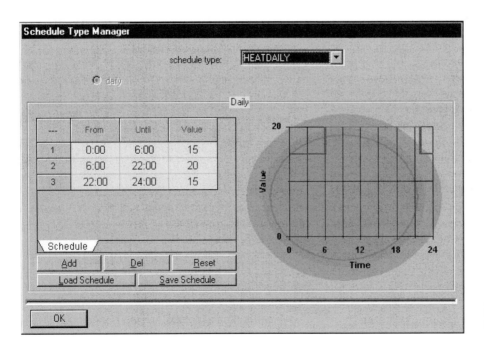

Figure G.3. Input window for schedule data

Appendix G. Trnsair – thermal simulation program for solar air systems 257

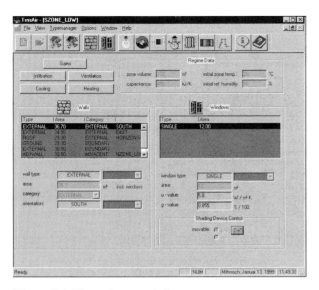

Figure G.4. The main zone window

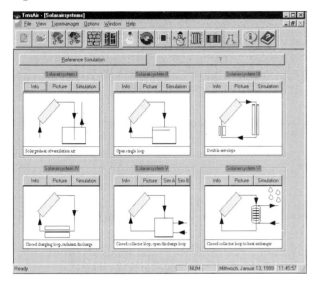

Figure G.5. Window for choosing the solar air system

CONCLUSION

Trnsair makes it possible to design a solar air heating system quite easily and with limited effort; and the results are quite good. Basic information on solar air systems and knowledge of how to operate a computer are necessary but easy to learn. Because user-friendly interface and a good manual are provided, it does not take very long to feel comfortable with the program.

The program is available from:

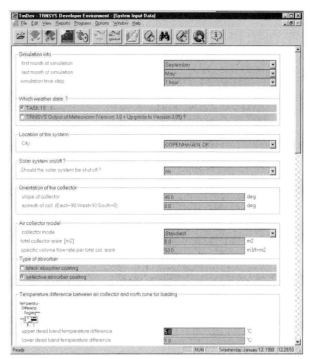

Figure G.6. Input window for relevant system parameters

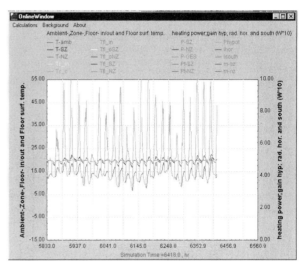

Figure G.7. On-line display of simulation results

TRANSSOLAR Energietechnik GmbH,
Nobelstrasse 15,
D 70567 Stuttgart,
Germany.
Tel. +49 711/67976-0; Fax +49 711/67976-11;
Email trnsair@transsolar.com

Prepared by Alex Knirsch

Appendix H: Nomogram assumptions

INTRODUCTION

A solar air heating system is installed to cover a part of the energy demand of a building and therefore the system performance is influenced by that energy demand. The system and the building and their interaction therefore have to be modelled in a thermal system simulation.

In order that the simulation results for different systems can be compared, the assumptions about the thermal behaviour of the building have been set up in a way that the results for different solar air heating systems have the same base, even though the different systems themselves influence the energy balance of the building. Nevertheless, comparison is difficult because of some assumptions or restrictions on the model, which influence the result (e.g. the heating demand).

As a first step, a so-called reference building was defined and is the basis of all the results presented in this book. System-relevant parameters that influence the thermal behaviour of the reference building were chosen in such a way that the energy consumption in the simulation of the building is the same when the system is switched off. For example, the double-envelope system will change the insulation level of the building as a result of the air gap created in front of the external walls. To produce the same energy demand for the building when the system is switched off, the wall of the reference building that would have had a double envelope has to be different.

For each system different assumptions are made in Trnsair. The following section describes the two sets of assumptions, general assumptions and system-dependent assumptions that influence system performance and energy consumption.

GENERAL ASSUMPTIONS

Short description of the general assumptions

The following reference cases have been used:

- Reference building I: dwelling; office;
- Reference building II: large volume building.

For most systems described in this book the reference building for dwellings and offices has been used. Therefore this building is called the reference building (Figure H.1) and the second type is called the large-volume building (lvb – Figure H.2).

Reference building I (dwelling)
The general assumptions for the reference building I for dwellings are:

- three different insulation levels, and therefore three building loss coefficients, 3.8 W/m²K down to 1 W/m²K;
- heating set point 20°C when occupied; 15°C during night set back;
- occupation time 6 a.m. until 10 p.m..;
- internal heat gain 3.5 W/m² (continuous);
- air change rate depending on the insulation level (0.8, 0.6 and 0.5 1/h) (better insulation level results in a higher air tightness);
- heated area 100 m² (whole building);
- air volume 300 m³ (whole building).

Reference building I (office)
The general assumptions for the reference building I for offices are:

- occupation time 8 a.m. until 5 p.m.;
- internal heat gain 14 (3) W/m² when occupied;
- air change rate depending on the insulation level and period of occupation: 0.4, 0.2 and 0.2 1/h if unoccupied; 1.0, 0.8 and 0.8 1itres/h when occupied.

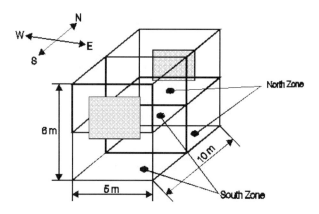

Figure H.1. Reference building

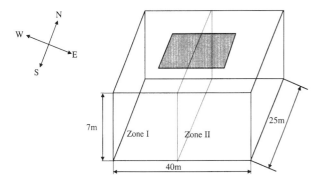

Figure H.2. Large-volume building

Reference building II (large-volume building)
For the reference building II, the large-volume building, the assumptions are:

- two insulation levels available (poor and good);
- heating set point 20°C when occupied; 15°C during night set back;
- occupation time 6 a.m. until 10 p.m.;
- internal heat gain 9 W/m² when occupied; 30 W/m² all the time;
- air change rate depending on the insulation level and period of occupation; 0.5, 0.3 1/h if unoccupied and an additional 0.5 (3.0) 1/h when occupied;
- heated area 1000 m² (whole building);
- air volume 7000 m³ (whole building).

Air collector model
The air collectors used are typical collectors and have different efficiencies. The collectors cover most types available. The four air collectors are

- standard;
- double-glazed;
- optimized;
- porous.

The air collector model is identical for all calculations. Depending on the system, the standard or the optimized collector is the basis of the nomograms.

DETAILED DESCRIPTION OF THE REFERENCE BUILDING

The reference building represents a two-apartment building with four rooms. It is a part of a building block. The reference building has an inside length of 10 m, an inside width of 5 m and an inside height of 6 m. The total air volume is 300 m³ and total building surface is 280 m². The outside measurements depend on the insulation standard. The net heated floor area is 100 m². The window area of the south wall is set to 12 m² (40% of the wall area). The glazing area of the window is set to 9.6 m² and the frame area is set to 2.4 m². The window area of the north wall is set to 5 m² (16.7% of the wall area). The glazing area of the north windows amounts to 4 m² and the frame area to 1 m².

As the thermal conditions are the same for both floors, the building is divided into two thermal zones, the south zone and the north zone (see Table H.2). The thermal zones are important for the simulation.

For Systems 2 and 3, it was necessary to divide the building into four thermal zones (ground, south first floor, north ground floor, north first floor). For System 2 the air gap (= envelope) was treated as two extra thermal zones.

The large-volume building has a floor area of 1000 m² and a volume of 7000 m³. As a result of its uniformity, it is divided into only two zones having roof windows.

The design of the walls

The composition of the wall materials defines five different insulation levels. The low insulation level only has insulation on the roof. Dependent on the thickness of insulation, buildings are referred to as middle- and high-insulated buildings.

Properties of the materials used
For the walls of the reference building the materials give in Table H.1 are used. The area of the walls is given in Table H.2.

The structure of the walls
For investigation of the thermal behaviour of different wall designs, two construction types are defined, heavy and light. The core of the heavy construction is made of bricks, while for the light one only wood is used.

Low-insulated reference building. For the low-insulated reference building only the heavy construction type is considered. The *external walls* are composed of 15 mm interior plaster, 240 mm air brick and 20 mm exterior plaster. This is therefore categorized as a heavy construction. The front convective heat transfer coefficient (h_{front}) of the external wall is set to 3 W/(m²K) and the back coefficient (h_{back}) is set to 25 W/(m²K). The surface absorptance for solar radiation of the front (a_{front}) and the back (a_{back}) are set to 0.6. This results in

Table H.1. Materials used

Material	Thermal conductivity (W/(mK))	Heat capacity (kJ/(kgK))	Density (kg/m³)
Screed	1.4	1.0	2000
Concrete	2.1	0.88	2400
Interior plaster	0.35	1.0	1200
Exterior plaster	0.7	1.0	1800
Roof seal	0.5	1.0	1200
Air brick	0.5	1.0	1200
Mineral wool	0.04	0.84	80
Polystyrene foam	0.04	1.25	30
Wood	0.15	2	600

Table H.2. Area of the walls (the values refer to the outside dimension)

Insulation level Construction style	Low Heavy	Middle Heavy	Middle Light	High Heavy	High Light
South zone					
Net heated area			50 m²		
Air volume			150 m³		
South wall	External 24.7 m²	External 26.2 m²	External 22.2 m²	External 29.9 m²	External 25.6 m²
Adjacent	Adjacent to north zone 30.0 m²	Adjacent to north zone 30.0 m²	Adjacent to north zone 30.0 m²	Adjacent to north zone 30.0 m²	Adjacent to north zone 30.0 m²
East wall	External 34.9 m²	External 36.1 m²	External 33.6 m²	External 38.5 m²	External 35.9 m²
West wall	Boundary identical (adiabatic) 34.9 m²	Boundary identical (adiabatic) 36.1 m²	Boundary identical (adiabatic) 33.6 m²	Boundary identical (adiabatic) 38.5 m²	Boundary identical (adiabatic) 35.9 m²
Floor	Boundary = 15°C 29.3 m²	Boundary = 15°C 29.9 m²	Boundary = 15°C 26.4 m²	Boundary = 15°C 32.6 m²	Boundary = 15°C 28.8 m²
Internal ceiling	Internal 25.0 m²	Internal 25.0 m²	Internal 25.0 m²	Internal 25.0 m²	Internal 25.0 m²
Roof	External 29.3 m²	External 29.9 m²	External 26.4 m²	External 32.6 m²	External 28.8 m²
Window	South oriented, single-glazed 12.0 m²	South oriented, double-glazed 12.0 m²	South oriented, double-glazed 12.0 m²	South oriented, low-*e* glazing 12.0 m²	South oriented, low-*e* glazing 12.0 m²
North zone					
Net heated area			50 m²		
Air volume			150 m³		
North wall	External 31.7 m²	External 33.2 m²	External 29.2 m²	External 36.9 m²	External 32.6 m²
Adjacent	Adjacent to south zone 30.0 m²	Adjacent to south zone 30.0 m²	Adjacent to south zone 30.0 m²	Adjacent to south zone 30.0 m²	Adjacent to south zone 30.0 m²
East wall	External 34.9 m²	External 36.1 m²	External 33.6 m²	External 38.5 m²	External 35.9 m²
West wall	Boundary identical (adiabatic) 34.9 m²	Boundary identical (adiabatic) 36.1 m²	Boundary identical (adiabatic) 33.6 m²	Boundary identical (adiabatic) 38.5 m²	Boundary identical (adiabatic) 35.9 m²
Floor	Boundary = 15°C 29.3 m²	Boundary = 15°C 29.9 m²	Boundary = 15°C 26.4 m²	Boundary = 15°C 32.6 m²	Boundary = 15°C 28.8 m²
Internal ceiling	Internal 25.0 m²	Internal 25.0 m²	Internal 25.0 m²	Internal 25.0 m²	Internal 25.0 m²
Roof	External 29.3 m²	External 29.9 m²	External 26.4 m²	External 32.6 m²	External 28.8 m²
Window	South oriented, single-glazed 5.0 m²	South oriented, double-glazed 5.0 m²	South oriented, double-glazed 5.0 m²	South oriented, low-*e* glazing 5.0 m²	South oriented, low-*e* glazing 5.0 m²

an overall heat transfer coefficient of the external walls of 1.7 W/(m²K).

The flat *roof* consists of four layers: 20 mm wood, 200 mm concrete: 50 mm mineral wool and a roof seal with a thickness of 20 mm. The overall heat transfer coefficient of the roof is therefore 0.63 W/(m²K) (h_{front} =3 W/(m²K), h_{back} = 25 W/(m²K); a_{front} = 0.1, a_{back} = 0.6).

The ground *floor* is made of 50 mm screed, 25 mm mineral wool and 250 mm concrete. The overall heat transfer coefficient of the floor is set to 1 W/(m²K) (h_{front}=3 W/(m²K), h_{back}= 0.001 » 0 W/(m²K); a_{front}= 0.8, a_{back}= 0.001 » 0).

The *ceiling* (internal floor) consists of 50 mm screed, 25 mm polyfoam and 200 mm concrete. The overall heat transfer coefficient of the ceiling is calculated as 1.1 W/(m²K) (h_{front}=3 W/(m²K), h_{back}= 3 W/(m²K); a_{front}= 0.7, a_{back}= 0.8).

The *internal wall* is also made of three layers. The interior finish on each side is 15 mm and the core is made of air brick with a thickness of 120 mm. The overall heat transfer coefficient of the internal wall is set to 2.1 W/(m²K) (h_{front}=3 W/(m²K), h_{back}= 3 W/(m²K); a_{front} = 0.6, a_{back}= 0.6).

The *windows* of the low-insulated reference building are single-glazed with an overall heat transfer coefficient of 5.2 W/(m²K). The heat transfer coefficient from the inside surface to the outside surface of the outer glazing is 33.3 W/(m²K). The convective heat transfer coefficient at the inside surface of the window is set to 3 W/(m²K) and the coefficient at the outside is set to 25 W/(m²K).

The heat capacitance of the air and the interior materials, such as furniture, plants etc., is assumed to be 250 kJ/K per zone.

Table H.3. *The structure of the walls and windows for different insulation levels and for the different construction types*

Insulation level Type of construction	Low Heavy	Middle Heavy	Middle Light	High Heavy	High Light
External wall					
Plaster			0.015 m		
Brick	0.240 m	0.240 m	None	0.240 m	None
Insulation	None	0.040 m	0.060 m	0.200 m	0.210 m
Plaster			0.020 m		
Total thickness	0.275 m	0.315 m	0.095 m	0.475 m	0.245 m
U-value	1.4 W/(m²K)	0.58 W/(m²K)		0.17 W/(m²K)	
Roof					
Wood			0.020 m		
Mineral wool	0.050 m	0.150 m	0.150 m	0.300 m	0.300 m
Concrete	0.200 m	0.200 m	None	0.200 m	None
Seal			0.020 m		
Total thickness	0.290 m	0.390 m	0.190 m	0.540 m	0.340 m
U-value	0.63 W/(m²K)	0.24 W/(m²K)		0.13 W/(m²K)	
Floor					
Screed			0.050 m		
Mineral wool	0.025 m	0.025 m	None	0.025 m	None
Concrete			0.250 m		
Polystyrene	None	0.075 m	0.100 m	0.175 m	0.200 m
Total thickness	0.325 m	0.400 m	0.400 m	0.500 m	0.500 m
U-value	1.05 W/(m²K)	0.35 W/(m²K)		0.19 W/(m²K)	
Internal wall					
Plaster			0.015 m		
Brick	0.12 m	0.12 m	None	0.12 m	None
Polystyrene	None	None	0.100 m	None	0.100 m
Plaster			0.015 m		
Total thickness	0.150 m	0.150 m	0.130 m	0.150 m	0.130 m
U-value			2.1 W/(m²K)		
Ceiling					
Mortar			0.050 m		
Mineral wool			0.025 m		
Concrete			0.200 m		
Total thickness			0.275 m		
U-value			1.1 W/(m²K)		
Window					
Glazing	Single	Double		Low-e	
U-value	5.8 W/(m²K)	2.8 W/(m²K)		1.4 W/(m²K)	
G-value	0.87	0.75		0.62	

More details are shown in Table H.3.

Middle-insulated reference building. The core structure of the walls of the middle construction is similar to the core construction of the reference building with low insulation. However, there are four differences between the low and the middle insulation standards; in the latter the external wall possesses an additional polystyrene layer of 40 mm and the insulation layer of the roof is 150 mm thick. Thus, the insulation layer is three times thicker than the insulation layer of the low-insulation standard. Even the floor is better insulated, with a polystyrene layer of 100 mm. There is no insulation between the screed layer and the concrete. Finally, the single-glazed windows are replaced by double-glazed windows with a heat transfer coefficient of 3.2W/(m²K).

The second type of the middle-insulated building is a lightweight wood construction, which is typical of the architecture of northern Europe. The ceiling and windows are the same as for the heavy-constructed building, but, instead of bricks, the external wall is made of an insulation construction between two plaster layers. For the light construction the 60 mm polystyrene layer is thicker than that for the heavy construction. The thickness of the insulation was chosen such that the U-value of the wall is the same for the two constructions. As with the external wall, the internal wall is made of a light structure, with a layer of 100 mm mineral wool between two plaster layers. The roof does not have a concrete layer but a light wood construction. The insulation of the floor is increased. The thickness of the polystyrene foam layer is now 75 mm, although the 25 mm polystyrene foam layer below the screed is the same.

Data for the large-volume building are given in Table H.4.

High-insulated reference building. In comparison with the middle-insulated reference building, the highly insulated building has no additional layers, but does have thicker insulation and better windows. The polystyrene foam of the external wall is twice as thick as that of the middle-insulated reference building (200 mm) and the thickness of the mineral wool for the roof is 300 mm and thus twice as thick as that of the middle-

Table H.4. Data for the large-volume building

Insulation level	Poor	Good
External wall	Brick 24 cm 1.39 W/(m²K)	Corrugated steel Mineral wool 8 cm 0.46 W/(m²K)
Roof	Polystyrene 3 cm, wood 0.92 W/(m²K)	Polystyrene 20 cm, wood 0.19 W/(m²K)
Floor	Concrete 25 cm, mineral wool 2.5 cm, screed 1.05 W/(m²K)	Concrete 25 cm, mineral wool 2.5 cm, screed 1.05 W/(m²K)
Windows	Double 2.8 W/(m²K)	Low-e 1.4 W/(m²K)

insulated building. The thickness of the polystyrene floor insulation is increased by 100 mm to 175 mm and the windows consist of double-glazed low-e glass with a heat transfer coefficient of 1.4 W/(m²K) (including the frame).

For the light wood construction, the external walls contain 220 mm polystyrene foam. The internal walls are made of a 100 mm polystyrene layer instead of bricks. The roof, floor, ceiling and windows are similar to those of the heavy-construction style. A summary of the five insulation standards is given in Table H.3.

Mass of the reference building

Depending on the type of construction and the insulation level, the standard reference building has a different mass. Table H.5 shows the total mass in tonnes for each insulation level and type of construction. The mass of the building in relation to the net heated floor area is also given.

Occupancy of the reference building

Two types of occupancy are defined for the reference building, an office building and a dwelling. The occupation schedule of the office is from 8 a.m. to 5 p.m. The dwelling is occupied from 6 a.m. to 10 p.m.
For the large-volume building, the time of occupation is in all cases 6 a.m. to 10 p.m.

Heating
The operation time of the conventional heating system can be chosen by the user. The room temperature for both types of occupancy of the reference building is set

Table H.5. Mass of the reference building

Insulation level	Low	Middle	Middle	High	High
Construction style	Heavy	Heavy	Light	Heavy	Light
Total mass (tonnes)	165.0	168.0	79.6	175.2	82.9
Mass/net area (kg/m²)	1650	1680	796	1752	829

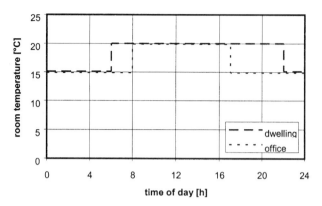

Figure H.3. Schedule for the room temperature dependent on the occupation

to 22°C when it is occupied and 15°C when it is unoccupied (see Figure H.3).

The room temperature in the large-volume building has been set to 15°C in case the building is used for sports or as a warehouse (night set back to 12.5°C). If the building is used for light industry it requires 20°C (15°C during the night).

Internal heat gains
Reference building. When the office is occupied, the internal gains are assumed to be 14 W/m². Otherwise, the internal gains are set to zero (see Figure H.4). The internal gains are made up of 75% convective and 25% radiative heat gains. For the dwelling, the internal gains are set to 2.3 W/m² constantly. The contribution of the convective heat gains is two-thirds and the contribution of the radiative heat gains is one-third.

Large-volume building. Internal gains are 9 W/m² in sports halls and warehouses, 30 W/m² in light industry. Other parameters can be taken from the reference building.

Air change and ventilation rate
Air change by infiltration for hygienic purposes. The amount of the natural and hygienic air change depends on the occupation of the reference building. For the dwelling, infiltration occurs throughout the

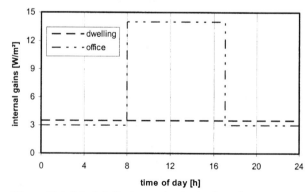

Figure H.4. Schedule for internal gains in the reference building

Table H.6. Boundary conditions for the control strategy in the reference building

Type of building	Office		Dwelling	
Occupation				
Schedule for occupation	8.00 a.m. to 5.00 p.m.		6.00 a.m. to 10.00 p.m.	
Heating				
Heating setpoint – occupied	20°C		20°C	
Heating setpoint – unoccupied	15°C		15°C	
Internal gains	Unoccupied	Occupied	Midnight–midnight (always)	
	3 W/m²	14 W/m²	3.5 W/m²	
Air change	Unoccupied	Occupied	Midnight–midnight (always)	
	Leakage	Hygenic ventilation + leakage	Leakage	Hygenic ventilaion + leakage
Low insulation	0.4	1.0*	(0.4)	0.8†
Medium insulation	0.2	0.8*	(0.3)	0.6†
High insulation	0.2	0.8*	(0.2)	0.5†
Ventilation				
Night flushing during: summertime room temperature > 24°C wintertime room temperature > 26°C	3.0 air changes per hour 900 m³/h		3.0 air changes per hour 900 m³/h	
Shading				
Shading of south-oriented windows: internal shading device factor	0.5		0.5	
Shading on	23°C		23°C	
Shading off	21°C		21°C	

* Total amount of air change per hour; mechanical ventilation.
† Total amount of air change per hour; natural ventilation.

whole year. For the office, a hygienic air change is only necessary when it is occupied. Table H.6 gives an overview of the air changes of the reference building.

In the large-volume building the infiltration is handled in the same way as for dwellings (Table H.7).

Air change by mechanical ventilation. During the summer, when the air heating is not used, an air change by ventilation is necessary in order to prevent overheating, especially of the south zone. When the room temperature is exceeds 24°C, a temperature controller signals a fan to blow fresh air from the outside into the building. During the heating period, the room temperature is set to 26°C and the air flow rate is set to 450 m³/h for one zone; thus the air change rate is three times per hour.

Shading
If the room temperature of south zone reaches 23°C, the shading device will be activated. The shading device has an internal shading device factor of 0.5. If the room temperature drops below 21°C the shading device will be lifted. In the large-volume building there are no shading devices.

Building loss coefficients

The building loss coefficient (BLC-value) represents a value that indicates the insulation quality of a construction. The BLC-value is calculated as the sum of the overall heat transfer coefficients (U-values) for the whole building multiplied by the building surface (A_{bui}) and the infiltration losses (Q_{inf}):

$$\text{BLC} = \frac{U \times A_{bui}}{A_{fl,net}} + \frac{Q_{inf}}{A_{fl,net}}.$$

Both the overall heat transfer coefficient and the infiltration losses are related to the net heated floor area ($A_{fl,net}$). Tables H.8 and H.9 show the building loss coefficients of the office and the dwelling for different insulation levels. The BLC-value of the office and the dwelling at the same insulation level are approximately equal.

Energy balance of the building

Each building has heat gains and heat losses. The energy delivered by an auxiliary heating device, electri-

Table H.7. Data for the large-volume building

Type of building	Sports hall/ Warehouse	Light industry
Occupation		
Schedule for occupation	6.00 a.m. to 10.00 p.m.	6.00 a.m. to 10.00 p.m.
Heating		
Heating set point – occupied	15°C	20°C
Set back	12.5°C	15°C
Internal gains	9 W/m²	30 W/m²
Air change	Leakage	Leakage
Poor insulation	0.5	0.5
Good insulation	0.3	0.3
Ventilation	0.5 air changes per hour 3500 m³/h	3.0 air changes per hour 21,000 m³/h
Shading	None	None

Table H.8. Building loss coefficients for the office

Insulation level	Construction type	Transmission losses (W/(m²K))	Infiltration losses (W/(m²K))	$BLC_{tot,office}$ (W/(m²K))
Low		2.94	0.79	3.731
Middle	Heavy	1.28	0.587	1.867
Middle	Light	1.19	0.587	1.777
High	Heavy	0.574	0.587	1.161
High	Light	0.537	0.587	1.123

Table H.9. Building loss coefficients for the dwelling

Insulation level	Construction type	Transmission losses (W/(m²K))	Infiltration losses (W/(m²K))	$BLC_{tot,office}$ (W/(m²K))
Low		2.94	0.816	3.757
Middle	Heavy	1.28	0.612	1.893
Middle	Light	1.19	0.612	1.800
High	Heavy	0.574	0.51	1.084
High	Light	0.537	0.51	1.047

cal devices and the solar radiation through the windows are gains by the reference building. The heat transmission through the walls and windows and the heat transfer by the exchange of air are energy losses. There are two kinds of air flows: air change by natural infiltration through open windows and air change through leakage of the building. The energy balance is given by:

$$Q_{aux} + Q_{sol} + Q_{int} = Q_{trans} + Q_{vent} + Q_{inf},$$

where:

Q_{aux} = energy delivered by auxiliary device;
Q_{trans} = transmission losses through the walls;
Q_{sys} = energy delivered by the solar system without window gains;
Q_{int} = internal heat gains from electrical devices and people;
Q_{sol} = solar gains by windows (passive);
Q_{inf} = heat losses by hygienic ventilation and leakage;
Q_{vent} = heat losses by mechanical ventilation to prevent overheating in the zones.

The total auxiliary energy is given by the heat demand in the south zone (Q_{sz}) and the north zone (Q_{nz}):

$$Q_{sz} + Q_{nz} = Q_{aux}.$$

The absolute heat demand of both building zones (Q_{load}) is calculated as the sum of the transmission losses and the infiltration losses:

$$Q_{trans} + Q_{inf} = Q_{load}.$$

SYSTEM-DEPENDENT ASSUMPTIONS

In the following sections concerning the system-dependent changes in the building, both the system description and the working points will be described.

System 1 assumptions – preheating of air

Figure H.5 shows System 1. A fan draws air from the outside into the building. If there is radiation, the air is preheated and reduces the heating demand.

For System 1 the building needs no change. The only modification that is necessary is to distinguish between leakage and hygienic ventilation because for this system only the hygienic part of the air will go through the air collector. Air that enters the building through leakage (lack of air tightness, open windows etc.) cannot be used for preheating by the collector.

The leakage rates that were specified can be found in the description of the reference building given above.

Controller settings
Starting the fan depends on different control strategies. For System 1 three control strategies are available:

1. The system is always running except when room temperature exceeds a certain value.

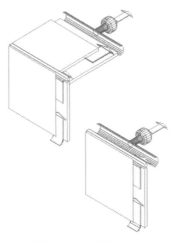

Figure H.5. Scheme of System 1

2. Controlled by the difference between the collector temperature and room temperature; the system is stopped when $T_{room} > T_{max}$.
3. Controlled by solar radiation; the system is stopped when $T_{room} > T_{max}$.

During the heating period the switch-off temperature is set to 26°C for control strategy 1.

Usage of system energy
The energy produced by the system is transferred equally to the north and south zones.

System 2 assumptions – natural convection with radiant discharge (Barra–Costantini)

Figure H.6 shows a possible application of System 2. Natural convection is used to drive the solar heating system and no fan is necessary. The heat gain of the system is transferred into the building by radiation through the ceiling hypocaust and by air which enters the north zone.

To model this system accurately, the reference building has to be modified. The building is therefore split into four thermal zones (Figure H.7):

- ground floor south;
- ground floor north;
- first floor south;
- first floor north.

The air collector is build into the facade (tilt 90°). Losses from the rear side of the collector are not taken into account.

Controller settings
The inlet temperature into the collector is room temperature. It is assumed that a temperature stratification of 2 K exists in the south zone. If there is enough radiation, the air collector raises the temperature and starts the natural circulation. The resulting mass flow then passes through the hypocaust, which is located in the slab that separates upper and lower floors. The air outlet to the room is located in the north zone.

The same loop is installed on the first floor. There the hypocaust is integrated into the roof (see Figure H.7).

The controller will stop the mass flow (in reality by closing a damper) if the temperature difference between the outlet temperature of the air collector and the average north-zone temperature is less than 1 K.

Usage of system energy
Most of the energy produced by the system is transferred by long-wave radiation into the room, while the remaining energy enters the north zone as a convective gain. For more details see Chapter II.2.

System 3 assumptions – double envelope

The building of System 3 is also modelled with four thermal zones. The envelope is built in front of the existing north and south wall, as shown in Figure H.8.

The gap between the exterior wall and the internal is set to 6 cm. The most important assumption is that the wall construction and also some layers of the wall composition have been adapted in some way, so that the reference case without the solar air system gives the same result as the other buildings.

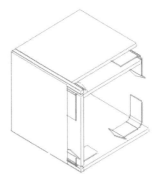

Figure H.6. Scheme of System 2

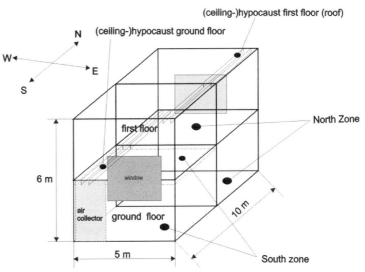

Figure H.7. Scheme of reference building with the four zones for System 2

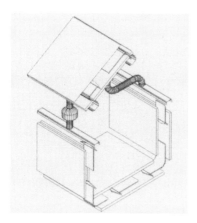

Figure H.8. Scheme of System 3

Controller settings

The fan starts if the outlet temperature of the air collector is 5 K above the temperature of the north envelope. The system stops if either the temperature difference decreases to 1 K or room temperature exceeds 26°C.

Usage of system energy

The energy produced by the system is transferred into the thermal mass of the inner wall of the envelope. This energy will then be transmitted, with a time delay, to the inside zones by long-wave radiation.

System 4 assumptions – hypocaust system

Figure H.9 gives an overall view of System 4. The air heated by the collector is force-driven through a hypocaust. The energy is discharged by long-wave radiation with a time delay that varies with the chosen hypocaust.

The most important assumption is that the hypocaust is only located in the floor/ceiling that separates the storeys. Thus, the energy losses that appear if the basement is also used as a hypocaust are avoided. In addition, discharge is possible on both sides of the slab.

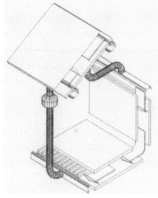

Figure H.9. Scheme of System 4

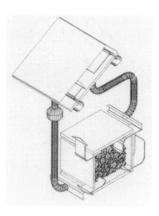

Figure H.10. Scheme of System 5a

Controller settings

The fan starts if the outlet temperature of the air collector is 5 K above the temperature of the north zone. In addition, the program checks whether the collector outlet temperature is at least 3 K above the average outlet temperature of the hypocaust (the average outlet temperature is the weighted outlet temperature of south and north hypocausts).

The system stops if either the temperature difference between the collector outlet and the north zone decreases to 1 K or the room temperature exceeds 26°C. It will also stop if the temperature difference between the collector outlet and the hypocaust outlet drops to 1 K.

To avoid overheating and because of the higher passive solar gains of the south zone, the mass flow rates for the south and north zones can be set. The default values for diverting the mass flow are 40% south and 60% north.

Usage of system energy

The energy produced by the system is transferred by long-wave radiation (from the slab) into the north and south zones. Because of the design of the system model, the energy can be radiated from both sides of the slab.

System 5a assumptions – rockbed storage

Figure H.10 gives an overall view of System 5a. The air heated by the collector is force-driven through a rockbed storage. The energy is also discharged by a fan.

The most important assumption of the control strategy is that the heated air enters the north zone without charging the rockbed storage if possible. If there is enough energy that cannot be used directly, the rockbed storage will be charged.

Controller settings

The fan starts if the temperature difference between the air collector outlet is 2 K above the temperature of the north zone. It will be shut off if the difference drops to 1 K.

For charging the rockbed, the outlet temperature of the air collector has to be 10 K above the rockbed

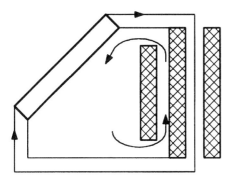

Figure H.11. Scheme of System 5b

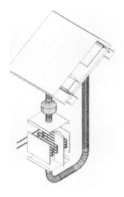

Figure H.12. Scheme of System 6

storage temperature. Charging will stop if the difference drops below 3 K. Discharging the rockbed depends on the temperature difference between the north zone and the rockbed temperature. Default values are 5 K for starting the fan and 1 K for switching it off.

The whole control strategy will be over-ridden if the temperature of the north zone exceeds the maximum temperature, which is set to 26°C.

Usage of system energy
The energy produced by the system is transferred by convection into the north zone. The energy losses can be used – if specified – to a certain extent, if the pipes are inside the building. This is also valid for rockbed storage. The surface of the rockbed faces the north zone (where there is a lack of passive gain). The amount of usable *thermal losses* from the rockbed storage can be specified (default 50%). To make sure that the south zone also benefits from these system gains, there must be an air exchange between north and south zones. The value of the air exchange rate depends on the specified mass flow rate of the collector.

System 5b assumptions – active discharge of murocaust wall zone

This system uses a murocaust instead of a rockbed storage (Figure H.11). The difference from System 5a is that the storage can be charged and discharged at the same time. In the model an extra zone is implemented in front of the murocaust. This heated murocaust zone is then vented by an extra discharge fan. The electric energy used by the fan is not taken into account, because it could cause too high system performance if the values are not set carefully.

The losses of the rear side of the murocaust have been neglected because it is assumed that an identical temperature exists on the other side of the murocaust.

Controller settings
The fan starts if the temperature difference between the air collector and the temperature of the north zone is greater than 5 K. It will be shut off if the difference drops to 1 K. Another controller, which is combined with the first one, turns on the fan if the outlet temperature of the air collector is more than 3 K above the murocaust outlet temperature. It will stop the fan if the difference is less than 1 K. This control is necessary because otherwise the thermal energy gain of the system is too small and the electrical energy for the fan is wasted.

Discharging the murocaust zone depends on a second controller that switches on the discharge fan. The murocaust zone temperature must be at least 3 K higher than the temperature of the north zone. It will stop the fan if the difference between these zones drops to 0.5 K.

The whole control strategy will be over-ridden if the temperature of the north zone exceeds a specified chosen maximum temperature, which is set to 26°C.

Usage of system energy
The energy is transferred by convection and, to a small extent, by radiation only to the north zone. To make sure that the south zone also benefits from system gains, there is an air exchange between north and south zones.

System 6 assumptions – solar-heated water

System 6 (Figure H.12) does not need any changes in the modelling of the building, because it is a water-based heating system that is not affected by, and does not affect, infiltration, ventilation or other data for the building.

The system is modelled with two storage systems, a domestic hot-water storage for usage of hot tap water and a buffer storage to store additional heat from the collector loop that is supplied to a water-based (low-temperature) heating system.

For the nomograms of System 6, the use of DHW was disabled (tap water = 0, together with a very high temperature difference between collector outlet and upper storage temperature) so that only the buffer storage is charged. The system output can then be compared with the other systems.

Controller settings
This system has different controllers. The following three points describe the control strategy introduced:

1. The system starts when there is a certain temperature difference between outlet collector temperature and the bottom temperature of the domestic hot-water storage (DHW) (e.g. 15°C = upper dead band). If the temperature difference drops below, for example, 5 K (lower dead band), the collector loop stops.
2. If possible, the buffer storage will be loaded:
   – The DHW storage will not be loaded if the top temperature is above a certain temperature (default 65°C).
   – The buffer storage will not be loaded if the top temperature is above a certain temperature (default 95°C).
3. If the collector loop (and the load loop) are not in operation, it is possible that the DHW storage will be loaded by the buffer storage. This happens only if the top temperature of the buffer storage is above the chosen hot-water temperature + 5 K. The loading stops if the temperature of the buffer storage drops to less than 2 K higher than the hot-water temperature.

As described above, charging the DHW storage can be disabled.

Usage of energy
The energy contained in the buffer storage is used if a heating demand occurs. For the heating demand, a resulting mass flow rate (depending on the design temperature of the heating system) is calculated. If the outlet buffer storage temperature is lower than the necessary supply temperature a boiler supplies the additional energy.

Prepared by Alex Knirsch

Appendix I: Survey of PC programs

MODELLING STRATEGY IN DIFFERENT SIMULATION TOOLS

This overview presents the modelling strategy used in some well-known programs (such as Sunned, Doe-2, Blast, Trnsys, etc.) to calculate the properties of solar air systems. The weakness and strengths of the different tools related to this topic are also documented.

Requirements

What is required is a detailed component description as a result of interaction with a building model, for example a building wall as a storage element.

Table I.1 gives the scope of some common programs.

OVERVIEW OF THE PROGRAMS

Trnsair

This program was developed especially for the purpose of modelling solar air heating systems together with a two-zone building model. The program is capable of simulating six predefined solar air heating systems representing the most common applications. The building is modelled in detail using the well-known Trnsys program, but with an easy-to-use interface. Any of the six solar air heating systems can be specified in detail as a result of entering only a few parameters. Dependent variables, e.g. storage volume per collector area, are pre-calculated to help the user get a good first result in designing the system, although the user can modify these values. A small limitation is that overhangs or wingwalls cannot be defined to simulate the shading effects of surroundings. For more details see Appendix G.

Trnsys

Trnsys is a modular simulation program. It recognizes a system description language in which the user specifies the components that constitute the system and the manner in which they are connected. The Trnsys library includes many of the components found in thermal (particularly solar) energy systems, as well as component routines to handle input of weather data or other time-dependent forcing functions and output of the simulation results. Trnsys has been configured to facilitate users adding their own components to the library. During Task 19 some additional components were developed, such as a detailed air collector model and a hypocaust model.

The intended users of this program are energy consultants, engineers and researchers. The user needs to know the different Trnsys components from which a system may be built and know how to deal with modular

Table I.1. Scope of some common programs

Program	Cross-ventilation	Whole building analysis	Convective heat transfer coefficient	Storage	Collector	Reporter*
Trnsair	Fan	Yes	$f(T)$	Yes	Yes	A. Knirsch
Trnsys	$f(\Delta T, \Delta p)$	Yes	$f(T)$	Yes	Yes	Matthias Schuler
Tas	Internal program	Yes	Constant	No	No	Frank Heidt
Blast	Fan	Yes	Constant	No	No	Ben Corpening
Doe-2	?	Yes	?	?	?	Gerhard Zweifel
Emgp3	Constant	(Yes)	Constant	Yes	Yes	Søren Ostergard
Apache	Open loop only	Yes	Constant	No	No	Frank Heidt
Esp-r	$f(\Delta T, \Delta p)$	Yes	$f(T)$	Yes	Yes	Søren Ostergard Jensen
Watsun	Fan	No	Constant	Yes	Yes	Matthias Schuler
Ida	Open	Yes	$f(T)$	Yes	Yes	Ove Mørck
Tsbi3	Fan	Yes	Approximated	No	Yes	Søren Ostergard Jensen
Suncode	Fan	Yes	Constant	Yes	No	Frank Heidt
Swift	Fan	Yes	Constant	Yes	No	John Hollick

* See the IEA Participants list in Appendix K for addresses.

simulation programs. A pre-processor (Presim or Iisiabt) helps less experienced users.

Tas

The construction element database of A-Tas contains no collector or storage type. The user therefore has to construct the model of the collector and the storage him- or herself and has to treat them as extra thermal zones.

The intended users of this program are engineers and experienced modellers working with Unix workstations (HP–Apollo). Because of the need for the user to model the collector, the storage and the control strategy, the program is not suited for standard users. In addition, the user has to set several operational parameters, which requires experience.

A-Tas prepares a detailed output of all important daily values. For monthly or annual output the user has to program the 'Report Generation'. This is like programming in an operating system and quite uncomfortable. The Report Generation only prepares simple text and tabular output. To create more impressive results and diagrams, the user has to transfer the data into a spreadsheet program like Excel. The temperature output of A-Tas includes only the mean zone temperatures (sensible and radiant) and surface temperatures. To calculate the air stratification, the user has to use an additional program called Ambient.

For the simulation of the whole HVAC system, the user has to work with the program B-Tas. This program uses the A-Tas output and is a tool for the optimization of the conditioning facilities and of the control strategy. Expert knowledge in heating/cooling and air-conditioning engineering is needed to use B-Tas.

Blast

Some of the strengths of Blast (Building Loads Analysis and System Thermodynamics) are valuable for modelling solar air systems, although Blast does not have any pre-programmed solar systems.

Blast can perform simulations using annual hourly weather data for specific locations, which provide better annual energy use predictions than estimations based only on design day data. Blast uses conduction transfer functions to model the transient heat flow through walls, showing the proper effects of thermal mass. This is important in designing passive solar houses. The actual distribution of sunlight on interior wall surfaces can be calculated if the geometry of the building is described in detail. This includes the effects of shading and self-shading by the building, wings and overhangs, as well as shading by objects not attached to the building (such as trees and nearby structures). Wingwalls, overhangs and detached shading objects can be modelled with varying transmittance based on the time of year. This is helpful for modelling shading by trees.

Modelling solar air systems can be done with normal Blast models, although there are some limitations. A collector can be modelled as a thermal zone; however, the zones in Blast are assumed to be isothermal. This restricts the peak air temperatures generated by the collector. Rockbed storage can be simulated in Blast by thermal mass in a zone. Transient storage can be modelled well, but convection heat transfer coefficients are not increased for forced convection. Thermal storage by the walls is automatically handled by Blast, as is convective and radiative heat transport from them. The greatest difficulty in modelling a solar air system is controlling the movement of air from one thermal zone to another. Blast allows air movement from zone to zone to follow a time schedule and to be restricted by the temperature difference between the zones; however, Blast cannot control air flow in terms of the temperature of a specific thermal zone. Therefore, a scheduled fan or natural convection could be modelled, but not a thermostatically controlled air flow.

Emgp3 (European Modelling Group Program 3)

The intended users of this program are energy consultants, engineers and researchers. The user needs to know the different Emgp3 elements from which a system may be built, and know how to work with modular simulation programs. A pre-processor helps the less experienced user. The manual contains examples showing how to model different systems, e.g. a system with rockbed storage, PCM storage or a hypocaust.

The building model is very simple. It is described as an element with an overall U-value, a heat capacity and hour-by-hour modelling of the interaction between several of the above-mentioned elements.

It is easy to model a solar collector, especially if the efficiency equation is known. Alternatively, it may be described by the number of covers, $\tau\alpha$-product, heat capacity, rear-side heat loss, emittance of the cover and absorber and the collector efficiency factor F'. Sun spaces can only be handled in a very simple way.

Energy and room temperatures are easily obtained, but the program does not report surface temperatures. Both active and passive interaction between a system and the building are possible. It is possible to model the six systems of Task 19; however, tricking may be needed in some cases.

Apache

Apache runs on an Intel-based platform (DOS version). No information was found on running Apache on Windows 95 or Windows NT.

The program has some interesting and 'modern' features:

- The whole program has a pull-down-menu interface.
- Input data can be entered interactively in a spreadsheet.
- While the simulation is running, results (e.g. zone air temperatures) are displayed on the screen and can easily be printed out.

Apache dynamically models the building and HVAC system using the CIBSE guide admittance method. Because the program has no predefined air collectors, all collector types must be modelled as the usual zones with solar gain. Rockbed, hypocaust and building mass storage are described via zones, but it is not possible to model PCM. Hourly room loads and room temperatures are written to a plot file and/or an output file; it is not possible to obtain the surface temperatures. Apache is designed to model common HVAC systems. It therefore assumes an open loop with one air inlet and one outlet.

Because the air exchange between the collector and the storage must be modelled as an HVAC system, it is not possible to model closed-loop systems. To achieve numerical convergence, it is necessary to choose very small time steps (typically one minute). Because the program has many operational modes, only skilled modellers should consider using it.

Esp-r

The Esp-r system allows simultaneous detailed modelling of a building and plant systems, which may include fluid-flow networks. It is therefore possible to model all the solar air systems of Task 19 at various levels of detail. The tool requires experience in modelling and simulation. For solar air systems, which are particularly sensitive to control and flow assumptions, the user should be knowledgeable about the systems being modelled.

Currently, modelling a building is well supported by a graphically-based 'project manager', with numerous examples. Similar support is currently being installed for plant systems, and a graphical plant and flow component network is under development. Currently users must be expert to model plant systems. However, most of the solar air heating systems do not require a plant network. If collectors were modelled as zones, they would normally be represented by multiple zones, giving greater accuracy. Alternatively a single-node air heater component can be used.

Rockbed, hypocaust and phase change materials (PCM) can be modelled as plant components.

All aspects of the building energy system are modelled simultaneously and dynamically. Each heat and mass transfer subsystem is modelled with its own state-of-the-art algorithm. Each of these systems is straightforward to model in Esp-r. They can be modelled either as building zones or as plant components. The simulation time step can be adjusted between a fraction of a minute and an hour, as required.

All temperatures (including temperatures within a construction) and heat fluxes are output as tables (time step by time step or integrated) or as graphs.

Ida and Nmf (integrated platform for model development and simulation)

Ida (from the Institute of Applied Mathematics, Stockholm) is a general simulation platform for modular differential algebraic equation systems. A graphical interface is available for certain applications. For general use, the solver can be used without the graphical interface. Ida solves the equations in residual form by solving the system matrixes. It uses an implicit solution, which improves the accuracy for many configurations.

Nmf (Neutral Model Format) is a model definition language for arbitrary components described in terms of algebraic and dynamical equations. No specific solution of the equation system has to be specified. The final choice of the input and output variables will vary, dependent on the role of the model in a complete system of individual components.

Nmf was recently approved by ASHRAE as a standard model definition format. An ASHRAE translator for multiple platforms will be available after the year 2000; Trnsys is its first target.

Ida has used Nmf as its standard model definition for a long time. Its solution methods are well suited to easy implementation of models described by Nmf. A translator (Neutran) is available. An integrated modelling environment based upon Neutran, Ida and Gnuplot features easy modelling and post processing with no need for programming. Special output for Matlab can also be produced. If new component models are to be created, a compiler (Lahey F77L3 or Microsoft Fortran) is required. The translator will automatically create the Fortran code and update the makefile.

At present, there are several Nmf libraries available in various fields, for example:

- zone and building construction;
- air exchange in buildings;
- fire-related smoke propagation;
- the Annex 10 component library;
- the ASHRAE HV AC2 toolkit;
- district heating subcentral.

Tsbi3 (Thermal Simulation of Buildings and System 3 generation)

The building model gives results very close to the average of results from a large variety of models compared in the framework of IEA Task 12.

A beta version of tsbi3 can model different solar walls (mass walls, trombe walls, internally ventilated solar walls and solar walls for preheating of ventilation air). Other aspects are as follows:

- rockbed: possible with care, modelled as building mass;
- hypocaust: possible with tricking, modelled as small zones;
- PCM: impossible;
- building mass: can be done.

All the output is available in both tabular and graphic form. The solar wall and the storage are part of the building and are thereby simulated together with the building performance (passive transfer and tricked active transfer).

There is only one temperature node in each zone, but a room can be modelled as several zones with air exchange between the zones. The program is designed for use by skilled engineering consultants.

Suncode/Sunrel

Sunrel is a public version of the proprietary program Suncode. It is under development by NREL (USA). Modelling a solar air system can be done in several ways.

A trombe wall as collector can easily be modelled, because Suncode has a trombe-wall type. This trombe-wall type models only buoyancy-driven ventilation. For modelling solar air systems with forced ventilation, the collector has to be modelled as a separate zone. For storage modelling, a rockbed type is available, but other storage types (e.g. hypocaust) must also be modelled as separate zones. Collectors and storage are connected with temperature-controlled fans, but they may not be connected in a closed loop. It is therefore not possible to model closed-loop systems (with the exception of a buoyancy-driven trombe-wall system).

Heat flows, room and surface temperatures can be output.

Because the program has many possibilities, a skilled engineer is needed to handle it.

Swift

The Swift program is a Windows-95-based program that simulates the heat output of a perforated unglazed collector. It utilizes hourly or monthly weather data and solar design details to estimate accurately the savings and return on investment for all types and colours of unglazed air collector. It was written by Enermodal Engineering Inc. of Kitchner, Canada, with funding from Natural Resources Canada. The Swift program calculations are based on actual test data accumulated over the past ten years. The program is suitable for all types of buildings and applications and it will calculate the solar savings, wall heat-loss recapture and destratification savings (where applicable). The program was released in March 1999.

Prepared by Matthias Schuler

Appendix J: Bibliography

BOOKS

1. **Collier JG, Dias RWM:** *Active Solar Heating Systems Design Manual* (1990). ASHRAE, Atlanta ISBN 0-910110-54-9.
   This manual offers solar designers a means to improve the quality and energy efficiency of solar systems. The design process has four parts:

   1. A conceptual analysis to determine whether a full feasibility study is warranted.
   2. A feasibility study for estimating the cost-effectiveness of the proposed project.
   3. A detailed design, which includes system performance calculations, economic evaluation, piping, structural, mechanical, thermal and other design analyses, subsystem drawings and component requirements.
   4. A design and construction package, into which the design analyses, design and installation drawings, cost estimate, economic evaluation and the solar energy system are compiled.

   At the end, in 'Lessons learned', problems commonly found in solar energy system design, potential effects of these problems and ways to avoid them are presented.

2. **Beckman W, Klein S, Duffie J:** *Solar Heating Design by the f-Chart Method* (1977). Solar Energy Laboratory, University of Wisconsin, Madison. John Wiley & Sons, New York. ISBN 0-471-03406-1.
   This book describes a practical method of sizing solar space and water heating systems, either based on liquid or air. Each of the six chapters addresses a particular aspect of the design of solar heating systems. Each chapter is nearly complete in itself, containing introductory material, references to pertinent literature and examples that illustrate the major points. The first four chapters present methods of estimating the various quantities that are needed in Chapter 5 to determine the thermal performance of systems and then in Chapter 6 to determine economical designs. The design procedure presented here, referred to as the '*f*-chart' method was developed by S. A. Klein in his PhD programme at the University of Wisconsin. The approach is to identify the important dimensionless variables of solar heating systems and to use detailed computer simulations to develop correlations between these variables and the long-term performance of these systems. The correlations developed for the heating systems are presented in graphical and equation form. The result is a simple method, requiring only monthly average meteorological data, which can be used to estimate the long-term thermal performance of solar heating systems as a function of the major system design parameters.

3. **Böer KW (ed.):** *Advances in Solar Energy* (1992) Vol. 7. American Solar Energy Society. ISBN 0-89553-250-6.
   One full chapter (approximately 25 pages) is dedicated to unglazed transpired solar (air) collectors with emphasis on free and forced convection and collector efficiency. Full-scale installations including Solarwall® projects are briefly discussed.

4. **de Winter F (ed.):** *Solar Collectors, Energy Storage, and Materials* (1990). MIT Press, Massachusetts. ISBN 0-262-04104-9
   The book is one of twelve volumes that summarize research in and development and implementation of solar thermal energy conversion technologies carried out under federal sponsorship during the last 11 years of the National Solar Energy Program in the USA. It deals with solar collectors, their concepts and design, the theory of collectors, the research on them and the cost issues. It also gives an overview of energy storage for solar systems such as electrochemical, chemical, mechanical, superconducting magnetic or thermal storages and materials for solar technologies.

5. **Duffie J, Beckman W:** *Solar Engineering of Thermal Processes* (1991). Solar Energy Laboratory. University of Wisconsin-Madison. Wiley-Interscience, New York. ISBN 0-471-51056-4.
   In the first two chapters of this book solar radiation, radiation data and the processing of the data to get it in forms needed for calculation of process performance are treated. The next three chapters re-

view heat transfer and radiation properties of opaque and transparent materials. Chapters 6 through 9 go into detail on collectors and storage. Chapters 10 and 11 are on system concepts and economics. They serve as an introduction to the balance of the book, which addresses applications and design methods. Chapter 15 is a discussion of passive heating. It uses many of the same concepts and calculation methods for estimating solar gains that are developed and used in active heating systems. Design procedures based on simulations are described in Chapters 14 and 18. The book also contains comparisons of predicted and measured performance of solar energy used for heating.

6. **Duffie JA, Beckman WA:** *Sonnenenergie – Thermische Prozesse* (1976). U. Pfriemer Verlag, Munich. ISBN 3-7906-0065-2.
The German translation of the first edition of the classical book by Duffie and Beckman.

7. **Fisch NM:** *Systemuntersuchungen zur Nutzung der Sonnenenergie bei der Beheizung von Wohngebäuden mit Luft als Wärmeträger* (1984). N. Fisch, TU-Braunschweig, Inst. f. Gebaüde u. Solartechnik, Mühlenpfadstr. 23, D-38106 Braunschweig. ISBN 3-922-429-10-6.
This doctoral thesis investigates solar air systems and, in particular, air-cooled flat-plate collectors and rockbeds. The steady and transient thermal behaviour of the components is determined in experiments; the temperature and heat flow distribution, the efficiency and time constant, as well as the pressure loss in the collector and the rockbeds, are calculated as functions of the air flow rate.
   The program Simul is developed to obtain knowledge on the most essential parameters and to provide a tool for optimization of heating systems. Computer models are developed for the system components and the collector, a two-dimensional temperature distribution in the collector and a thermal stratification in the heat storage are taken into account. Good agreement is found between the computed and the measured results for the air collector, the rockbed and a model system.

8. **Fricke J, Borst W:** *Energie. Ein Lehrbuch der physikalischen Grundlagen. Regenerative Energiequellen, Energiespeicherung, Energietransport, Energiekonservierung* (1981). R. Oldenburg Verlag GmbH, Munich. ISBN 3-486-24971-1.
Notes of physics lectures at the University of Würzburg and the University of Illinois in Carbondale are the basis of this book. As well as fundamentals, it reviews regenerative sources of energy and the physical principles of transformation, transport and storage of energy. One chapter deals specially with solar flat collectors.

9. **Garg HP:** *Advances in Solar Energy Technology*. Vol. 1. Collection and Storage Systems (1990). D. Reidel Publishing Company, Dordrecht. ISBN 90-277-2430-X.
Although this book deals mostly with advanced flat-plate collectors, concentrating collectors, solar ponds, solar water heating and the design process, there is also a chapter on storage of solar energy. Thermal storage and thermochemical storage are described, as also are phase change materials (salt hydrates, paraffins, etc.). A subchapter covers sensible heat storage in liquid, solid and dual media, i.e. hot-water storage systems, rockbeds, long-term energy storage, high-temperature energy storage and energy storage for passive solar architecture. The advantages and disadvantages of each are explained.

10. **Gillett WB:** *Solar Collectors, Test Methods and Design Guidelines* (1985). Kluwer Academic Publishers, Dordrecht. ISBN 90-277-2052-5.
This book is a guide to the design of solar air and water collectors. The test methods are more interesting from a scientific point of view, but there are also good explanations of the different test methods used to evaluate collector data.

11. **Hastings, R. (ed.):** *Passive Solar Commercial and Institutional Buildings. A Sourcebook of Examples and Design Insights* (1993). International Energy Agency, Paris, France and John Wiley & Sons, Chichester, UK. ISBN 0-471-93943-9. → Systems 4, 5.
The results of research performed by solar experts from 12 countries during the years 1986 to 1991 are presented here. Forty-five building case studies are featured. The buildings range from a large university complex with multiple atria in Norway to a sports hall employing a mass wall in Spain. Insights gained from these projects have been expanded by conducting parametric studies using computer models, some of which were developed specifically for this research. One part of the book deals with solar heating and includes guidelines and monitored example buildings with air collector and storage systems.

12. **Hollick J, Aislin E:** *Conserval Solarwall® – Air Heating System Design Manual* (1990). Conserval Engineering Inc., 200 Wildcat Road, Downsview, Ontario M3J, 2N5, Canada.
This design manual details the design parameters, illustrates numerous typical situations found in industrial buildings and explains the use of basic ventilation principles necessary to achieve a comfortable indoor environment. Adequate fresh air can be provided at similar capital costs to a fuel-burning system with significant energy savings and energy cost avoidances using the Solarwall® system.

13. **Khartchenko N:** *Thermische Solaranlagen. Grundlagen, Planung und Auslegung* (1995). Springer-Verlag Berlin, Heidelberg, New York. ISBN 3-540-58300-9.
    Here is a book for students and beginners in the field of solar technology. It shares knowledge of thermal dynamics, heat transfer and hydrodynamics so as to give a better understanding of the processes in the components of solar systems. The book describes the state of the art of modern collectors, storage systems and other equipment. Accepted methods for calculating and planning active and passive solar systems are presented, as are also methods for economicl analysis and optimization of solar systems.

14. **Kok HWM, Andrews S:** *Passive and hybrid solar low energy buildings – Construction issues*. Design information booklet number five. Internal Energy agency: Solar heating and cooling program, Task VIII (1989). Available from the Superintendent of Documents, U.S. Government Printing Office, Washington DC 20402, USA.
    Construction problems unique to the use of passive and hybrid solar systems are identified in this booklet and several proven solutions are offered for different countries; representative construction details are provided. The intention is to define where construction detailing is crucial to the performance of low-energy passive solar homes and to provide some ideas on how these detailing problems can be solved for a range of construction technologies. Topics include windows, sun spaces and thermal storage walls.

15. **Kornher S, Zaugg A:** The complete handbook of solar heating systems. (1984). Rodale Press, Emmaus, Pennsylvania. ISBN 0-87857-442-5.
    Skilled manual workers are the audience for this book, which includes a wealth of construction details. It leads both beginners and experienced do-it-yourselfers and tradespeople through a variety of systems and components.

16. **Kreider JF, Hoogendoorn ChJ, Kreith F:** *Solar Design – Components, Systems, Economics* (1989). Hemisphere, Bristol, Pennsylvania. ISBN 0-89116-406-5.
    This book emphasizes system and component design and includes a dozen pages on air-based mechanical solar heating systems.

17. **Kreith F, Kreider J:** *Principles of Solar Engineering* (1978). Series in thermal and fluids engineering. Hemisphere, McGraw-Hill, New York. ISBN 0-07-035476-6.
    The objective of this book is to update and expand the classic works on the use of solar energy and to present all aspects in a teachable framework for engineers. The level of presentation in this book assumes that the reader has had a course in basic thermodynamics, has some background in chemistry and physics and has a knowledge of calculus and ordinary differential equations. Given these prerequisites, the book presents all the information necessary for the design and analysis of solar energy conversion systems. Basic, fundamental approaches for an engineer to evaluate specific systems are emphasized. Chapters also address methods of solar collection and thermal conversion, system analysis, components and economics of solar systems, solar heating systems, solar cooling and dehumidification, solar electric power and process heat and natural solar conversion systems.

18. **Labhard E, Binz A, Zanoni T:** *Erneuerbare Energien und Architektur. Fragestellungen im Entwurfsprozess – Ein Leitfaden* (1995). Bundesamt für Konjunkturfragen, Impulsprogramm PACER. Bern, Switzerland. ISBN 3-905232-48-0.
    This guide addresses architects in the early process of planning. It seeks to establish whether renewable energy, for example the use of air collectors, is sensible. It questions whether a building is suitable for the use of air collectors, regarding orientation and climate, if it is possible to situate enough air collectors and storage very close to them, if the air duct can be short but have a big sectional area and if there will be no overheating. It is a checklist for the planner.
    The manual is also available in Italian as *Energie rinnovabili e architettura* and is one of many documents of the Swiss PACER Programme.

19. **Levy ME, Evans D, Gardstein C:** *Passive Solar Construction Handbook*. Stephen Winter Association, Inc., Rodale Press, Emmaus, Pennsylvania.
    Basic concepts in the design of passive homes are discussed here and general design guidelines are offered. Many construction details are included and materials specifications are given. Special issues are thermal storage walls, attached sun spaces and convective loops. Rules of thumb are mentioned concerning collectors, storage, shading, etc. Building examples are used to explain concepts. A list of suppliers of some special materials (e.g. PCM) is attached as well as a simplified rockbed design guide and a recommended analysis tool.

20. **Löf G (ed.):** *Active Solar Systems* (1993). MIT Press, Massachusetts. ISBN 0-262-12167-0.
    This book is also one of a series of twelve volumes on solar thermal energy conversion technologies. Covered here are the design, the analysis and control methods of active solar systems. The book also gives an overview of modelling and simulation.

Active solar systems are divided into solar water heating, solar space heating and solar cooling. Both performance and costs are addressed.

21. **Morhenne J, Langensiepen B:** *Planungsgrundlagen für solarbeheizte Hypokausten* (1995). Ingenieurbüro Morhenne, Schülkestrasse 10, D-42277 Wuppertal, Germany.
A design guide for solar air heated hypocausts, based on numerical analysis and dynamic simulation results, has been developed. It helps the planner to dimension hypocausts and murocausts for the heating of dwellings. The heat flux of radiation and convection to the room, as well as the necessary energy coming from the collector, can be estimated by using nomograms. Results are the collector area and the dimension of the hypocaust (length, channel diameter, spacing of the tubes and the distance to surface). Available construction types are symmetric and asymmetric murocausts and hypocausts. The accuracy of the nomograms is sufficient for dimensioning purposes. The tool can be used in the preliminary design phase as well.

22. **Røstvik H:** *The Sunshine Revolution* (1992). SunLab Publishers, Stavanger, Norway. ISBN 82-91052-03-4.
Solar energy is suggested as a solution to the world's present energy situation. The principles and the components of solar heating are explained in a way that is easy to understand. Examples of houses with solar heating, either with water or air as the medium, are illustrated with text, photographs and diagrams from all over the world. Alternative uses of solar energy, such as solar cars, photovoltaic systems for electrical power etc., are also discussed.

23. **Solar Energy Research Institute (SERI):** *Engineering Principles and Concepts for Active Solar Systems* (1988). Taylor & Francis Inc., Philadelphia. ISBN 0-89116-855-9.
This book is divided into four parts: advanced passive solar materials research; collector technology research; cooling systems research; and systems analysis and applications research. Part one reviews work on new aperture concepts for controlling solar heat gains. It also encompasses work on low-cost thermal storage materials that have high thermal storage capacity and can be integrated with conventional building elements, and work on materials and methods to transport heat efficiently between any building exterior surface and the building's interior by non-mechanical means. Part two encompasses advanced low-to-medium-temperature flat-plate collectors, and medium-to-high-temperature evacuated tube/concentrating collectors. The focus is on design innovations using new materials and fabrication techniques. The last part deals with experimental testing, analysis and evaluation of solar heating or cooling.

REPORTS

1. **Blatter-Spalinger, M.:** *Messung eines Fensterkollektorsystems. Schlussbericht* (1997). HBT Solararchitektur, ETH-Hönggerberg. CH-8093 Zurich, switzerland/BFE, Bundesamt für Energie. → System 4.
A single-family house with a window air collector system and a ground-level floor used as storage is reported. Measurements from October 1996 to September 1997 show that the house is a low-energy house. The storage floor is able to hold heat even through very cold periods, providing that there are very few thermal bridges to the ambient. The rooms behind the air collector do not overheat. The control of the fans should be improved.

2. **Bosanac M, Østergaard Jensen S:** *In-situ Solar Air Collector Array Test* (1998). Solar Energy Laboratory, DTI Energy. → Systems 1, 6.
A method for in situ testing of solar air collector arrays is reported, including guidelines on the type and accuracy of the measuring equipment and a method for identification of the array parameters. The test had to be a short-term test method, in order to reduce the measurement costs and because of the infrequency of long periods with high irradiance in many European regions. The method was validated by in situ measurements on two representative systems, both having an air-to-water heat exchanger. An important conclusion of the work is that air heating systems equipped with an air-to-liquid heat exchanger may be successfully tested by measurements on the liquid side of heat exchanger only. This simplifies the test procedure and significantly reduces costs.

3. **Brühwiler D:** *Messprojekt EFH Gwadt, St. Gallenkappel. Kombination von Fensterkollektoren und Warmluft-Holzofen, beide mit Thermozirkulation angetrieben* (1995). BFE, Bundesamt für Energie, CH-3003 Bern, Switzerland. → System 4.
The well-insulated house Gwadt has a south-oriented window–collector facade and walls and floors used as storage. A wood-burning stove is installed to provide back-up heating and is tied into the storage. The air movement is not fan driven but by thermal circulation. The heating energy demand is 163 MJ/m^2a. Thermal comfort is good and the system proved to be user-friendly. The collector gains were reduced as a result of using thermal circulation because the collector and the storage were at the same elevation.

4. **Eckerlin P, Klopfer A, Schröder J:** *Untersuchung und Entwicklung von Systemen zu Speicherung*

thermischer Energie im Temperaturbereich zwischen – 25°C und +150°C (1980). Kommission der europäischen Gemeinschaften, Generaldirektion Informationsmarkt und Innovation.→Systems 3, 4, 5.
A detailed experimental investigation of materials suitable for latent heat storage is reported. A theoretical study of the possibilities of storing low-grade heat by means of reversible chemical reactions was conducted. On the basis of thermodynamic considerations, criteria are discussed for the selection of suitable reactions. Of the more than 30 systems investigated, all but one are based on gas/solid and gas/liquid interactions. In addition, chemical heat-pump systems, using heat from the ambient air or the ground, have been included in this study. The highest energy density of a storage unit could be obtained with a system working as a chemical heat pump extracting heat from the ground. The construction and the costs of a seasonal storage system, consisting of wet soil as the storage medium surrounded by a layer of dry earth or foam insulation, are discussed.

5. **Erhorn H, Franke A, Gertis K, Kiessl K, Rath J:** *Thermisches Verhalten luftdurchströmter Kiesspeicher* (1987). WB 10/87 des Fraunhofer-Instituts für Bauphysik, POB 800469, D-70504 Stuttgart, Germany.→System 5.
The exact determination of the thermal behaviour of rockbeds is decisive for their dimensioning and the choice of the material. To date, most investigations have been experimental, not allowing general results because of the unique aspects of each experiment. In this work an attempt was made to investigate mathematically the thermodynamic behaviour of such storage. The results of these calculations have been compared to experiments carried out at the University of Essen. The length of the rockbeds, the duration of the experiment, the amount of air passing through the pebble bed store and the entrance temperature were varied. The congruence between the calculation and the experiments is very good, even at extreme conditions such as a sinusoidal variation in entrance temperature. The results of the parametric study lead to practical conclusions:

- The rocks of the storage should have a small diameter and a high heat capacity, and should be strongly compressed. Their thermal conductivity has almost no influence.
- The length of the rockbeds is not only determined by the needed amount of storage.

6. **Erhorn H, Gertis K, Rath J, Wagner J:** *Ein einfaches Hybridsystem zur Heizenergieeinsparung mittels Kiesspeicher und verglastem Dachraum* (1987). WB 8/87 des Fraunhofer-Instituts für Bauphysik, POB 800469, D-70504 Stuttgart, Germany.→System 5.
Monitoring and conclusions from two unoccupied experimental buildings are reported. The cabins are of a passive solar design. One cabin (the hybrid cabin) has a roof inclined to the south with a 10 m^2 glazed area, a rockbed 3 m long, 1 m wide and 0.5 m high and two fans. If the solar radiation is transmitted through the window in the roof, it will be absorbed by the black surface of the insulated attic and transformed into heat. The solar-heated air goes either directly to the room or, if there is no heat demand, into the rockbed in the cellar (System 5). The rockbed may also be used for cooling in the summer.

During the night, cool air is blown through the rockbed. Then, during the day, if the room overheats, the room air is circulated through the cool rockbed and then back to the room.

The experiments showed that hybrid systems have to be carefully planned: The energy consumption (including the electric energy of the fans) in the hybrid cabin is about 70% of that in the reference building. Although it is very well insulated, the glazed attic transmits the heat to the room with a delay. The rockbed in the cellar with a length of 3 m was too big and placed wrongly. During the summer there was a higher temperature in the hybrid cabin than in the passive cabin. Cooling through the rockbed only moderated the temperature. Shutters for the skylight would improve this situation.

7. **Erhorn H, Reiss J, Szerman M:** *Hybride Heizsysteme, Phase II* (1994). WB 79/1994 des Fraunhofer-Instituts für Bauphysik, POB 800469, D-70504 Stuttgart, Germany.→System 5.
In this study the influence of hybrid heating systems on the auxiliary heating energy demand, the heating costs and the CO_2 emissions are calculated for a row (terrace) house, a duplex and a single-family house. Using the Suncode program, different kinds of systems, such as active or passive discharge and different thicknesses of insulation between the hypocaust ceiling and the room, were analysed. A reduction of the heating demand up to 15% was possible.

8. **Gütermann A, Krüsi P:** *Solarhaus Lenherr, Schwyz. Wintergarten-Fensterkollektor-Kombination mit Boden-/Wandspeicher* (1992). Bundesamt für Energie, CH-3003 Bern, Switzerland.→Systems 4, 5.
This house in Switzerland has a three-storey sunspace on the south side. Part of the sunspace facade is a double facade used as collector. With the help of fans, the gained heat is stored in the massive house ceilings and inner walls. The house and the system were monitored and computer parametric studies carried out of the storage insulation, the operation of the 'collector', the size of the storage and the insulation of the house.

9. **Gütermann A, Krüsi P:** *Thermische Messungen am Stahllager Kägi & Co. in Wintherthur* (1990). Schlussbericht. HBT Solararchitektur, ETH-Hönggerberg. CH-8093 Zürich, Switzerland and BFE, Bundesamt für Energie, Bern, Switzerland.
This unheated warehouse for steel beams had several comfort problems for the workers. The cellar, which is also used, was damp and cold, leading to condensation and corrosion on the surface of the steel and increased illness amongst the workers. The solution was to circulate warmed air from below the shed roof through a solar air collector into the cellar. Measurements were made to analyse and optimize the efficiency of the system. The control had some errors, which caused the ventilation to operate for too few hours. Subsequently, the control value was changed from relative humidity to dew-point temperature. After this change the average temperature of the cellar was not more than 3 K below that of the hall. It was more economical to build this system than to install an oil heating system.

10. **Haas K, Brühwiler D:** *Bürohaus Haas & Partner AG – Ergebnisse eines Messprojektes* (1990). Forschungsstelle für Solararchitektur, ETH Hönggerberg, Zürich, Switzerland and Bundesamt für Energie, Bern, Switzerland. Available from Amt für Umweltschutz, Sektion Energie, Linsebühlstrasse 91, 9001 St. Gallen, Switzerland. → System 5.
This Swiss office building with a window collector facade and a rockbed built is described in the first part of the report. The costs are also given. The monitoring results and their analysis are shown in the second part. Finally, results from computer simulation of passive use, active use and summertime use are reported.

11. **Kurer T, Filleux C, Lang R, Gasser H:** *Forschungsprojekt 'Solar Trap': Passive und Aktive Sonnenenergienutzung bei Gebäuden* (1982). Schlussbericht über das Forschungsprojekt Solar Trap zu Händen des Nationalen Energie-Forschungs-Fonds, Zürich. Available from C. Filleux, Basler & Hofmann, CH-8029 Zurich, Switzerland. → System 5.
Here you can find information on the efficiency of the 'solar trap' window collector and the energy balance for a heating season of a house built with a window collector, a murocaust wall and a vertical rockbed. Measurements from a Lausanne test cell are also reported. The computer program Soltrap is described. It was developed to optimize the solar trap system and its components.

12. **Kristiansen F, Østergaard Jensen S:** *Forøgelse af solvarmeanlægs ydelse.* Report No. 264, Laboratoriet for Varmeisolering, Danmarks Tekniske Universitet, Postbox 141, DK-2630 Taastrup, Denmark. (1994). ISSN 0905-1511. → System 6.
The efficiency of two different air/liquid collectors was measured under standard conditions in the solar simulator at the Thermal Insulation Laboratory, and efficiency formulae were derived. Based on the efficiency tests, computer simulations of the performance of solar systems with the two collector types were carried out, with two different programs. Combined systems with 5 m^2 solar collectors for single-family houses were investigated and compared with a similar system with normal liquid collectors. None of the simulation programs developed so far can handle combined air/liquid systems correctly. Therefore it was necessary to simplify the assumptions regarding the operation and control of the solar system. The different simplifications leading to different estimations of the performance are compared and discussed in the report. The simulations indicate that the performance of air/liquid systems will be 10–20% higher than the performance of similar liquid systems. Finally, a number of Danish air/liquid systems that have already been installed and the first user feedback are described.

13. **Kristiansen, F., Østergaard Jensen, S.:** *Luft- og væske-solvarmeanlæg med varmepumpe, vp-sol. Måling og beregning* (1998). Report No. R-015, Institut for bygninger og energi, Danmarks Tekniske Universitet, Postbox 141, DK-2630 Taastrup, Denmark. ISBN 87-7877-020-3. → System 6.
The annual thermal performance of an air/liquid solar heating system with a heat pump has been measured and compared with simulations carried out using the program Kviksol. The system performs as predicted. Some mistakes were found, such as: the heat pump overheats the tank; the sensor of the differential control of the liquid pump is placed inappropriately; and the air should be driven through the air collector only at times when the temperature difference between the air collector and the outside air exceeds 5 K. After changes to correct these problems have been made, it is expected that the use of the solar energy from the system can be increased by about 40%.

14. **Mørck O:** *Hybrid Solar Low-Energy Dwellings in Denmark* (1996). EU DG XVII for Energy. Thermie Programme, project no. SE 140/91 DK. Cenergia Energy Consultant. → Systems 5, 6.
This solar low-energy housing building project, comprising 50 new dwellings built near Hillerød in Denmark, demonstrates an energy-saving and renewable utilization concept employing the following principles:

- solar air collectors combined with an air-to-water heat exchanger;
- hydrologic floor heating with integrated hybrid solar storage;
- low-temperature district heating;
- domestic hot-water heating in the summer months by the air solar system.

This project demonstrates that this concept would result in a 55% reduction in energy consumption for heating and domestic hot water compared to the new dwellings constructed according to the Danish building regulations. The economic viability is satisfactorily, but further cost reductions are needed. The construction was completed in September 1994 and the monitoring period reported lasted from 1 November 1994 to 31 October 1995.

15. **Østergaard Jensen S:** *Luft-til-vand varmevekslere til tagrumssolfangere* (1991). Report No. 218, Laboratoriet for varmeisolering, Danmarks Tekniske Højskole, Postbox 141, DK-2630 Taastrup, Denmark. → System 6.
This report deals with the possibility of transferring energy from a roof-space collector to a water-based heating system by means of an air-to-water heat exchanger (System 6). The heat exchangers were originally developed to be used together with heat pumps.
The efficiency of the two most promising heat exchangers has been determined through experiments. The efficiencies have been determined for several combinations of air and water flows through the heat exchangers. On the basis of the measured efficiencies, the performance of a traditional solar heating system with a roof space collector and an air-to-water heat exchanger was compared with the performance of a traditional solar heating system with water-based absorbers in the collectors.

16. **Østergaard Jensen S:** *MF-demonstration project at Wewer's Brickyard in Helsinge* (1994). Report no. 267, Laboratoriet for varmeisolering, Danmarks Tekniske Højskole, Postbox 141, DK-2630 Taastrup, Denmark. ISSN 0905-1511. → System 1.
The Multi-Function Solar Energy panel was developed in a previous R&D project financed by the Danish Ministry of Energy. The outer part of the panel consists of a flat plate – either a metal sheet or a transparent plate – and behind this plate a metal sheet with a trapezium-corrugated metal sheet is located. The panel can therefore act both as a solar collector and as a heat exchanger between fresh air to the building and exhaust air from the building. If the exhaust air is led down between the insulated wall and the trapezium-shaped plate, the heat loss through the wall will be further reduced, as the temperature will be higher on the outside of the insulated wall during the heating season than the ambient air temperature. For design conditions the MF panel will reduce the heat loss through the wall or roof by 89%. Simulations have shown that the performance of the MF panel is very high. Depending on the area, the flow rate and the energy demand of the building, the annual performance will be between 250 and 1350 kWh/m^2.
An area of 126 m^2 of MF panels was installed on the south facing roof of the workshop at Wewer's Brickyard in Helsinge. The building was equipped with a new ventilation system connecting the MF panel to the rooms of the workshop. A comprehensive measuring system was installed in the workshop in order to obtain detailed information on the performance of the MF panel system. The performance of the MF panel as solar collector and as heat exchanger was higher than expected. The mean system efficiency of the MF panel acting as solar collector was measured to be approximately 39%, while the system efficiency of the panel used as heat exchanger was measured to be 44%. The performance of the MF panel for the heating season 1992–93 was measured to be 96 kWh/m^{2}a. If, however, the MF panel system had operated as intended with the actual low flow rates, the yearly performance would have been 353 kWh/m^2.

17. **Østergaard Jensen, S.:** *Results from test on a Multi-Function Solar Energy Panel* (1990). Report No. 213, Laboratoriet for varmeisolering, Danmarks Tekniske Højskole, Postbox 141, DK-2630 Taastrup, Denmark. → System 1.
This report deals with the experience and conclusions gained from tests carried out on two prototypes of a Multi-Function Solar Energy Panel (MF panel). Besides being the wall (or the roof) of a house, the panels may serve as an air-based solar collector for direct heat injection into the house and as a heat exchanger between fresh air to the house and exhaust air from the house. The panel has a smaller heat loss to the surroundings than a normal wall (or roof).
The efficiency of the two MF panels, operating only as solar collectors or only as heat exchangers, has been measured for different flows through the panel. Based on the measured efficiencies, simulations have been performed to establish knowledge of the annual performance of such panels mounted in a house under different operational conditions (tilt, orientation and air flow through the panel).
Some preliminary investigations were performed in order to define theoretical equations for calculating the efficiency of the MF panel used as solar collector and heat exchanger, to determine the influence of condensation in the MF panel and to quantify the reduction of the heat loss compared to a normal wall.

18. **Østergaard Jensen S:** *Roof Space Collector - Validation and simulations with Emgp2* (1987). Report No. 87-15, Laboratoriet for varmeisolering, Danmarks

Tekniske Højskole, Postbox 141, DK-2630 Taastrup, Denmark. → System 5.

In 1986 an experimental house with a roof-space collector connected to a rockbed was built at the outdoor test area of the Thermal Insulation Laboratory. The purpose of the project was to determine the solar fraction of this kind of solar heating. Because the measuring period lasted less than a year and the actual weather was abnormal, a model was needed to predict the performance of the solar heating system. Emgp2 (European Modeling Group Program 2) was chosen. By means of that program a model of the space heating system was developed and validated using the measured data from the experimental house. The report describes the development and validation of the model and the results obtained using the model.

19. **Østergaard Jensen S:** *Test of the Summer House Package from Aidt Miljø* (1994). Report No. 94-1, Laboratoriet for varmeisolering, Danmarks Tekniske Højskole, Postbox 141, DK-2630 Taastrup, Denmark. (1994). → System 1.

The experience from tests carried out on the Summer House Package from Aidt Miljø is documented. The tests have established the efficiency of the package, depending on the solar radiation. The knowledge gained is the necessary basis for any calculation of the driving effect the collector can provide.

20. **Reiss J, Erhorn H:** 'Comparative study of different hybrid systems in practical application, a demonstration project of the International Energy Agency (IEA)' (1996). *EuroSun '96 Proceedings*. DGS-Sonnenenergie Verlags-GmbH. → Systems 1, 4, 5.

This contribution to EuroSun '96 presents the development of hybrid systems in multi-storey buildings made by the Department of Heat Technology of the Fraunhofer Institute of Building Physics. In Berlin in 1986 a housing block was constructed with a passive discharging storage component (System 4). In this building heating energy consumption was less than 30 kWh/m²a. At the same site, controlled discharge of the storage component was tried in 1993 and 1995 (System 5). In the 'Ganghoferstrasse' project in Munich, six different hybrid systems are currently being tested in one building. This project is a model for city development, demonstrating that extremely low heating energy demand (< 20 kWh/m²a) is possible within the cost frame of public housing.

21. **Reis J, Erhorn H, Stricker R:** *Passive und hybride Solarenergienutzung im Mehrfamilienwohnhausbau, Messergebnisse und energetische Analyse des Deutschen IEA-Task VIII-Gebäudes in Berlin* (1994). WB 64/91 des Fraunhofer-Instituts für Bauphysik, POB 800469, D-70504 Stuttgart, Germany.→ System 4.

Seven of 31 apartments of a multi-storey building were monitored over two years to show how much heating energy could be reduced by the use of solar energy in passive and hybrid constructions. Four different solar apartments with passive and hybrid constructions (System 4) were compared with a reference apartment and with two atrium apartments with and without a beadwall system. The heating energy consumption of the solar apartments was measured to be 43% to 62% below the consumption of the reference apartments. During the coldest months in winter (November to March) 20% to 33% of the solar radiation on the collector could be delivered to the hypocaust ceilings and used for heating. The solar apartments did not overheat during the summer. The use of the beadwall system in the heating period was not possible in the planned way. In principle, passive solar use has to be given priority over hybrid or active use. The technology of hybrid systems must be absolutely failsafe regarding wrong manipulations by the occupants.

22. **Reiss J, Erhorn H:** *Solare Hybridsysteme in einer Reihenhaus-Wohnanlage am Weinmeisterhornweg in Berlin* (1997). WB 88/97 des Fraunhofer-Instituts für Bauphysik, POB 800469, D-70504 Stuttgart, Germany.→ System 5.

This project quantified heating-energy savings from a hybrid system. Detailed measurements were necessary to take into account the behaviour of the occupants. Two upper-storey apartments were monitored, one with a hybrid system and one without as a reference. The hybrid system consists of air collectors connected to a hypocaust wall right behind the collectors. Both charging and discharging is done using fans (System 5). The measurements started in December 1994 and ended in May 1996.

The hybrid system decreased the heating energy demand by 8.4 kWh/m²a, about 18%. The collector delivered 100 kWh/m². The active discharging proved to be a success.

23. **Ruppert P:** *Meteolabor Betriebsgebäude. Schlussbericht* (1990). HBT Solararchitektur, ETH-Hönggerberg. CH-8093 Zürich, Switzerland. → System 4.

The Meteolabor plant was built 1984/85 as a low-energy building with a solar air heating system using a sunspace and a facade air collector. The mass of the floor slab provides hypocaust storage. The thermal time constant of the building is about one month. In this report one year of measurements are analysed. The measured energy consumption is much lower than that of an ordinary building. The room temperature changes very little even on days with high radiation. The solar gains of the sunspace and the air collectors are compared. The amounts are similar but the collectors gain most in the winter and the sunspace in autumn and spring, which is explained by the seasonal sun angles.

24. **Schreck H, Hillmann G, Nagel J:** *Dokumentation passiver und hybrider Solarenergiehäuser - Ausgewählte Beispiele*. Prepared in the framework of the BMFT research project 'Solarenergienutzung mit passiven (baulichen) Mitteln, Az.' PLE-03E-4460A. Available from IBUS, Institut für Bau-, Umwelt- und Solarforschung GmbH, Berlin, Germany. → Systems 4, 5.
The construction of the houses in this survey was completed towards the end of 1983. In this survey 111 houses from all over the world have been documented in terms of their system of gains, distribution, storage and delivery, far-reaching ideas, integration of the system in the building and the availability of data. Included system types are: direct solar gain, indirect gain, isolated gain and hybrid systems.

25. **Stricker R:** *Bauteilintegrierte Heiz- und Haustechnik für zukünftige Baukonzepte (Hybride Heizungssysteme), Vorausberechnung des thermischen und energetischen Verhaltens von Hybridsystemen* (1991). WB 65/1991 des Fraunhofer-Instituts für Bauphysik, POB 800469, D-70504 Stuttgart, Germany. → System 4, 5.
A calculation model that assesses the thermal and energy behaviour of solar hybrid systems has been further developed. In order to verify the new model, a hypocaust ceiling was monitored and simulated. The influence of variations of the storage size and material was examined for a typical storage construction, a hypocaust ceiling. It was found that for a facade collector either a floor or wall storage is adequate and that its density should be more than 800 kg/m³. The following conclusions are presented:

- The solar system reduces the heating energy demand from 9 to 26% (absolute values 2 and 18 kWh/m²a), depending on the building type and insulation level.
- Active discharge is necessary for comfort.
- A hybrid system shortens the heating period.
- The hybrid system can also be used for cooling.

26. **Wettermark G, Carlsson B, Stimne H:** *Storage of Heat. A Survey of Efforts and Possibilities*. (1979). Report D 2, Swedish Council for Building Research.
Various methods of storing heat are surveyed. Different principles on how to gain, use and store heat are given. Storing heat as sensible heat, latent heat and by conversion to another form of energy (i.e. thermochemical) are explained and specified.

27. **Zweifel G:** *Messprojekt Schüpfen. Schlussbericht* (1989). Bundesamt für Energiewirtschaft/EMPA, Abteilung Haustechnik, CH-8600 Dübendorf, Switzerland. → System 1.
The south facade of a row house (terrace house), built with a massive construction, has a vertical air collector between two bands of large windows. Four partition walls between north and south rooms incorporate phase-change heat storage. The ventilation system is driven by PV cells. During the period February 1987 to June 1988 detailed measurements were made. The results are:

- The building has a total energy demand of 408 MJ/m² of which 270 MJ/m² is for space heating.
- The air collector contribution was 8% to 10%.
- The latent storage functions only in the upper part of the upper storey. In the other parts there is no phase change.
- The storage requires major reconstruction.

28. **Zweifel G, Koschenz M:** *Optimierung der Luftkollektoranlage mit Latentspeicher in Schüpfen. Schlussbericht* (1996). Eidgenössische Materialprüfungs- und Forschungsanstalt (EMPA). Available from HBT Solararchitektur, ETH Hönggerberg, CH-8093 Zürich, Switzerland. → System 5.
The aim of the project was to optimize the air-collector system in Schüpfen and provide general guidelines for latent storage. Laboratory tests with phase-change materials were carried out at the EMPA. The functionality of a passive discharged latent storage was proven and the test data were used to validate a computer model specially developed for this project (a module for Trnsys). A parametric study in Schüpfen was done, in which the size of the storage and the melting temperature were varyied. The reduction of the heating-energy demand was much higher than the use of electrical power for active discharging. The use of the air collector for producing domestic hot water during the summer was also considered. After that, four variants on the Schüpfen system were studied, one with passive and one with active discharge, one with a combined passive and active discharge and a larger storage and, finally, one without the latent storage in order to minimize the costs and the space used for the system. The variants were evaluated in terms of the costs. In the second phase of the project the air collector was changed to an air–water-system with a full year's use and this was monitored and optimized. The gains from the collector were lower than estimated and the air–water heat exchanger was less efficient than expected.

Prepared by Heike Kluttig

Appendix K: IEA Solar Heating and Cooling Programme

The International Energy Agency (IEA) was established in 1974 as an autonomous agency within the framework of the Organization for Economic Cooperation and Development (OECD) to carry out a comprehensive programme of energy cooperation among its 24 member countries and the Commission of the European Communities.

An important part of the Agency's programme involves collaboration in the research, development and demonstration of new energy technologies to reduce excessive reliance on imported oil, increase long-term energy security and reduce greenhouse gas emissions. The IEA's R&D activities are headed by the Committee on Energy Research and Technology (CERT) and supported by a small Secretariat staff, headquartered in Paris. In addition, three Working Parties are charged with monitoring the various collaborative energy agreements, identifying new areas for cooperation and advising the CERT on policy matters.

Collaborative programmes in the various energy technology areas are conducted under Implementing Agreements, which are signed by contracting parties (government agencies or entities designated by them). There are currently 40 Implementing Agreements covering fossil fuel technologies, renewable energy technologies, efficient energy end-use technologies, nuclear fusion science and technology, and energy technology information centres.

The Solar Heating and Cooling Programme was one of the first IEA Implementing Agreements to be established. Since 1977, its 20 members have been collaborating to advance active solar, passive solar and photovoltaic technologies and their application in buildings. The member countries are:

Australia	Finland	New Zealand
Austria	France	Norway
Belgium	Germany	Spain
Canada	Italy	Sweden
Denmark	Japan	Switzerland
European Commission	Mexico The Netherlands	UK USA

A total of 26 Tasks have been initiated, 19 of which have been completed. Each Task is managed by an Operating Agent from one of the participating countries. Overall control of the programme rests with an Executive Committee comprised of one representative from each contracting party to the Implementing Agreement. In addition, a number of special ad hoc activities – working groups, conferences and workshops – have been organized.

The Tasks of the IEA Solar Heating and Cooling Programme, both completed and current, are as follows:

Completed Tasks

Task 1	Investigation of the Performance of Solar Heating and Cooling Systems
Task 2	Coordination of Solar Heating and Cooling R&D
Task 3	Performance Testing of Solar Collectors
Task 4	Development of an Insolation Handbook and Instrument Package
Task 5	Use of Existing Meteorological Information for Solar Energy Application
Task 6	Performance of Solar Systems Using Evacuated Collectors
Task 7	Central Solar Heating Plants with Seasonal Storage
Task 8	Passive and Hybrid Solar Low Energy Buildings
Task 9	Solar Radiation and Pyranometry Studies
Task 10	Solar Materials R&D
Task 11	Passive and Hybrid Solar Commercial Buildings
Task 12	Building Energy Analysis and Design Tools for Solar Applications
Task 13	Advanced Solar Low Energy Buildings
Task 14	Advanced Active Solar Energy Systems
Task 16	Photovoltaics in Buildings
Task 17	Measuring and Modeling Spectral Radiation
Task 18	Advanced Glazing and Associated Materials for Solar and Building Applications
Task 19	Solar Air Systems
Task 20	Solar Energy in Building Renovation

Completed Working Groups

CSHPSS
ISOLDE
Materials in Solar Thermal Collectors

Current Tasks

Task 21 Daylight in Buildings
Task 22 Building Energy Analysis Tools
Task 23 Optimization of Solar Energy Use in Large Buildings
Task 24 Solar Procurement
Task 25 Solar Assisted Air Conditioning of Buildings
Task 26 Solar Combisystems
Task 27 Performance of Solar Facade Components
Task 28 Solar Sustainable Housing
Task 29 Solar Crop Drying
Task 30 Solar City (Task Definition Phase)

Current Working Groups

Evaluation of Task 13 Houses
PV/Thermal Systems

To receive a publications catalogue or learn more about the IEA Solar Heating and Cooling Programme visit the web site at http://www.iea-shc.org or contact the SHC Executive Secretary, Pamela Murphy, Morse Associates Inc., 1808 Corcoran Street, NW, Washington, DC 20009, USA. Tel. +1 202 483 2393; Fax +1 202 265 2248; Email pmurphy@MorseAssociatesInc.com.

TASK 19: SOLAR AIR SYSTEMS

Duration

October 1993 to April 1999.

Objectives

The goal of Task 19 is to facilitate the use of solar air systems for residential, institutional and industrial buildings by:

- documenting exemplary buildings to inspire designers and building clients with the reliability, performance and aesthetics of building-integrated systems (Lead country: Switzerland);
- writing a Systems Design Handbook to help engineers choose, dimension and detail a system, while assessing energy performance and non-energy issues (Lead country: Denmark);
- developing a computer tool to analyse key design variables by means of the Trnsys modules of Task 19 systems and a user-friendly interface (Lead country: Germany);
- compiling a catalogue of manufactured components to inform designers what is available 'off the shelf' for solar air systems and from where (Lead country: Norway);
- testing collectors under laboratory conditions to help manufacturers optimize performance and to provide standardized data for consumers (Lead country: Austria).

EXPERTS OF IEA TASK 19

Thomas Zelger
Kanzlei Dr. Bruck, Prinz-Eugen-Strasse 66, A-1040 Wien, Austria.
Tel. +43 1 503 5559; Fax +43 1 503 5558;
Email bruck@magnet.et

Hubert Fechner
Fach 8, Oest. Forsch. zentrum, Faradygasse 3, A-1031 Wien, Austria.
Tel. +43 1 797 47 299; Fax +43 1 797 47 594;
Email fechner@email.arsenal.ac.at

Sture Larsen
Architekt, Lindauerstrasse 33, A-6912 Hörbranz, Austria.
Tel. +43 5573 831 00; Fax +43 5573 841 90;
Email solarsen@computerhaus.at

John Hollick
Solar Wall International Ltd, 200 Wildcat Road, CDN-Toronto ONT M3J 2N5, Canada.
Tel. +1 416 661 70 57; Fax +1 416 661 71 46;
Email conserval@globalserve.net

Søren Østergaard Jensen
Solar Energy Laboratory, Danish Tech. Institute, Postbox 141, DK-2630 Taastrup, Denmark.
Tel. +45 43 50 45 59; Fax +45 43 50 72 22;
Email soren.o.jensen@dti.dk

Ove Mørck
Cenergia ApS, Sct. Jacobs Vej 4, DK-2750 Ballerup, Denmark.
Tel. +45 44 66 00 99; Fax +45 44 66 01 36;
Email ocm@cenergia.dk

Heike Kluttig
Fraunhofer-Institut für Bauphysik, Nobelstrasse 12, D-70569 Stuttgart, Germany.
Tel. +49 711 970 3322; Fax +49 711 970 3399;
Email hk@ibp.fhg.de

Alexander Knirsch
Transsolar, Nobelstrasse 15, D-70569 Stuttgart, Germany.
Tel. +49 711 679 76 23; Fax +49 711 679 76 11;
Email knirsch@transsolar.com

Helmut Meyer
Transsolar, Nobelstrasse 15, D-70569 Stuttgart, Germany.
Tel. +0711 679 76 35; Fax +0711 679 76 11;
Email meyer@transsolar.com

Johann Reiss
Fraunhofer-Institut für Bauphysik, Nobelstrasse 12, D-70569 Stuttgart, Germany.
Tel. +49 711 970 33 37; Fax +49 711 970 33 99;
Email reiss@ibp.fht.de

Joachim Morhenne
Ing. büro Morhenne GbR, Schülkestr. 10, D-42277 Wuppertal, Germany.
Tel. +49 202 264 0290; Fax +49 202 264 0298;
Email IBMorhenne@t-online.de

Frank Heidt
Fachbereich Bauphy. u. Solar., Universität GH Siegen, Adolf Reichweinsstrasse, D-57068 Siegen, Germany.
Tel. +49 271 740 41 81; Fax +49 271 740 23 79;
Email heidt@physik.uni-siegen.de

Hans Erhorn
Fraunhofer-Institut für Bauphysik, Nobelstrasse 12, D-70569 Stuttgart, Germany.
Tel. +49 711 970 33 80; Fax +49 711 970 33 99;
Email erh@ibp.fhg.de

Matthias Schuler
Transsolar, Ingenieurgesch. GmbH, Nobelstrasse 15, D-70569 Stuttgart, Germany.
Tel. +49 711 679 76 11; Fax +49 711 679 76 0;
Email schuler@transsolar.com

Siegfried Schröpf
Solar Env. Tech., Grammer KG, Wernher v. Braunstr. 6, D-92224 Amberg, Germany.
Tel. +49 962 160 1150; Fax +49 962 160 1151

Gianni Scudo
DI.Tec, Politecnico di Milano, Via Durando 10, I-20158 Milano, Italy.
Tel. +39 2 2349 5729; Fax +39 2 2399 5746

Giancarlo Rossi
I.U.A.V. Dept. Building Construction, Università di Venezia, S. Croce, 191, I-30135 Venezia, Italy.
Tel. +39 41 257 1303; Fax +39 41 257 1313;
Email rossigc@brezza.iuav.unive.it

Harald N. Rostvik
Sunlab/ABB, Alexander Kiellandsqt 2, N-4009 Stavanger, Norway.
Tel. +47 51 53 34 42; Fax +47 51 52 40 62

Torbjörn Jilar
Department of Biosystems & Technology, Swedish University of Agriculture, PO Box 30, S-23053 Alnarp, Sweden.
Tel. +46 40 41 54 74; Fax +46 40 41 54 75;
Email Torbjorn.Jilar@jbt.slu.se

Christer Nordström
Ch. Nordström Arkitektkontor AB, Asstigen 14, S-43645 Askim, Sweden.
Tel. +46 31 28 28 64; Fax +46 31 68 10 88;
Email cna@cna.se

Gerhard Zweifel
Abt. HLK, Zentralschw. Technikum Luz., CH 6048 Horw, Switzerland.
Tel. +41 41 349 33 49; Fax +41 41349 39 60;
Email GZweifel@ztl.ch

Peter Elste
Basler & Hofmann, Forchstrasse 395, CH-8029 Zürich, Switzerland.
Tel. +41 1 387 11 22; Fax +41 1 387 11 00;
Email pelste@bhz.ch

Charles Filleux
Basler & Hofmann, Forchstrasse 395, CH-8029 Zürich, Switzerland.
Tel. +41 1 387 11 22; Fax +41 1 387 11 00;
Email chfilleux@bhz.ch

Robert Hastings
Solararchitektur, ETH-Hönggerberg, CH-8093 Zürich, Switzerland.
Tel. +41 1 633 2988; Fax +41 1 633 1169;
Email hastings@orl.arch.ethz.ch

Karel Fort
Ing. Büro, Weiherweg 19, CH-8604 Volketswil, Switzerland.
Tel. +41 1 946 08 04; Fax +41 1 946 08 04;
Email k_fort@compuserve.com

Kevin Lomas
Institute of Energy and Sustainable Development, De Montfort University, The Gateway, Leicester LE1 9BH, UK.
Tel. +44 116 257 7445; Fax +44 116 257 7449;
Email klomas@dmu.ac.uk

Index

Acoustic criteria 204
Acoustics 11
Acrylic covers 141
Agricultural (purposes) 150
Air flow 26, 138
Annuity factors 248
Apartment buildings 150
Apartments 50
Assumptions 26
Atria 172
Attic space collectors 92

Bara Costantini 5, 265
BLC (building loss coefficient) 263
Buffer 159
Building envelope 5

Canopy 151
Capital 247
Climate(s) 7, 21, 229, 240
Closed loop (systems) 6, 10, 11, 69, 91, 135
CO_2 9
Collector efficiency
    in-situ determination 252
Collectors
    attic space 92
    perforated 150
    roof-integrated 91
    site-built 142
    solar air 250
    spatial 172
    sunspace 49, 50
    window 69, 71, 146
Comfort 10
Commissioning 11, 13
Condensation 121, 191
Containment 189
Controls 215
Cooling 120, 150
COP (coefficient of performance) 225

Degree days 225, 229, 243
Dehumidification 19
Design
    process 7, 13
    temperature 225
DHW (domestic hot water) 102, 115, 213

Double
    envelope 49, 265
    facade 158
    flow 42
    -shell facade 158
Drying 150
Dwellings 31, 34

Earth
coupled channel 121
temperatures 126
Economic evaluation 246
Efficiency 133
Electricity 66
Exhaust air facade 161

Face velocity 187
Fan 105, 201
    axial 201
    cross-flow 201
    radial 201
Filter 10, 143
Fire protection 221
Flat-plate
    air collector 141
    collectors 133
Flow rates 252
Freeze protection 106, 214
Friction coefficient 203

Glauber's salt 193
Glazed balconies 172
Global Warming Potential (GWP) 9

Heat exchange 6, 102, 104, 106, 116
Heat exchangers
    air-to-water 212
    fin-tube 212
History 3
Hospitals 150
Hybrid system 225
Hydraulic
    aspects 142, 148, 213,
    design 180, 182
Hygiene 10
Hypocaust 38, 69, 70, 74, 91, 96, 121, 126, 175, 225, 266

IEA 283
Industrial buildings 150
Industry 25, 26, 50
Insulation 141, 190
    levels 258
Intake air facade 161
Interest 247
Investment 246

Large (volume) buildings 25, 26, 29, 34
    internal heat gains 262
Latent heat 193
Latitude 230-4, 243
Leakage 142
Life Cycle Assessment (LCA) 9
Light industry 31
Liquid systems 102
Longitude 243

Maintenance buildings 150
Manifolds 141
Materials, properties 227
Meteonorm 236
Meteorological data 229
Microprocessor controls 216
Murocaust(s) 69, 225, 267
    storage 175
    controller settings 267

Natural
    convection 265
    ventilation 158
Night ventilation 121
Nomogram assumptions 258

Offices 19, 50, 150, 164
Open discharge loop 91
Open loop 10, 38, 135
Orientation 7, 33, 73

Parallel connection 139
Parking garages 150
Payback 246, 248
PC programs 269
Perforated
    collectors 150
    unglazed collectors 150
Phase change materials (pcm) 193
Photovoltaic (PV) 24
    cells 128 207
    solar panels 128
Preheating 25
Pressure
    drop 140, 142, 182, 187
    losses 202, 209
Products 144

Reference BuildingI
    internal heat gains 262
    materials used 259
    occupancy 262
    office 258
Reference systems 19, 70, 185
Rock bed 11, 91
    storage 96, 266
Roof-integrated collectors 91
Roof space 93
    collector 253

Schools 19, 50, 150
Sensors 216
Serial connection 139
Set points 105, 215
Shoebox 225
Simulation program 255
Site-built collectors 142
Smoke detectors 221
Snow 143
Solar
    chimney 39
    protection 149
SOLARWALL 150, 156
Spatial collectors 172
Sports halls 19, 26, 31, 150
SSI (Solar Similarity Index) 229, 230, 243
Storage 18, 187, 193
Stratification 198
Sunspace collector 49, 50
Sunspaces 172
Swimming halls 19

Testing 250
Thermostats 215
Thermosyphon 38, 255
Throttling 206
Tilt 33, 73
Timers 216
Toxicity 194
TRNSAIR 236, 255, 269
TRNSYS 255, 269
Two loop systems 70

Venetian blind 69
Ventilation 5, 17
    buffer 160

Warehouses 25, 26, 31, 50
Water 102
    heating 115
    solar-heated 267
Window collector(s) 69, 71, 146
Worksheet 67